Evolution's Fatal Flaw

The Inevitable Consequence of the Need to Ensure Species Survival

For JoANNE + Dutch

Beware the Flaw

[signature] Wood

Lawrence Wood

iUniverse, Inc.
New York   Bloomington

Evolution's Fatal Flaw
The Inevitable Consequence of the Need to Ensure Species Survival

iUniverse books may be ordered through booksellers or by contacting:

iUniverse
1663 Liberty Drive
Bloomington, IN 47403
www.iuniverse.com
1-800-Authors (1-800-288-4677)

Because of the dynamic nature of the Internet, any Web addresses or links contained in this book may have changed since publication and may no longer be valid.

ISBN: 978-1-4401-7151-2 (sc)
ISBN: 978-1-4401-7153-6 (dj)
ISBN: 978-1-4401-7152-9 (ebk)

Printed in the United States of America

iUniverse rev. date: 9/28/2009

CONTENTS

FIGURES

TABLES

ACKNOWLEDGEMENTS

I have discussed this book with many people; however, I would like to paricularly acknowledge the assistance of my friends Bert Coates and Ken Fialkowski, who read and commented on the many versions of the book as it evolved. I would also like to thank my wife Mary for her patience in listening as I monopolized dinner table conversations with my latest inspirations as the book evolved.

DEDICATION

This book is dedicated to those persons whom I designate as "Closet Questioners." Those individuals, some of whom are mentioned in this book, who have been raised in religiously conservative families in which they were taught that the Book of Genesis is literally true; however, as they matured, found that a literal interpretation of the Bible is inconsistent with what they observe and learn; BUT, are afraid to voice any questions for fear of retribution or ostracism from their family and friends. This book may answer many of their questions.

PROLOGUE.

The primary objective of this book is to validate at least 2 controversial claims: the first claim is that evolution is the correct explanation for our origins, while the second claim is that evolution was the inevitable consequence of the need to ensure species survival; however, evolution's success in ensuring species survival has been achieved at potentially great cost, which is "Evolution's Fatal Flaw," disastrous population growth, especially in our species. Fully understanding how the fatal flaw developed, and to appreciate its significance, requires the majority of the material in this book; hence, a discussion of evolution's fatal flaw and potential solutions must be deferred to later chapters.

Clearly, the complexity and controversy attendant with evolution requires an inordinate amount of evidence to accomplish this objective. The evidence I will present is scientific, pragmatic and has a basis of a Millennia of scientific observation and investigation coupled with a few timely accidental discoveries.

As the evidence for evolution accumulated and a unique approach to the presentation of this evidence occurred to me, I occasionally considered writing a book; however, the decision to finally write this book was prompted by two chance encounters at sea. The first encounter occurred on a trans-Atlantic voyage aboard the Queen Mary, while the other occurred one lovely summer day while I was on a whale watching excursion on Puget Sound.

Boat rides on Puget Sound to observe the magnificent black-and-white killer whales, properly known as Orcinus orca, or simply orca swim gracefully through the water in search of salmon and surface porpoise-like to breathe, are a prime attraction in Puget Sound. Our ship's captain was an expert at finding orcas, so we had a great time chasing them around the Sound, always at a safe distance, and observing the magnificent mammals porpoising through the water. As I watched the orcas, I noted that they never opened their mouths to breathe, even though they came far enough out of the water to do so. It seemed odd because most mammals breathe through their mouths.

Onboard the vessel was a charming lady with a degree in marine biology. She was apparently friends with the captain and enjoyed answering questions about orcas. I asked her about orcas' breathing. "It's all about the nostril," she explained. "The nostril was originally in front, but gradually over time, it evolved." At that point she suddenly stopped her explanation and said, "Oh, I can't use that word. Many people don't like it."

I was astounded! Here we were in the middle of Puget Sound, listening to a marine biologist explain a simple evolutionary transition, and the ugly head of anti-evolution appeared.

A little while later, I encountered the biologist again. This time she was seated between two women. I looked at the biologist and said something to the effect of, "I was rather surprised and a bit bothered by your comment about not being able to discuss evolution."

Before the biologist could respond, the woman seated to her right said, "I suppose you believe in evolution!" Her tone was a bit strident but seemed to indicate that she was a "believer."

I replied to her, "Just as I don't use the term 'I believe in electricity,' I don't use the term 'I believe in evolution.'"

Before she could respond, the woman seated to the biologist's left drew herself up and said haughtily, "Well, I don't care. I was created." She got up and departed. With this as the obvious conversation stopper, I departed also but would have loved to ask what she meant by the statement, "I was created."

Later on, I encountered the biologist once more. She stood alone at the ship's rail. I asked her how people reacted when a subject came up that involved an evolution-oriented explanation. She said people asked, "Can't you explain it another way?" to which she usually says, "Sorry, I can't," and the subject is dropped. She seemed rather troubled by this but didn't seem to have a way to thwart it. She merely told me, in a rather dejected manner, "They just don't understand."

We must have spent five hours traveling around the Sound that beautiful afternoon, and by the end of it, we had lost track of the number of orca sightings. It was a wonderful day.

As we drove home from the voyage, snatches of the conversations kept echoing in my mind: "Can't you explain it another way?" "I suppose you believe in Evolution!" "Well, I don't care, I was created!"

I had been following the evolution controversy in a cursory manner for several years, noting that a small but vocal and assertive minority continually persisted in attempts to prevent the teaching of evolution in public schools. But here, in a small group of people on a whale watching boat in Puget Sound, presumably representing a random selection from the general population, the dichotomy of views associated with the controversy was dramatically presented to me "up close and personal."

BOOK IDEA BEGINS TO FORM.

Farther along on our journey home, I recalled the other chance encounter at sea aboard the Queen Mary. Purely by chance, our dinner companions included a charming lady who was an author. We got to talking about her work and I mentioned that occasionally the "itch to write a book" struck me. She said that writing a book is a wonderful experience and that for a relatively unknown person such as myself, self publishing offers a way to get started. One of the publishers she had suggested was I Universe. As I mulled that encounter over in my mind, I recalled a series of Gallup polls which succinctly captured the divergent beliefs related to evolution in one three-part question, and I promised myself I would review these surveys since I knew Gallup had an excellent reputation.[1] Perhaps I could build a book around the surveys.

Upon arriving home, I dug out my copies of the surveys. There were eight, beginning in 1982, with the last in 2006. In each survey, the Gallup organization posed the following question, or something very close to it, to representative groups of Americans:

> "Which of the following statements comes closest to your views on the origin and development of human

beings? 1) God created human beings pretty much in their present form at one time within the last 10,000 years or so, 2) Human beings have developed over millions of years from other forms of life, but God guided this process, 3) Human beings have evolved over millions of years from other forms of life, but God had <u>no part</u> in this process"[2].

Table P-1 summarizes the Gallup poll results. The numbers represent the percent agreeing with the indicated statement.

	God created humans in present form: Creationism	Humans developed, with God guiding: ID	Humans developed, but God had no part in process: Evolution	No Opinion
1982	44%	38%	9%	9%
1993, June	47	35	11	7
1997, Nov	44	39	10	7
1999, Aug	47	40	9	4
2001, Feb	45	37	12	6
2004, Nov	45	38	13	4
2005, Sept	53	31	12	4
2006, May	46	36	13	5
Average	**46**	**37**	**11**	**6**

Table P-1. Summary of Gallup poll results

As indicated in the table, option 1 is generally termed "creationism." Creationism applies to those people who believe in a strictly literal interpretation of the Bible, especially the book of Genesis. Option 2 is generally termed "intelligent design," or ID, which applies to those who believe in the Bible but feel creationism is too restrictive. Option 3 is a general definition of evolution; however, as will become apparent, a more appropriate definition is "Evolution is the gradual improvement of life forms over billions of years controlled by Natural

Selection's filter which selects the most fit from the variations produced by reproduction."

Of particular interest is that the responses have been surprisingly uniform. Moreover, these responses seemed to reflect the opinions revealed on the whale watching boat as well.

BROADER IMPLICATIONS OF THESE GALLUP POLLS.

As I reflected on the responses to these polls, it struck me that these surveys raised two additional and extremely important questions that had much broader implications than the three questions originally posed in the Gallup polls:

1. Why are there three mutually exclusive origin views extant?

2. Since these views are mutually exclusive, which one is the true one?

While the Gallup poll only specified human origins, the basis for views expressed in option 1—a literal interpretation of the book of Genesis—begins with the origin of the earth; therefore I believe the Gallop question is a bit limited. In addition, I believe the phrase "your views" is rather ambiguous and should be replaced with the more precise phrase "your understanding of explanations" because understanding should be interpreted as the ability to answer one or more of the six little friends questions[2]: what, when, where, who, how, and why?

How and *why* are the more important[3] questions, particularly *why* because many useful explanations do not answer the all-import question of why? With these modifications, the Gallup poll questionnaire would become:

> "Which of the following statements comes closest to your *understanding of the explanations* of the origin and development of human beings and the earth? 1) God created human beings and the earth pretty much in their present form at one time within the last 10,000 years or so; 2) Human beings and the earth have evolved over millions of years and God guided this process; 3) Human beings and the earth have evolved over millions of years, but God had no part in this process."

And the two general questions now become:

1. Why are there three mutually exclusive origin *explanations*?

2. Since these *explanations* are mutually exclusive, which one is the true?

To my knowledge, the plethora of books attempting to demonstrate that neither creationism nor ID are correct explanations of our origins do not satisfactorily, if at all, address the question: "Why are there three origins explanations?" I believe the answer to this question is one of the keys to explaining the Gallup poll results. My hunch about using the Gallup polls as a basis for a book was correct.

As will be demonstrated, answering the "Why three origins explanations? question is best accomplished by an in-depth review of the history of the development of our understanding of ourselves and the world around us. In addition, an examination of the "Why three origins explanations?" question will shed considerable light on one of the more contentious issues confronting America today, the conflict between science and religion.

It should be noted, that the information gathered to answer the "Why three origins explanations?" question will also establish the correct answers to the other two questions of "Which of the three mutually exclusive options posed by the Gallup poll is correct?" and "Is evolution the consequence of the need to ensure species survival?"

In summary, accomplishing the book's primary objective of validating the claims that evolution is the correct explanation of human origins and is a consequence of the need for species survival, requires the accomplishment of three "supporting" objectives:

1 Answering the question "Why are there three mutually exclusive origin explanations?"

2. Answering the question "Since these explanations are mutually exclusive, which one is true?"

3. Establishing unequivocally that the reason for the existence of evolution is the need for species survival

Accomplishing these three objectives will require considerable supporting information. **As introduced above,** due to the controversy surrounding the issue of evolution and the complexity of the process of evolution, accomplishing these objectives will require the presentation of as much supporting information as is feasible. Accordingly, several chapters of background information are provided to assure the best possible understanding of this complex subject.[4]

A secondary purpose of providing considerable background information is to parry statements such as this one, which can be found on a Web site called Just what is creationism trying to say?: "There is *no reason not to believe* that *God created* our universe, earth, plants, animals, and people just as described in the book of Genesis!" [emphasis in original] [3]

POSSIBLE APPROACH TO ANSWERING WHY THERE ARE THREE EXPLANATION OF OUR ORIGINS.

As I reflected upon the replacement of the word "view" with "explanation" and the extension to the earth (and by implication, the universe), I suddenly realized a "background" project I had been pursuing that began a few years after finishing graduate school in 1967 might provide a possible approach to answering the first question, from which the answer to the second question should then become clear.

While in graduate school, I engaged in some interesting discussions regarding how the explanations of ourselves and the world around us had developed. It was noted that textbooks and lectures organize material into nice logical patterns, but that was not how the original explanations were developed. Development of original explanations often involved blind alleys, serendipitous discoveries, and audacious hypotheses. This being an intriguing sidelight, I decided to pursue it to the extent I had time after graduate school.

A central aspect of this endeavor is the fundamental issue: what do we know and how we can have confidence in this knowledge? After a while, it became apparent that this would be most easily established by examining the historical development of any explanation.

It should be noted that the general process of obtaining a correct explanation that one can have confidence in is similar to the process of solving any mystery[5]. The essential task is to make observations and then provide explanations for those observations—a process that, if done properly, is usually called the scientific method. In general, the process tends to be circular, with improved observations requiring improved explanations, which then often suggest new observation possibilities, etc. In this way, the process is never ending. Of utmost importance regarding observations, an observation must obey these criteria:

- The observation is performed by a recognized, competent observer.
- The observer employs generally accepted observational instruments and techniques.
- If the observation involves a planned experiment, as they usually do, the experiment must be repeatable.

Regarding explanations, an explanation must obey these criteria:

- It must satisfactorily explain the observations without recourse to unverifiable causes.
- It must be understandable by persons recognized to be competent in the subject being explained.

Investigations that violate either of these criteria are suspect. For example, experiments performed by an incompetent observer or experiments that cannot be repeated will generally be discredited or not take seriously. More importantly, these types of experiments may lead competent observers down blind alleys. Similarly, explanations that resort to unverifiable causes such as "The Devil made me do it" are not useful

EXPLANATIONS—TWO PHASES OF DEVELOPMENT:

After a few years of exploration, I discovered something that was never explicitly mentioned in any of my courses or textbooks: development of explanations in almost any field has usually evolved in two relatively distinct phases:

I. First Phase: the earliest explanations were usually incorrect because they were based upon limited and/or incorrect observations and incorrect interpretations/explanations of those observations.

II. Second Phase: Improved observational techniques were eventually developed, which raised significant questions regarding the early explanations. This resulted in the development of correct explanations based upon the improved observations.

This second phase usually developed in two parts, beginning with empirically based explanations, which typically involved laborious examination of large numbers of observations from which explanatory information was "distilled."[6] Empirical explanations were then usually followed by improved "theoretical" explanations[7] based upon an understanding of the underlying fundamental principals regarding the phenomenon being explained. Unfortunately, these improved explanations often involved complex mathematics.

While empirical explanations were generally a significant advance in the development of understanding, properly answering three of the six little friends questions (what, when, and where), they rarely answered the important *how* and *why* because they didn't provide sufficient understanding of fundamentals. The answers to *how* and *why* required an understanding of the fundamental principles which were embodied in the theoretical explanations. It is important to note that normally a "theoretical" explanation is just an improvement of an empirical explanation and rarely negates it

Interestingly, the empirical or theoretical explanations rarely answer the *who* part of the six little friends. The reason for this will be explained later.

EXPLANATIONS DEVELOPED IN PHASES DUE TO IMPEDIMENTS.

Further study of the development of explanations revealed that the principal reason for the "phased" development was the existence of at least five impediments to developing a correct explanation:

1. Illusions—things are not what they seem.

2. The unaided human eye has limitations.

3. Explanations often conflict with "common sense" and "intuition."

4. Explanations sometimes conflict with authority, especially family and religious authority.

5. Unrealistic expectations—things are not always what we desire.

The principal impediment is number 1, the illusions, of which there are many. The major ones are:

1. The apparently solid earth;

2. The apparent motion of the sun and planets around the earth;

3. The apparent same size of the sun and moon and the apparent closeness of both the sun and moon;

4. The apparent motion of the stars around the earth, and the apparent closeness of the stars; and

5. The apparently unchanging physical and biological features of the earth.

Illusions 2, 3, and 4 are called the "astronomical illusions" because they involve phenomena that occur above the earth. While they are related, dividing them as shown will facilitate an understanding of how they were resolved:

- The resolution of illusion 2—the apparent motion of the sun and planets around the earth—led to the correct explanation of the shape of the solar system.

- The resolution of illusion 3—the apparent same size and closeness of the sun and moon—led to the correct explanation of the size of the solar system.

- The determination of the correct size of the solar system aided in the resolution illusion 4—the apparent motion of the stars around the earth, and the apparent closeness of the stars— which led to a correct explanation of the size and age of the universe.

Regarding illusion 5, it is important to distinguish between the physical and biological features that are apparently unchanging. Most physical features, such as mountains, do not appear to change because erosion and mountain building involve glacially slow processes. However, only one biological feature does not appear to change: animal and plant structures. The process which alters these biological features is also glacially slow. In contrast, other biological processes such as breathing and eating are readily observable because they occur within the relatively short animal and plant lifetimes.

It is equally important to note that, with the exception of the astronomical illusions, almost nothing was available to the casual observer, who was equipped only with the unaided eye, to suspect that any of these were indeed illusions. This alone would make the illusions difficult to resolve, but the other impediments greatly exacerbated the problem and still cause difficulty today.

PRODUCTION OF ILLUSIONS.

The illusions are produced principally by impediment 2, the limitation of the human eye.

> **Illusion 1**, the apparently solid earth, is produced by the inability of the human eye to "see" the atomic structure of matter; therefore resolution of the illusion had to wait for the development of sophisticated observational instruments.

> **Illusions 2 and 4**, the apparent motion of the sun and planets and stars around the earth, are produced by the rotation of the earth, something that is totally imperceptible unless you know how and where to look.

> **Illusion 3**, the apparent same size and closeness of the sun and moon, is produced the fact that we perceive distances by reference to known objects (e.g., a car more distant than another appears farther away because it looks smaller). When viewing objects like the sun and moon, we have no reference.

As mentioned above, illusion 5—Earth's apparently unchanging physical features and biological animal and plant structures—is produced by the glacial slowness of the physical and biological processes which shape them when compared with human lifetimes. At one to two inches per century, it takes millions of years to erode a mountain; accordingly, this erosion is imperceptible, even over centuries.

CONSEQUENCES OF THESE ILLUSIONS.

The astronomical illusions 2, 3, and 4 led to the *incorrect* belief that humans were at the center of a rather small universe that revolved around the earth.

The apparently unchanging earth features led to the *incorrect* belief that the earth was probably not very old, which then led logically to two questions: how did the earth, sun, moon, and stars originate (i.e., where did they come from?), and what or who controls the various earthly phenomena, such as the cycle of the seasons, and the sun and moon?

RESOLVING THE ILLUSIONS.

Understanding how these incorrect beliefs were corrected is best accomplished by examining, in sufficient detail, how the illusions were resolved. This will then provide the answers to the questions of why there are three explanations of origin and which one is correct, thereby accomplishing objectives 1 and 2. Accomplishing objective 3 will require an application of the information provided in the explanations of the illusions.

Originally, my investigations into the development of our understanding of ourselves and our surroundings had concentrated largely on how the five illusions were resolved from a scientific viewpoint, essentially by improved observational instruments and techniques coupled with improved explanations. As I pursued these investigations into the historical development of our understanding of ourselves and the world around us, I was not actively involved in any religious organization; therefore, I did not fully realize the extent of the organized resistance to the current understandings, especially evolution, until I encountered the Gallup poll information. Pondering the Gallup poll results, I realized that the answer to the "Why three origin explanations?" question would be found via the same approach: an examination of the historical development of explanations via the resolution of the illusions, with emphasis on the source of the three different explanations.

As I mentioned above, I believe a significant reason for the continuing controversy regarding the three explanations of our origins is the complexity associated with the resolution of the five illusions mentioned earlier. Hence, I am convinced that a modest understanding of how the improved understandings of ourselves and our surroundings developed, which led to the resolution of the illusions, is essential to an understanding and acceptance of which of the three origins explanations is correct. Accordingly, the first twelve chapters of this book are devoted to the resolution of the five illusions. In particular, chapter 12 will complete the explanation of how evolution works.

Once the material necessary to resolve the five illusions has been presented, a correct explanation of how the earth was formed and how life developed on it will be easily understood and is presented in chapter 13. Tracing the development of life on Earth will vividly illustrate how the process of evolution functions, particularly natural selection.

The material presented in the first thirteen chapters will then provide a sound basis for answering the question of why there are three explanations, which will be presented in chapter 14. This will also satisfy the first objective the book, by conclusively demonstrating that option 3 in the Gallup poll—evolution—is the correct explanation of our origins

Chapters 15 will present a short overview of the religious reaction to evolution, which has seriously impeded acceptance of evolution in America and is a significant reason for the Gallup results. Chapter 16 will extend the discussion by examining why, in spite of a vast quantity of evidence against creationism and ID, many people still believe in them.

Finally, chapter 17 will provide a contrast to the creationism-based Web site, "Answers in Genesis" by demonstrating that "Answers of Evolution?" are superior. Besides establishing the validity of Answers in Evolution, chapter 17 will discuss such topics as why we must die, what life's purpose is, and why we have such a strong need for sex. Answering the last "obvious" question, will lead to the demonstration of why evolution is the inevitable consequence of the need to ensure species survival and an explanation why this results in Evolution's Fatal Flaw.

Regarding the resolution of the illusions, we will begin with the solid earth illusion. While the subject may not seem germane to this book, a cursory understanding of matter will provide an important foundation for an understanding of why we can categorically state that the earth is much more than 6,000 years old and how *invisible* X-rays were discovered—a discovery that led ultimately to the determination of DNA structure, which is central to an understanding of how evolution works.

For example, chapter 1 describes the development of our understanding of electricity and magnetism, which definitely seem outside the purview of this book; however, it is common knowledge today that nerve impulses, which control all of our activities including mental activity and reproduction, are carried by electric currents—a fact discovered almost serendipitously. Accordingly, it is an important component to understanding evolution because later chapters explain in detail how reproduction plays a key role in evolution.

Regarding the first part of this book, the book of Genesis is the principal source book for creationism and much of ID; accordingly, the contribution provided by Genesis to the resolution of each illusion will be included.

Finally, the information provided in these first thirteen chapters will provide insight into the tortuous paths normally followed in search of correct explanations. Moreover, it's an interesting story that may provide you with useful insights into the world we inhabit.

Since one of the prime objectives of this book is to demonstrate that evolution is the correct explanation of our origins, emphasis will be placed upon understanding developments which apply to creationism and ID.

Beginning with creationism, its main pillars are: Earth's age is less the 10,000 years, the earth was created by God, and Earth's geological features are the result of the biblical flood. ID's main pillar is evidence of an intelligent designer directing evolution. Emphasis will therefore be placed upon these issues.

WHY IS THIS BOOK UNIQUE?

There are numerous books, articles, and Web pages that either defend or refute creationism, ID, or evolution, several of which will be discussed in this book. However, none appear to take the approach presented in this book, which examines the Gallup poll's discoveries regarding the three explanations of our origins and addresses two questions I've never seen articulated that derive from the Gallup Polls: "Why are there three explanations of our origins?" which becomes the key to answering the second question, "Which origins explanation is correct?"

In addition to answering these two questions by explaining how evolution works and conclusively demonstrating that it is the correct answer, this book explains that demonstrating the correctness of evolution as an explanation of our origins should be augmented by explaining why the process of evolution developed, i.e., evolution developed because of the need to ensure species survival, which unfortunately resulted in Evolution's Fatal Flaw, explained near the end of the book.

Finally, this book demonstrates that these questions are best answered by examining the history of the development of a correct understanding of ourselves and the universe we inhabit, which has been impeded by illusions, conflicts with common sense, conflicts with authority, and our reluctance to accept that things are not always as we desire.

A Few Comments on Reading this Book.

By now it should be apparent that this book's subject is complex. Explaining the resolution of the illusions requires the examination of material from many fields of inquiry, including but not limited to anthropology, astronomy, biology, geology, paleontology, and physics. This is, of course, a rather wide range of subjects. However, as I mentioned above, I believe that to adequately satisfy the objectives of this book, an in-depth examination of these materials is necessary to counter attacks on and statements against evolution.

For example, Texas School Board President Dr. McLeroy, DDS is leading a crusade against evolution because he believes the earth was created 10,000 years ago (see chapter 17).

Or there are statements like:

- "There is *no reason not to believe* that *God created* our universe, earth, plants, animals, and people just as described in the book of Genesis!"[emphasis in original] [Ch 9, ref 6].

- "The 'geologic column', which is cited as *physical evidence* of evolution occurring in the past, is *better explained* as the result of a *devastating global flood* which happened about 5,000 years ago, as described in the Bible" [Ch 9, ref 9].

- "The fossils, and the sedimentary deposits they were entombed within, have *simply been misinterpreted by the scientific community*. The fossil record is instead a recording of a devastating global scale flood" [Ch 9, ref 13].

- It is evident, when all the facts are examined that *there is no scientific evidence* that the biblical account of Noah's ark is a myth or fable [Ch 9, ref 17].

People like Dr. McLeroy and statements like those listed above and others sprinkled about in the text cannot just be dismissed out of hand as the ravings of misinformed ignoramuses as some scientists seem to prefer doing. These attacks must be met with a solid set of verifiable explanations backed by sufficient scientific evidence that can withstand assaults from those who argue *there is no scientific evidence.* This book provides the needed material. In addition, the book will explain in chapter 16 why persons such as Dr. McLeroy still believe in creationism.

On the other hand, this book is not intended to be a detailed science book: there is just not enough room. However, ample references are provided in the back of this book for those wanting more information. In addition, much that has been learned within the last 200 years relies heavily on relatively complex mathematics, which I will generally avoid except where an explanation is impossible without some reference to the math behind it. In these cases, only the most essential and accessible aspects will be included in the discussion.

Accordingly, to facilitate the presentation and understanding of this material, especially for those not familiar with it, I have divided the material into several chapters. In addition, summaries of the important points are provided at strategic break points within some of the chapters and at the end of each chapter. Finally, to improve continuity between chapters, at the end of each chapter you will find a "Looking Ahead" section, which encapsulates the material to be provided in the following chapter.

In view of this organization, you have two options on how to read through the book: (1) read straight through, or (2) begin with just the paragraphs that seem most useful and read the summaries in order to get the "gist" of the material.

RESOLVING ILLUSIONS RESEMBLES A GOOD MURDER MYSTERY.

As a final note, one of the interesting aspects of resolving these illusions is the remarkable resemblance to the resolution of a good murder mystery. Consider a typical mystery story: A person is found dead, slumped over his or her desk, inside a locked room, gun in hand, with a bullet wound in head—obviously a suicide. But person's spouse believes the suicide was a clever illusion and it was actually a murder. "My wife/husband would never do such a thing," he or she proclaims vehemently. The spouse seeks out a detective, the most famous of which is Sherlock Holmes. Entering Holmes's 222-B Baker St. flat, the spouse exclaims, "My wife/husband couldn't/wouldn't have killed him/herself, Mr. Holmes, because of [insert reason]. Please, Mr. Holmes, you've got to find out who did it."

Holmes reluctantly agrees and thus begins an exhaustive investigation including a thorough examination of the murder scene and interviews with possible suspects. Eventually, Holmes resolves the locked room illusion and deduces the correct explanation of the murder, which is always arrived at because Holmes was more observant than everyone else and found clues others had overlooked and/or misinterpreted. No one clue tells much, but the combination and proper organization of them clearly explains what happened. Eventually at the end of book Holmes solves the locked window. The murderer had arranged in some way to lock the window from the outside. When asked how he solved the mystery, Holmes often intoned the immortal lines, first uttered in *The Sign of Four*, to his constant companion, Dr. Watson, "How often have I said to you that when you have eliminated the impossible, whatever remains, *however improbable*, must be the truth" [4].

It is doubtful that Conan Doyle realized how generally applicable Holmes's famous saying was to scientific investigation, but we will see that it has been invoked, perhaps not consciously but often, as a significant aid to scientific investigation. We will term it "Sherlock's theorem," and we will notice that it plays an important role in answering the question of which answer to the origins question is correct.

There is debate about whether some scientists have doubts about evolution, but this is irrelevant because this book is directed toward a more general audience, an audience that I hope enjoys a good detective mystery because this book describes the solution of one of the most complex mysteries known—the explanation of ourselves and our surroundings. This book recounts the efforts of hundreds of "detectives," some who worked together and others independently to unravel this greatest mystery of all. It's an exciting story that I'm sure you will enjoy.

A FINAL NOTE:

In the preface to his monumental "History of Western Philosophy" [2, ch 2], Bertram Russell acknowledged the limitations faced by a person attempting to cover a broad subject involving many disciplines. Since this book falls into this category, I believe a few lines from Russell's preface, adapted to this book are appropriate.

"I owe a word of explanation and apology to specialists on any part of my enormous project. It is obviously impossible to know as much about every subject covered in this book as can be known about the subject by a person whose field is the subject. I have no doubt that every single subject I have discussed, is better known to many than to me; therefore, I ask the indulgence of those readers who find my knowledge of this or that portion of these subjects less adequate than it would have been if there had been no need to remember "times winged chariot" [2, ch 2, p-x]

LOOKING AHEAD.

The first two chapters will discuss early investigations into electricity and magnetism (E&M)[8] phenomena in the first chapter and "solid" matter in the second chapter, since early observers believed they were separate. This is due to the fact that "solid" matter can be "observed" directly, whereas E&M phenomena cannot. We can only observe the effects of E&M; hence the initial development of a proper explanation of the structure of matter followed two paths.

CHAPTER 1:
ELECTROMAGNETIC PHENOMENON
INVESTIGATIONS PRIOR TO 1897

EARLIEST OBSERVATIONS AND EXPLANATIONS OF THE EFFECTS OF ELECTRICITY.

It is reasonable to assume that the first observation of electrical phenomena was the observation of lightning—jagged streaks of bright light followed by a loud bang and usually accompanied by hard rain. Early humans, of course, had no idea what lightning was, often attributing it to the anger of gods. In the *Aeneid*, Virgil explains how Jupiter, king of the Roman gods, throws lightning bolts as weapons to kill humans or cause destruction [1]. As discussed below, beliefs that God was involved with lightning persisted into the 1800s, often with disastrous consequences [2].

In addition to lightning, which is actually "current" or "flowing" electricity, another form of electricity called static electricity is produced when cloth or fur rubs a solid object. This form of electricity was also undoubtedly known to early humans.

Early observers might have detected two "forms" of static electricity, one of which was associated with amber. Amber, a fossilized resin, can be an item of beauty. It was often used as jewelry for millennia [3]. We can never know the identity of the first person to rub amber with fur and observe that small bits of material such as straw were then attracted to the amber, but the Greek philosopher Thales (625–546)

BCE[1] was apparently one of the first to record the effect [4]. But, as with lightning, no one understood what caused the phenomenon.

Another "form" of static electricity can be made by rubbing a glass rod with silk. Because the earliest man-made glass objects, mainly non-transparent glass beads, are thought to date back to around 3500 BCE [5], this form of static electricity was probably known by at least the time of Thales, although no record of it seems to exist. This second form would have been detected if the observer brought the fur-rubbed amber close to silk-rubbed glass because one would have attracted the other. Two pieces of rubbed amber or two pieces of rubbed glass would repel each other.

While early investigators had no idea what exactly was happening when a rubbed object acquired the ability to attract or repel, they probably reasoned that something was being transferred between the fur or silk and the rubbed object. They eventually called this something a "charge." Therefore, when rubbed with fur or silk, an amber piece or glass rod became "charged."

The electrostatic effect produced by rubbing dissimilar materials together is now called tribo-electricity, from the Greek word for friction.

EARLIEST OBSERVATION OF THE EFFECTS OF MAGNETISM.

Just as lightning and static electricity were probably the first observations of electrical effects, naturally formed magnets were probably known for thousands of years. Natural magnets are created primarily by the iron-based mineral magnetite, the most common mineral to display magnetic properties [6]. Pieces of this material, which is often called lodestone [7], must have been found from time to time. Careful observers noted that, depending upon orientation, pieces of lodestone would either repel or attract each other.

Early records dating to the second century CE, show that the Chinese discovered that when placed on a piece of wood floating in a bowl of water, lodestone would always point to the pole star, essentially creating the first practical compass [8].

While a number of interesting phenomena had been observed (e.g., the existence of static electricity and magnetism), there were few realistic explanations of these phenomena (i.e., how the effects were produced). Serious investigation into E&M began with Girolamo Cardano.

Girolamo Cardano (1501–1576) was an Italian Renaissance physician, astrologer, and gambler He apparently led quite a life [9] While early observers had undoubtedly "discovered" electricity and magnetism, Cardano was the first to write a formal paper distinguishing the two phenomena from one another. His most famous work was *De subtilitate rerum* published in 1550 in Nuremberg by Johann Petreius [9]. This…

> "… was his great encyclopedia of scientific knowledge. This was the most advanced presentation of physical knowledge up to its time. It contains many remarkable observations and ideas, including Cardano's distinction between the attractive powers of rubbed amber (electric) and the lodestone (magnetic), his pre-evolutionary belief in creation as progressive development, and the premise that natural law was unified and could be known through observation and experiment²"[9].

As stated previously, the fact that lodestone had two "magnetic poles," one at each end of a magnet, whereas rubbed amber has only one "pole" must have been known long before Cardano, but he got credit for being the first person to formalize this knowledge.

English physician, **William Gilbert (1544–1603)** performed the first organized investigations of electricity and magnetism. He and Sir Thomas Browne are credited with the origination of the English word "electricity." Gilbert coined the term "electricus" from the Latin *electricus*, meaning "amber-like attractive properties," which Browne then anglicized into electricity [10].

Because the fact that two pieces of lodestone had the tendency to attract or repel each other depending upon their orientations must have been known for thousands of years, Gilbert would also have known of this property of magnetic materials. His experiments must have easily led him to conclude that the earth is a giant magnet, which correctly explains the functioning of a compass. It was then natural for Gilbert to apply the terms "North and South Poles" to the ends of a magnet.

Gilbert observed that like poles, (N-N or S-S) repelled each other while unlike poles, (N-S) attracted. One of the more mysterious aspects of the magnetic attraction/repulsion force was that it exerts its effect when magnets are merely brought next to each other. Touching is not required. The same is true of the force generated when amber is rubbed.

Gilbert studied these electrostatic effects and employed the term "electric force" to them. He summarized his work in *De Magnete* [10], a thorough exposition that was highly influential. A copy of the book is available at this site [11].

Unfortunately, Cardano and Gilbert were at a disadvantage in their studies. They lacked practical means for creating and storing electricity, which is imperative for practical investigation. As mentioned above, a significant impediment to establishing a correct explanation of electricity was that, unlike solid matter, electricity is invisible and only detectable by its effects. Compounding the difficulty, one of the principal detectable effects—forces generated by electrical entities—act at a distance. Thus, in order to study electricity, practical means for laboratory creation of electricity as well a practical means for detecting it had to be developed.

It should be noted that our inability to "see" electricity has not changed. As anyone who has accidentally touched a "live" wire knows, there is nothing to distinguish a wire attached to a source of electricity from a normal piece of wire just lying around.

The development of practical methods for storing and detecting electricity played significant roles in the establishment of a correct explanation of electricity. Moreover, since moving electricity generates magnetism, practical methods for storing and detecting electricity were instrumental in the development of a correct explanation for magnetism.

In 1650, German scientist and inventor **Otto von Guericke (1602–1686)** [12] invented the first instrument capable of generating electrostatic energy. The electrostatic generator consisted of a rotating disc with two "brushes" rubbing against it, thereby duplicating the rubbing of amber in a more efficient manner. Electrostatic energy could then be extracted from one of the brushes for experimentation [12].

French chemist **Charles François du Fay (1698–1739)** further refined the explanation of static electricity. He formalized the effects of rubbing amber with fur and a glass rod with silk. The former he termed "resinous electricity" (amber is a resin) and the latter he called "vitreous electricity" (*vitreous* being the Latin term for glass).

One of Du Fay's more important contributions was the summarization of all existing electrical knowledge up to that time [13] [14]:

1. All bodies can be electrically charged by heating and rubbing, except metals and soft/liquid bodies.

2. All bodies, including metal and liquid, can be charged by influence (induction).

3. The electrical properties of an object unique to color are affected by the dye, not the color itself.

4. Glass is as satisfactory as silk as an insulator.

5. Thread conducts are better wet than dry.

6. There are two states of electrification, vitreous and resinous.

7. Bodies electrified (charged) with vitreous electricity attract bodies electrified with resinous electricity and repel other bodies electrified with vitreous electricity.

One of the most important electricity investigating instruments, the **Leyden jar**, was invented in the city of Leyden in the Netherlands in 1745 by Pieter van Musschenbroek. The Leyden jar solved the electricity storage problem by allowing the storage of large amounts of static electricity inside a compact object [15]. It made significant experiments using the stored electricity possible. However, the problem of detection still needed to be addressed.

Because electricity cannot be observed directly, the second element of an electrostatic experiment is a means for detecting the presence of electrical charge. Although Du Fay must have devised some detection method, as discussed above, the first formal detection instrument was the electroscope. Two forms were developed.

In 1754, British weaver's apprentice **John Canton (1718–1772)** invented the pith ball electroscope [16]. A pith ball can be any lightweight, nonconducting substance. Canton's device employed silk thread hung from an insulated hook. When an object bearing an electric charge, such as rubbed amber rod, is brought near the pith ball, the ball will be deflected.

In 1786, a more sensitive instrument called the gold leaf electroscope was invented by **Abraham Bennet (1750–1799)** [17]. The Bennet electroscope employs two gold leaves suspended by a thin, nonconductive wire in glass cylinder. The two leaves are connected to a conductive knob at the top of the jar. When an electrically charged object is brought near the knob, the leaves separate.

However, while these devices were useful for detecting the presence of electrical charge (and still are to this day), actually measuring the amount of force generated by an electrical charge required a more sensitive instrument. Such an instrument was invented by **Charles de Coulomb (1736–1806)**.

Coulomb made two important contributions to the concept of electricity: he invented the torsional balance, an extremely sensitive force measuring device [18] that enabled him to measure the minute forces exerted by electric charges, and he then used these measurements to empirically deduce his famous "Coulombs law," which states that "the force generated between two electrical charges is proportional to the product of the amount of electricity carried by the two charges, Q_1 and Q_2, and inversely proportional to the square of the distance separating them." Coulomb published his results in 1790 [19].

Sir William Watson (1715–1787), a British physicist and physician, was one of the earliest experimenters with discharges from a Leyden jar and is known for discovering current electricity. He was one of the first to observe that a flash of light accompanies a discharge from a Leyden charge. This presumably led him conclude that electric current flows from the static charge in the Leyden jar. Beginning in 1745, he published a series of articles titled "Experiments on the Nature of Electricity." In assembling his experimental apparatus, Watson coined the term "circuit" [20].

Italian physiologist **Luigi Galvani's** (1737 - 1798) famous frog leg twitch experiment led to a source of electricity considerably superior to the Leyden jar. There is some confusion regarding when and how Galvani observed the twitching of a frog leg connected to a metallic conductor, but Galvani was a physician who performed anatomical studies on frogs (as well as other animals) with a metal scalpel. Thus, his metal scalpel must have accidentally touched a muscle in a dissected frogs leg, thereby causing a twitch and enabling Galvani to be the first person to observe the famous twitch and deduce that electrical action causes muscles to function[3] [21] [22].

In 1780, Galvani carried his frog experiments further by discovering, probably through trial and error with many metals, that when two different metals are connected together and then touched to different parts of a frog's leg nerve, the leg contracted [23].

Galvani's associate, **Alessandro Volta (1745–1827)** expanded upon the frog experiments and discovered an improved source of electricity. Count Alessandro Giuseppe Antonio Anastasio Volta was born in Como in the Italian province of Lombardy. While Volta made many contributions to the explanation of electricity, his most famous was occasioned by his association with Galvani. Volta deduced that the moist frog tissues could be replaced by water-soaked cardboard and that the muscular contraction of the frog's leg, which demonstrated the generation of electrical energy, could be replaced by another detector [24].

Building on Galvani's work with dissimilar metals, Volta created a vertical stack containing copper and zinc discs (he had experimentally determined these to be the best materials) separated by cloth or cardboard, soaked in salt water. He connected all of the copper discs and all the zinc discs together and terminated in a heavy piece of wire (now called a terminal). This arrangement, known as a Voltaic pile— or, more commonly, a Voltaic cell (see figure 1-1)—will produce an electrical current when the two terminals are connected by a conductor; thus the cell is a source of electrical energy [25].

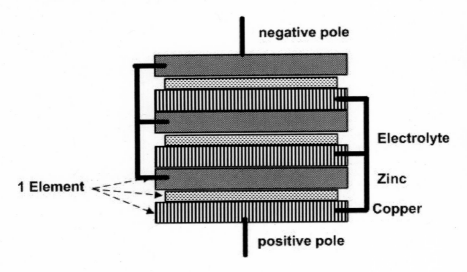

negative pole

Electrolyte

Zinc

1 Element

Copper

positive pole

Figure 1-1. Voltaic cell

Volta announced his invention to the London Royal Society on March 20, 1800. Volta's invention, now called a battery, revolutionized electrical investigations because it provided a continuous source of electricity instead of the short duration source available from a Leyden jar.

Experiments with electricity were not limited to Europe. History records that American inventor **Ben Franklin (1706–1790)** conducted numerous electricity experiments. In 1750, Franklin published a proposed a method for determining that lightning was the same form of electricity as the forms produced through other means. The method he proposed was flying a kite in an electrical storm [2]. Franklin stated:

> "When rain has wet the kite twine so that it can conduct the electric fire freely, you will find it streams out plentifully from the key at the approach of your knuckle, and with this key a phial, or Leiden jar [*sic*], maybe charged: and from electric fire [Franklin must have thought of lightning as a type of fire] thus obtained spirits may be kindled, and all other electric experiments [may be] performed which are usually done by the help of a rubber glass globe or tube; and

therefore the sameness of the electrical matter with that of lightning completely demonstrated"[2].

Franklin had a close association with French scientists, and in 1752, French scientist **Thomas François Dalibard** (1709-1799) conducted Franklin's experiment using a 40-foot-tall iron rod instead of a kite. He extracted electrical sparks from a cloud, which he stored in a Leyden jar. Dalbard determined the electricity in the Leyden jar was identical to other forms, just as Franklin predicted [26].

It's not clear whether Franklin actually performed the kite experiment since he cautioned it was dangerous; however, some anecdotal evidence suggests he actually carried it out. On the other hand, the fact that he conceived of and published the method is as important as actually having carried it out.

Franklin is also credited with replacing the terms "resinous" and "vitreous" with the more general terms "positive" and "negative" to apply to any form of electricity and not just electrostatic. Unfortunately, Franklin deduced that electricity flows from the positive electrode of a source of electricity to the negative electrode. This has caused electrical engineers no small grief because electricity actually flows the other way, but one cannot blame Franklin because no one in Franklin's time knew what the "charges" that carried electricity were [2].

Another significant Franklin contribution was the invention of the lightning rod. In one of his more astute observations, he observed the sharp pointed conductors were more effective than rounded pointed conductors. This observation led him to hypothesize that a conductor with a sharp point, placed next to a building with the point a few feet higher than the building with the other end firmly connected to the ground, might draw the electrical discharges from a lighting storm and conduct them harmlessly to the ground. His hypothesis was verified by several experiments on his own house. Following these successful demonstrations, the building in Philadelphia we now know as Independence Hall was outfitted with lightning rods in 1752 [2].

As a sad commentary on the continuing conflict between science and religion, there was resistance to the implementation of lightning rods on some churches. Religiously oriented individuals believed that lightning was the work of supernatural forces [27]. Some European Christians in the 1700s even believed lightning had a diabolical origin, based apparently on an interpretation of the phrase from Ephesians 2:2 [28]:"in which you used to live when you followed the ways of this world and of the ruler of the kingdom of the air, the spirit who is now at work in those who are disobedient." As a consequence, several wooden churches were destroyed by fire, one on three different occasions [28].

While considerable progress in the development of methods for generating and storing electricity were being made, little progress was being made toward an understanding of E&M. This gap was filled by **Michael Faraday (1791–1867)** who is considered to be one of history's foremost electrical experimenters.

Farday made three fundamental contributions and revealed relationships between E&M by continuing experiments of **Hans Christian Ørsted (1777–1851)**. In April 1820, Ørsted serendipitously placed a compass near a current-carrying wire and observed that the compass needle was deflected, thus confirming a direct relationship between electricity and magnetism [29].

Faraday took Ørsted's experiments one step further by placing a wire between two magnets (see figure 1-2). Then, when a current was passed through the wire, the magnetic field produced by the wire interacted with the magnetic field produced by the magnets and generated a force in the wire, resulting in what is the basis of an electric motor.

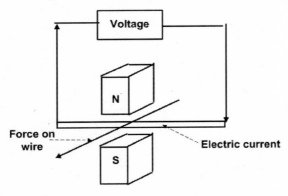

Figure 1-2. Demonstration of the electric motor principle

Secondly, Faraday demonstrated the inverse effect by passing a wire between two magnets and noting that an electrical current was produced, which illustrates the basis of an electric generator (see figure 1-3).

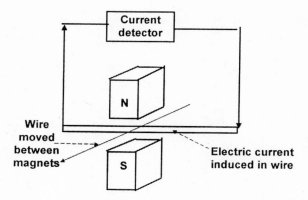

Figure 1-3. Demonstration of the electric generator principle

Finally, Faraday reasoned that a wire moving through a magnetic field was equivalent to a changing magnetic field in an adjacent wire. Thus, he concluded that if two wires are placed parallel to each other, changing the electric current through one by switching the current on and off would produce a time-varying magnetic field that would "induce" a current in the adjacent wire, as shown figure 1-4.

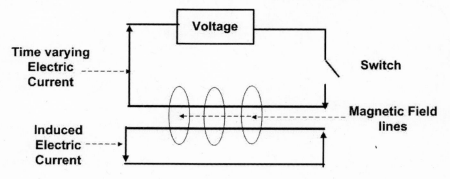

Figure 1-4. Demonstration of the electric current induction

Faraday verified this hypothesis by some tests using an apparatus similar to this simple diagram. This phenomenon is termed "electromagnetic induction" and led to the invention of the induction coil and the discovery of electromagnetic radiation. The induction of a current in wire by a time-varying current in an adjacent wire is also the basis for radiofrequency transmission, as discovered by German physicist Hertz Heinrich, which we will discuss shortly.

Through these experiments, Faraday also solved the problem of forces acting at a distance by demonstrating the existence of a "force field" associated with either a flowing current (as shown figure 1-4) or a magnet [30]. Faraday's three discoveries were among history's most important technical advances. Thus, his seemingly simple discoveries became the basis for much of modern life.

English physicist and inventor **William Sturgeon (1783–1850)**, presumably learning of Faraday's work, deduced that a current passing through a coil of wire would act like a magnet. Moreover, if a piece of soft iron was placed inside the coil, it would result in a much stronger electromagnet. Sturgeon demonstrated his new invention in 1825 by lifting nine pounds with the current from a single battery [31].

Reflecting on Sturgeon's electromagnet and Faraday's demonstration of induction between two wires, Irish scientist **Nicholas Callan (1799–1864)** realized in 1836 that if he wound a few turns of thick wire around a cylindrical form such as a cardboard tube, then wound many turns of a thinner wire on top of the first wire, and finally connected a battery to the thicker coil, a voltage would be induced into the second coil when the battery connection was switched on and off. He also believed that the larger number of turns on the second coil, the greater the voltage in the second coil. A surmise he verified experimentally [32]. Callan's invention is known as an "induction coil," which has many uses including the "ignition" coil in an automobile.

Next, through a brilliant set of deductions rivaling those of Sir Isaac Newton (see chapter 2), **James Clerk Maxwell (1831–1879)**, perhaps the nineteenth century's most influential scientist, organized all of the previous experimental data and formulae associated with electricity and magnetism—especially those of Faraday and Coulomb—into an integrated set of equations that bear his name [33]. These equations demonstrated that electricity, magnetism, and even light are manifestations of the same phenomenon. Thus, all the "laws of electricity and magnetism" are merely special cases of Maxwell's equations.

Maxwell's equations are a set of second order differential equations[4], a subject beyond the scope of this book except to point out that solutions to equations of this type often represent waves. The characteristics of a typical wave are shown in figure 1-5.

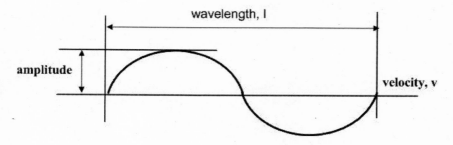

Figure 1-5. Typical E/M wave parameters and shape

Of particular interest is the fact that the frequency of an E/M is directly proportional to the velocity of the wave and inversely proportional to the wavelength of the wave. For an E/M wave, the velocity is the velocity of light, c, which is constant. Thus, the shorter the wavelength of an E/M wave, the higher the frequency.

In the case of the solutions of Maxwell's equations, the waves are a mixture of electric and magnetic fields. In addition, Maxwell was able to use the solutions to calculate the velocity of electromagnetic radiation, and it was the speed of light[5] as mentioned above. Because of this result, Maxwell commented,

"The agreement of the results [solutions of his equations yield a wave traveling at the speed of light] seems to show that light and magnetism are affections of the same substance, and that light is an electromagnetic disturbance propagated through the field according to electromagnetic laws" [33].

During the Middle Ages, it was believed that we could "see" an object because something was projected from the eye [34]. We now understand that we actually see because light is reflected from an object and into our eye, and is the basis of many experiments.

Heinrich Hertz (1857–1894) was a German physicist who built upon Maxwell's equations and Faraday's work on induction and was the first to experimentally demonstrate the existence of Maxwell's predicted electromagnetic (E/M) waves [35]. To generate E/M waves, Hertz employed the apparatus shown in figure 1-6.

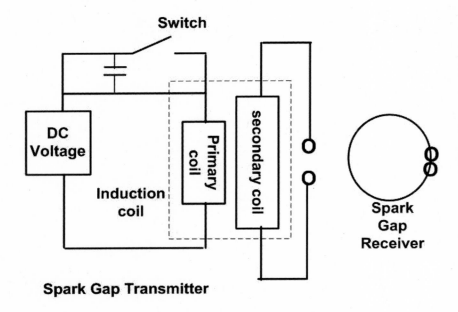

Figure 1-6 Demonstration of E/M wave propagation

The figure shows that Hertz's basic equipment was an induction coil configured as a "spark-gap transmitter" and a receiving coil configured as a "spark gap receiver." The addition of the capacitor produces an oscillating current when the switch contacts are closed (for information regarding this type of circuit see [36]). When the contacts are closed, an oscillating high voltage is induced in the secondary coil, which then "radiates" away from the spark gap transmitter and induces an oscillating current in the receiver, which can be detected by sparks across the spark gap [36].

Hertz's experiment conclusively demonstrated the existence of the electromagnetic waves predicted by Maxwell. In addition to confirming Maxwell's predictions, Hertz's efforts led to the modern radio. And, as will be explained below, electromagnetic radiation also plays an important part in the establishment of a correct explanation of solid matter.

Otto von Guericke (1602–1686) is another key player in the advancement of knowledge of electricity and magnetism. Guericke had many interests, and besides inventing the first electrostatic generator, in 1650 Guericke invented a vacuum pump consisting of a piston and an air gun cylinder with two-way flaps designed to pull air out of whatever vessel it was connected to. This device became valuable because it enabled the fabrication of evacuated glass jars [12]. Guericke's pump enabled others such as Heinrich Geissler to develop evacuated glass spheres and cylinders. Evacuated glass spheres and cylinders can conduct electric currents which ultimately led to the discovery of the electron

Heinrich Geissler (1814–1879) is known for using induction coils to study gases in vacuum tubes. Geissler had two skills that enabled him to invent an important piece of experimental apparatus now known as the Geissler tube: the ability to blow glass and a knowledge of physics. The Geissler tube is fabricated from blown glass and has a port that allows the withdrawal of air and insertion of gases. The tube also has two electrodes to which an induction coil is connected. The high voltage generated by the induction coil causes gases to glow, a principle used, for example, in neon display tubes [37].

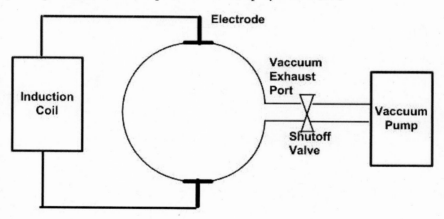

Figure 1-7. Sketch of Geissler tube

Of particular interest is the fact that the light created by the discharge in the tube has a characteristic color, a very important property, as discussed below.

Sir William Crookes (1832–1919) expanded on the concept of the Geissler tube by creating an evacuated glass cone containing three electrodes. This creation is called the Crookes tube (see figure 1-8).

anode

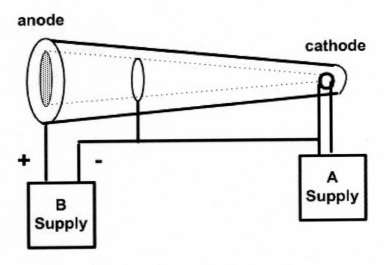

Figure 1-8. Sketch of Crookes tube

One electrode called the "anode" is connected to the wide end of the tube, and a second electrode called the "cathode" is connected at the narrow end. The third electrode, a thin piece of sheet metal, is placed in the middle. A low voltage battery (A) is connected to the cathode, and a high voltage battery (B) is connected to the anode [38].

Crookes's principal objective was the investigation of phosphorescent effects that had been observed in Geissler tubes. Phosphorescent material applied to the anode end of the tube would glow when voltages were applied as shown in figure 1-8. However, the middle electrode cast a shadow, as shown by the gray oval at the end of the tube. Crookes explained this by positing the existence of rays of particles that were emitted by the cathode. He called the rays "cathode rays" and said they were striking the phosphorescent material and causing the glow [37].

In 1891, prescient physicist George Johnstone Stoney (1826–1911) proposed the term "electron" as the fundamental unit of electric charge [39]. As we shall see in chapter 3, the "rays" Crookes observed were actually electrons.

SUMMARY OF CHAPTER 1.

If we include the Greeks, twenty-four investigators of electromagnetic phenomenon labored over a period in excess of 2000 years, trying to explain electromagnetic phenomena. By 1897, these efforts had yielded a collection of many facts:

- Electricity comes in two types of charges: positive and negative. Two objects of the same charge repel each other, but a positively charged object attracts a negatively charged object.

- Electricity can have two forms: static and flowing.

- Electrically charged objects generate a force on objects, which acts at a distance (i.e., the effect of a charged object is observed when the charged object is only brought near another object). The "action at a distance" effect is explained by an electric "field" created by the charged object.

- William Gilbert confirmed magnetism comes in one type, but each magnet has two so-called "poles," a north (N) and south (S).

- N poles repel each other, S poles repel each other, and N and S poles attract each other.

- Magnets generate a force on each other which "acts at a distance" (i.e., the effect of a magnet is observed when the magnet is bought near another magnet). As with electricity, the "action at a distance" effect is explained by a magnetic "field" created by the magnet.

- Von Guericke invented first electrostatic generator.

- Pieter van Musschenbroek invented the Leyden jar, a container for static electricity.

- Canton and Bennet invented devices to detect electrostatic effects.

- Volta invented the Voltaic cell, a continuous source of electricity.

- Ben Franklin invented the lightning rod, which was initially opposed by some religious conservatives.

- Michael Faraday demonstrated that:

  - Electricity and magnetism interact.

- A current-carrying conductor generates a magnetic field perpendicular to the conductor with N and S poles dependent upon direction of flow.

- A force perpendicular to a current-carrying wire is produced when the wire is placed in a magnetic field.

- If two conductors are placed parallel and a current is varying in one conductor, a varying magnetic field is created which induces a current in the other.

- Maxwell developed set of equations that predicted E/M waves and E/M wave velocity.

- Hertz's experiments confirmed the existence of E/M waves.

- Geissler invented electrostatic discharge globe.

- Crookes extended the Geissler tube to a more practical tubular device.

All of this, of course, was extremely useful. However, to this point no one knew *what* an electric charge is or *how* and *why* it produced the effects it did. Similarly, no one knew *what* magnetism was or *how* and *why* it produced the effect it did.

LOOKING AHEAD.

Having followed the path of electricity and magnetism through to 1897, we now return to the beginning of "solid" matter investigations.

CHAPTER 2:
"SOLID" EARTH INVESTIGATIONS
PRIOR TO 1897

CREATING THE ILLUSION.

To early observers and even casual observers today, the earth and all material objects such as animals, plants, rocks, etc., appear to be solid. The "solid" earth illusion is created by the inability of the human eye to "see" atoms, which on average are about 10^{-8} inches, or about 0.000000001 inches in diameter [1].

One of the earliest books to provide an explanation of ourselves and our surroundings is the Bible. However, regarding the "solid" earth illusion and whether atoms exist, the Bible only offers these few pieces of information in the book of Genesis (see chapter 5 for more regarding Genesis):

- "In the beginning God created the heaven and the earth (Gen 1:1)"

- "And God said, Let the waters under the heaven be gathered together unto one place, and let the dry land appear: and it was so. And God called the dry land Earth; and the gathering together of the waters called the Seas: and God saw that it was good (Gen 1:9–10)."

Based upon those few lines, it is reasonable to believe that the authors of Genesis did not realize that apparently solid "dry land" consisted of atoms; hence Genesis provides no guidance in this area.

EARLIEST OBSERVATIONS OF SOLID MATTER.

Because atoms are invisible, it is reasonable to ask, "What led anyone to suspect that atoms exist?" The answer to this question has at least two parts:

1. Greek curiosity

2. Alchemists' desire to turn baser metals into gold

An explanation of the underlying structure of matter began with the study of the observable properties of matter, which were undoubtedly known since ancient times. Using their unaided eyes, the earliest observers noted matter seemed to exist in four basic forms:

1. Liquid material (mainly water)

2. Gaseous material (mainly the atmosphere)

3. Fire (caused by lightning before humans developed methods of creating fire)

4. Solid material (mainly the earth)

However, as this table 2-1 indicates, several elements must have also been known to the ancients.

Element	Earliest Use
Copper	Ca. 9000 BCE
Gold	Before 6000 BCE
Lead	7000 BCE
Silver	Before 5000 BCE
Iron	Before 5000 BCE
Carbon	3750 BCE

Table 2-1. Some of the oldest known elements

GREEK EXPLANATION OF MATTER.

Some of the early Greek philosophers were led to suspect the existence of atoms when they observed the disappearance of water when it evaporated[1]. They asked the obvious question: "Where did the water go?" Others observed that water could squeeze through tiny openings. In fifth century BCE, these and presumably other observations led the Greek philosopher Leucippus to formulate the concept that matter is composed of an infinite number of "atoms" [2, pg 65. The word "atom" is derived from the Greek word for "indivisible." Very little of Leucippus's writings have survived, so most of our knowledge of his work is found through his student Democritus (460–370 BCE). Both felt that atoms were physically but not geometrically indivisible and so small as to be invisible. They also believed there was empty space between the atoms [2], pg 65.

On the other hand, other Greek philosophers did not agree with Leucippus and Democritus. Instead they believed in the four manifestations of matter: water, air, fire, and earth. However, they could not agree on which of the four manifestations was the most basic. Greek philosopher Thales (625–546 BCE) stated that water was the most basic constituent of matter [3]. On the other hand, another Greek philosopher, Anaximenes (ca. 650–528 BCE), thought that air was the primitive element [4], and yet another Greek philosopher, Heraclitus (535–475 BCE), preferred fire [5].

Finally, a fourth Greek philosopher, Empedocles (ca. 492–432 BCE), ended the squabble by agreeing with everyone and proposing matter was composed of all four of the classic "elements" [6].

Thus the Greeks were able to answer the question, "*What* is matter?" However, they were unable to answer the question, "*How* is matter constructed from these basic elements?" Moreover, it is not clear why they did not include some of the elements from table 1-1 in their list. The concept that the basic elements were water, air, fire, and earth prevailed. The atomistic view of Democritus posed even more difficulties to explain, especially because atoms were deemed invisible and thus their existence was difficult to prove at the time because it violated the common sense impediment (recall that the common sense impediment introduced above causes people to reject a correct explanation since it violates "common sense").

Thus, water, air, fire, and earth became the accepted explanation for the constituents of matter for centuries. The concept acquired considerable authority until the Arabic alchemists began their search for the magical philosopher's stone, which was believed to have the power to convert baser metals such as lead into gold [7]

ALCHEMISTS DISCOVER THE FIRST "NEW" ELEMENTS.

The origins of Western alchemy are traceable back to ancient Egypt. It is a part of the occult tradition [7]. The term "occult" refers to a number of magical organizations and the teachings and practices as taught by them [8].

In the eighth century CE, an Arabian alchemist whose Latinized name is Geber analyzed the classic elements, which included the four Greek elements, an Indian element called aether, sulphur, mercury, and salt [9]. Geber's studies of the qualities of an element, such as its hotness or coldness, led him to believe that one could transmute one element into another by rearranging its basic qualities. These changes would be mediated presumably by the legendary philosopher's stone [9].

Supposedly, belief in the existence of the philosopher's stone developed along with other magical formulas and instruments, aided and abetted by the knowledge that lead and gold have similar weights. Therefore, they believed there ought to be some way to convert lead into gold. Of course, the philosopher's stone was never shown to exist, but circa 800 CE, Geber's investigations into various chemical combinations led to the discovery of three new "real" elements: antimony, arsenic, and bismuth. Not only that, but "the practical aspect of alchemy generated the basics of modern inorganic chemistry, namely concerning procedures, equipment and the identification and use of many current substances

HISTORICAL DISCOVERY OF THE ELEMENTS.

A delightful article in the online Wikipedia encyclopedia provides a time line for the discovery of the elements [10], which can be divided into four periods as exhibited in this table

Time Period of Discovery	Number of Elements Discovered
Known since antiquity (see table 2-1)	9
Discovered between 1250 CE and 1741 CE	6
Discovered between 1741 and 1900	72
Discovered between 1900 and 1925	5

Table 2-2. Time periods when elements were discovered

As this information shows, the preponderance of discoveries occurred in the relatively short time between 1741 and 1900. Table 2-3 provides a list of the years in which many of the elements of interest to this book were identified.

Element	Symbol	Year identified
Carbon	*C*	*Antiquity*
Argon	Ar	1895
Calcium	Ca	1808
Hydrogen	*H*	*1766*
Lead	Pb	Antiquity
Nitrogen	N	1772
Oxygen	***O***	***1774***
Phosphorus	Ph	1669
Potassium	K	1807
Radium	Ra	1898
Rubidium	Rb	1861
Sodium	Ma	1807
Strontium	Sr	1787
Uranium	U	1789

Table 2-3. Elements of particular interest to this book

Those elements of particular importance, such as carbon, hydrogen, and oxygen—the basic building blocks of biological matter—are highlighted in italics. Others are listed as they are central to such important functions as radioactive dating.

From a biological viewpoint, carbon is the most important element because its structure allows the construction of extremely complex biological molecules. Compounds that are formed with carbon are called "organic," and the importance of carbon is attested by a branch of chemistry, organic chemistry, which is devoted exclusively to carbon compounds. Carbon's unique properties will be discussed further below.

This description of the elements provides enough information to understand how they were discovered and what they do, but not enough to understand how they were originally created. The answer to that question will be presented in chapter 8.

Empirical Investigations of Solid Matter before 1897.

The modern study of the behavior of matter is the province of chemistry which is defined as "The science of the composition, structure, properties, and reactions of matter, especially of atomic and molecular systems" [11]. As mentioned previously, chemistry was preceded by alchemy, which had been initiated by the followers of Islam. While some of the alchemists' goals were outlandish, early alchemists like Geber succeeded in doing some useful work, and their efforts were the main source of knowledge pertaining to matter until about the 1600s.

Interestingly, much of the work now performed by chemists was conducted by physicians during the 1600s as they sought new substances to cure their patients. One physician in particular, Irishman **Robert Boyle (1627–1691)**, became one of the important representatives of the transition from alchemy to chemistry [12]. Boyle expanded the efforts of early alchemist/chemists van Helmont and Paracelsus. Van Helmont was a Flemish chemist/alchemist and is considered the "father" of pneumatic chemistry [13]. Van Helmont, who coined the term "gas," was also a bit of a mystic and a follower of another alchemist/ chemist named Paracelsus, a.k.a. Philippus Aureolus (1493–1541). Paracelsus made great contributions to the development of practical medicines and eschewed the barbaric practice of bloodletting. However, his fascinations with mysticism blunted the overall effectiveness of his work[2] [13].

Building upon these predecessors, Boyle deduced an empirical relationship between the pressure, volume, and temperature of a gas, which is now known as Boyle's law. It states, "For a fixed amount of an ideal gas kept at a fixed temperature, P [pressure] and V [volume] are inversely proportional (while one increases, the other decreases)" [12]. Thus if one decreases the volume of a gas, one also increases the pressure (as anyone who has ever used a bicycle pump knows).

He also proposed the first modern definition of an element: "I now mean by an element, certain primitive and simple if perfectly unmingled bodies, or of one another, immediately compounded, and into which they are ultimately resolved" [14] pg 208.

1. Boyle summed up his work in the treatise *The Skeptical Chymist*, which was published in 1661. Boyle employed the dialogue form (perhaps imitating Plato) and postulated that matter consists of atoms and clusters of atoms in motion. His treatise essentially destroyed medieval alchemy [8, p. 209]. However, there are two important points to remember regarding Boyle's contributions.. He had help. He didn't work out the laws of gases by himself.

2. His formulations were arrived at empirically through many observations.

Antoine Lavoisier (1734-1794) advanced the field of chemistry even further. Before Lavoisier, the prevailing view held that burning was caused by the release of a gas called phlogiston. But he conducted a brilliant series of experiments that conclusively demonstrated that fire was a process whereby an atmospheric gas, which he named oxygen (derived from the Greek word for "acid former"), combined with a gas released by a burning substance [9]. In his 1783 publication, *Reflexions sur le Phlogistique* (Reflections on Phlogiston), Lavoisier argued that phlogiston was an inconsistent explanation for combustion [15].

Lavoisier also demonstrated that rust was caused by the combination of oxygen with iron. Thus Lavoisier was one of the first persons to demonstrate that elementary forms of matter combine to form compounds. These and other of Lavoisier's experiments represented the transition from alchemy to chemistry, which was capped by his publication of *Traité Élémentaire de Chimie* (Elementary Treatise of Chemistry) in 1789, which earned Lavoisier the title of "Father of Chemistry" [15].

Just as Boyle's *Skeptical Chymist* signaled the end of alchemy, Lavoisier's *Elementary Treatise on Chemistry* created modern chemical language [14, p. 209]. It should be noted that, as with Boyle, Lavoisier arrived at his conclusions empirically via many painstaking experiments.

Unfortunately, Lavoisier's brilliance did not extend to his political affiliations. He sided with Louis VI and joined him at the guillotine during the French Revolution.

Another notable advancement in the 1700s was the confirmation of the atomic hypothesis. While Democritus had proposed the atomic theory for the structure of matter in the fourth century BCE, this hypothesis was controversial until British school teacher **John Dalton (1766–1844)** [16] made a number of astute observations that confirmed it. Dalton had access to the several newly discovered elements mentioned above, many of which had been revealed by examining compounds formed by the elements. In his Papers and Extracts [14, p. 277], Dalton states:

> "There are three distinctions in the kinds of bodies, or states which have more especially claimed attention of philosophical chemists; namely, those which are marked by the terms elastic fluids [gases], liquids, and solids ...
>
> These observations have tacitly led to the conclusion, which seems universally adopted, that all bodies of sensible magnitude, whether liquid of solid are constituted of a vast number of extremely small particles or atoms, of matter bound together according to circumstances, and which as it endeavors to prevent their separation ...
>
> Whether the ultimate particles of a body, such as water, are all alike, that is of the same figure, weight, etc., is a question of some importance....

Therefore we may conclude that the ultimate particles of all homogeneous bodies are perfectly alike in weight, figure, etc. In other words, every particle of water is like every other particle of water.... Chemical analysis and synthesis go no further than to the separation of particles from one another ... we might as well attempt to introduce a new planet into the solar system as to destroy one atom of hydrogen. [Dalton was, of course, unaware of the fusion process, which was discovered over a century after his death, and in which hydrogen is converted into helium as is described below] ...

In all chemical investigations, it has justly been considered an important object to ascertain the relative weights of the samples which constitute a compound... Now it is one great object of this work, to show the importance and advantage of ascertaining the relative weights of the ultimate particles, both of simple and compound bodies, the number of simple elementary particles which constitute one compound particle, and the number of less compound particles which enter into the formation of one more compound particles." [14, p. 278]

The last sentence is generally known as Dalton's law of definite proportions, which essentially states that atoms combine to form compounds in definite (repeatable) proportions. Thus, water is always a combination of two atoms of hydrogen and one atom of oxygen, a fact demonstrated by the formula H_2O.

The exact timing of Dalton's discoveries is a bit uncertain, but Dalton read a paper in November 1802 in which the law of multiple proportions appears to be anticipated in the words: "The elements of oxygen may combine with a certain portion of nitrous gas or with twice that portion, but with no intermediate quantity" [16]

As with previous explanation developments, it should be noted that Dalton's law is an empirical formula deduced from a large number of experimental observations rather than from a more fundamental explanation. Dalton hadn't a clue *why* elements behaved as they did, which as mentioned is a fundamental problem with empirical explanations: they merely answer the question *what* and not *how* or *why*.

As previously mentioned, chemists in the 1800s discovered and examined the properties of about half of the presently known elements. By 1869, sixty-one elements had been identified and their properties sufficiently investigated enough for Russian scientist **Dimitri Mendeleev (1834–1907)** and **Julius Lother Meyer (1830–1895)** of Germany to notice a "periodic" relationship between the elements. Based on these relationships, they independently proposed the periodic law of the elements which states:

"The properties of an element depend upon the weight of the element."

Of particular significance, they observed that similar properties appeared at regular periodic intervals in a list of the elements arranged by weight. This led them to an ingenious arrangement of the elements in a periodic "table," first published in 1869 [17].

You have probably have seen a periodic table, but a simple example of the periodicity of elements is provided in the small portion of the periodic table shown in table 2-3.

First Column		Next to last Column
Lithium		*Fluorine*
Sodium		Chlorine
Potassium		Bromine
Rubidium		Iodine

Table 2-4. Sample from periodic table of the elements

Table 2-4 shows the elements lithium, sodium, potassium, and rubidium, which are in the first column of the table. These elements form strong compounds with fluorine, chlorine, bromine, and iodine, which are in the next to last column of the table [1], pg 79. Fluorine is italicized because it was not discovered until 1886. However, the existence of fluorine was predicted by noting that a lighter element with similar properties to the other elements in the next to last column should exist. The remarkable achievement of the periodic table is that it allowed prediction of the properties of the missing elements, which greatly assisted in finding them. Again, it must be emphasized that the periodic table was empirically derived. No one at the time had the foggiest notion *why* elements behaved in a periodic manner.

In addition to the investigations of Boyle, Dalton, Mendeleev, and others of similar inclination, investigations into the nature of matter proceeded along another path in the mid to late 1800s—a path that was initiated by Newton through his interest in the behavior of light. As will be seen, light and the structure of matter are intimately connected.

Everyone has seen a rainbow or looked at white light spreading into the colors of the rainbow when it passes through a glass prism. The separation of light by a prism into various colors is termed the "spectrum of white light" and is due to the fact that light consists of many waves and each color has a unique wavelength.

The fact that light consists of waves was conclusively demonstrated in the early 1800s by **Thomas Young (1773–1829)** via an ingenious "double-slit" experiment [19]. To appreciate Young's experiment, the ability of waves to interfere (i.e., to either cancel or reinforce each other) must be understood. The two sketches in figure 2-1 illustrate what happens when light waves cancel and when waves reinforce.

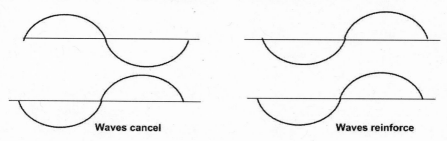

Waves cancel   **Waves reinforce**

Figure 2-1. Wave cancellation and reinforcement

In the case of wave cancellation, no light is seen, and in the case of reinforcement, light is enhanced.

Young created a deceptively simple piece of apparatus: a cardboard sheet with two narrow slits cut into it, a source of light, and a screen. He placed a source of light on one side and a screen on the other. The apparatus and results are shown in figure 2-2.

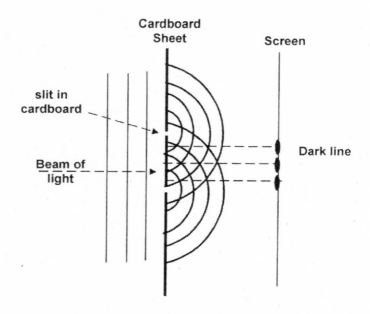

Figure 2-2. Young's double slit experiment

When a beam of light strikes two slits in the cardboard, Young reasoned each slit would act as an origin point, and if light was a wave, it would exit each slit in curved wave form. The waves from the two slits would overlap and interfere with each other, resulting in light (reinforced waves) and dark (canceled waves) bands on the screen. This is exactly what he observed.

Young extended his crude apparatus by forming hundreds of closely spaced lines on a glass plate, thereby creating a scientific instrument known as a "diffraction grating." The fine lines act as a "super prism" that can spread a beam of light much farther than a glass prism, revealing that light is not continuous as it appears in the spectrum produced by a glass prism, but that it actually consists of distinct lines.

The property of light spreading by a diffraction grating is particularly useful in the study of gaseous discharges such as those generated in a Geissler tube. As mentioned in the previous section, placing a gas in an evacuated "electric discharge" tube and passing an electric current through it will cause the gas to glow, and the glowing light can be studied with a diffraction grating. For example, if hydrogen gas is placed in a discharge tube and then light from the glowing hydrogen gas is passed through a diffraction grating, a pattern—part of the Hydrogen spectrum that falls in the visible spectrum, 400 nm to 700 nm—will be obtained (see figure 2-3).

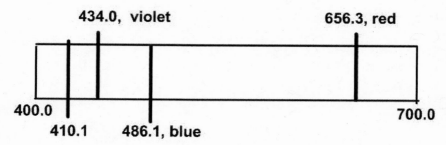

Figure 2-3. Hydrogen spectrum superimposed on visible spectrum, wavelength in nano-meters

It should be noted 1 nm = 10^{-9} meters, which is a very small number. Hence, the wavelength of visible light is very small. It should also be noted that spacing between spectral lines diminishes as the wavelength approaches the blue end of the spectrum. The significance of this will be explained shortly.

As explained in [20], the wavelengths of E/M energy span a much greater range than visible light, as shown in the simple pictorial in figure 2-4.

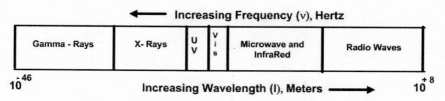

Figure 2-4. Simplified depiction of the electromagnetic spectrum

This figure demonstrates that increasing frequency f is opposite to increasing wavelength l

Noting that the spectrum lines tend to become closer together at shorter wavelengths l, or increasing frequency f, and it had occurred to some, that a formula of the form 1/f might fit the spectral lines since, as **f** increases, 1/f decreases. A number of investigators attempted to apply this reasoning to develop a formula for the patterns, but it was not until 1885 that **Johann Jakob Balmer (1825–1898)** succeeded. He successfully demonstrated that the spectrum of hydrogen could be expressed as a simple empirical formula of the form: $f = R^*(1/ n_1^2 - 1/ n_2^2)$.

Where n_1 and n_2 are constants, f is the frequency and R is an experimental constant employed to make values balance[21], R was named after **Johannes Rydberg (1854–1919)**, who developed the most general series formula in 1896 [22]. Please note that it is not important to understand this equation. Just remember its form because it will appear again.

Balmer achieved the best agreement with the experimental lines by allowing $n_1 = 2$ and letting $n_2 = 3, 4, 5, 6 \ldots$ The experimental value of R is equal to 1.097373×10^7. Keep this number in mind also. You will see it again later.

SUMMARY OF CHAPTER 2.

As with the E/M investigations, if we include the Greeks, thirteen investigators into the mystery of apparently "solid" matter labored over a period in excess of 2000 years, trying to explain matter. By 1897, these efforts had yielded a collection of many facts:

- Democritus's belief in the existence of the atom was confirmed by the discovery of elements.

- By 1897, most of the elements (eighty-seven out of ninety-two naturally occurring elements) were identified.

- Robert Boyle, a physician, showed medieval alchemy's attempts to turn led into gold were impossible.

- Lavoisier created modern chemistry.

- In 1802, Dalton established the law of proportions, which accounted for the configuration of molecules (e.g., a water molecule is formed from two hydrogen atoms and one oxygen atom).

- In 1869, Mendeleev showed elements with similar properties can be arranged in a periodic table.

- Thomas Young demonstrated the phenomenon of light diffraction, which led to the diffraction grating, a basic tool for examining atomic spectra.

- Experiments with gases in electrostatic discharge bottles and diffraction gratings demonstrated the existence of unique patterns in the light emitted from the gas atoms—the atoms spectrum or "fingerprint."

- In 1885, Balmer succeeded in deriving a formula for the spectrum of the simplest atom—hydrogen.

But, as with the E&M investigations, no one knew *how* or *why* something like Dalton's law occurred or *how* the spectral patterns were generated. And perhaps most importantly, no one realized that E&M phenomenon and "solid" matter were related.

LOOKING AHEAD.

In a series of groundbreaking experiments, investigators finally discover that E&M and "solid" matter are related.

CHAPTER 3:
EMPIRICAL RESOLUTION OF THE "SOLID" EARTH ILLUSION BETWEEN 1897 AND 1911

The last two chapters have demonstrated that by 1897 a large amount of empirical data pertaining to electricity, magnetism, and the structure of "solid" matter had been accumulated. However, these findings raised as many questions as they answered, and no one realized these investigative paths were related. The connection and some clarification of electricity were provided by three definitive experiments using a modified Crookes tube performed in 1897 by Sir J. J. Thompson [Ch1, 37].

This chapter will explain the investigations of "solid" matter, including E&M phenomena between 1897 and 1911. This time period was chosen because most of the investigation between 1897 and 1911 were essentially empirical and didn't effectively answer the questions *how* and *why*.

THOMPSON DEMONSTRATES E&M PHENOMENON ARE A MANIFESTATION OF MATTER.

In his first experiment, **Sir J.J. Thompson (1856–1940)** examined the effect of a magnetic field on the apparently negatively charged cathode rays produced in Crookes tube. To conduct this experiment, he constructed a Crookes tube with two "detector" electrodes in the anode end. The two electrodes were in turn connected to an electrometer, an instrument that can measure small quantities of charge. When Thompson connected the voltages to the tube, the electrometer detected electric charge. However, when two magnets were placed against the tube to create a magnetic field, no electric charge was detected, therefore proving cathode rays are deflected by a magnetic field [1].

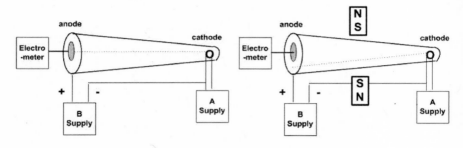

Figure 3-1. Demonstration that cathode rays are deflected by a magnetic field

For his second experiment, Thompson employed a Crookes tube with phosphorescent material on the anode end. When voltage was applied, the cathode rays struck the center of the anode end, but when an electric field was applied using two plates connected to a voltaic cell, the rays were deflected toward the positive plate, conclusively demonstrating that the cathode rays were negatively charged [1].

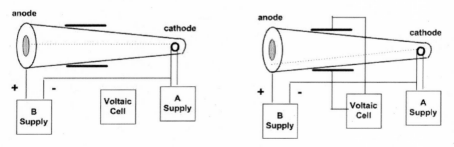

Figure 3-2. Demonstration that cathode rays are deflected by electric field

In his last experiment, Thompson constructed a tube in which he could measure the ratio between the amount of electric charge carried by a cathode ray and the mass of the cathode ray [2]. Figure 3-3 is a simple sketch of the apparatus.

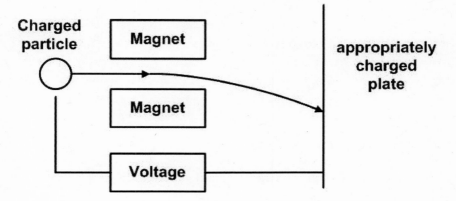

Figure 3-3. Measurement of charge-to-mass ratio of cathode rays

If the voltage and magnets are properly adjusted, deflection will be proportional to the ratio between the amount of charge and the mass of the charged particle, which allows the ratio to be measured.

After many careful measurements, Thompson concluded that the charge-to-mass ratio of the deflected cathode rays was much greater than the charge-to-mass ratio of the hydrogen ion. Therefore either a cathode ray carried a charge much greater than the hydrogen ion, or a cathode ray must be much lighter.

Summarizing his work, Thompson concluded that the cathode rays must not only be coming from the direction of the cathode but also from the material of the cathode in the Crookes tube. Moreover, the deflection experiments and charge-to-mass measurements convinced him the cathode rays were the electrons (a theory previously predicted by George Stoney [Ch1, 38]). Sir J. J. Thompson had discovered the electron!

THOMPSON'S ATOM MODEL.

The discovery of the negative electron created quite a problem for Thompson. Normal matter is obviously electrically neutral, thus while Thompson couldn't see an atom, he was had to conclude that the atom must contain sufficient positive charge to balance the negative electrons. Still believing atoms were indivisible, Thompson was forced to invoke Sherlock's theorem and postulate a sort of a "raisin muffin" model of the atom, with negatively charged electrons interspersed in the positively charged mass of the atom.

It is important to note that Thompson did not know what this positive mass was; he merely knew it had to exist in order to balance the negative electron charge. His concept of a raisin muffin in which the positive mass was a contiguous structure was reasonable at the time. Although a modest advance, Thompson's model was the first revision to atomic structure in 2200 years.

THOMPSON'S DISCOVERY EXPLAINS ELECTRICITY.

The discovery of the electron explained some of the mysteries of electricity and magnetism, in particular the fact that electricity and magnetism are not separate items but rather merely manifestations of effects produced by the electron. The basic unanswered questions regarding E&M were

1. What is an electric charge?
2. How is static electricity and electric current produced?
3. What is magnetism?
4. How and why is its effects produced?

Thompson's electron discovery answered questions 1 and 2: There is only one "type" of electric charge—the negative electron. Negative static electricity is produced by a surplus of electrons while positive static electricity is produced by a deficit of electrons. Electric current is produced by a flow of negative electrons from a negative electrode to a positive electrode.

But in this case, as in others that came before him, Thompson could not answer *how* and *why* the effects were produced. This will be explained below.

As for the questions regarding magnetism, magnetism is produced by moving electrons, which explains magnetic effects caused by electric charge flowing through a wire, but the existence of the electron could not fully answer questions 3 and 4.

THE SERENDIPITOUS DISCOVERY OF X-RAYS.

Other persons besides Thompson were experimenting with electric currents in vacuum tubes. In 1895, German physicist **Wilhelm Conrad Röntgen (1845–1923)** experimented with various types of electric discharge tubes. He performed some experiments with tubes in which the accelerating voltage was very high. Röntgen had intended to place a fluorescent material, barium platinocyanide, inside a tube to examine the effects of electrons on fluorescent material. In order to preclude anything emanating from the discharge tube from producing false data, he covered the tube with cardboard.

However, before he could perform the desired experiment, he observed that a piece cardboard he coated with the fluorescent material and laid a few feet from the discharge tube in preparation for his next experiment was fluorescing. After examining his apparatus, Röntgen was forced by Sherlock's theorem to conclude that something that could pass through the glass and the cardboard was emanating from the electric discharge tube and causing the external fluorescent-coated cardboard to fluoresce. Not knowing what the mysterious radiation was, he initially termed them "X-rays" [3].

Röntgen expanded his experiments and determined the X-rays would pass through most material except lead. The fluorescent-coated cardboard allowed him the take "pictures" of items like his hand. With that, the diagnostic benefits of X-rays had been launched [3].

The establishment of the proper explanation of X-rays, that they are very high frequency E/M radiation, required many years and a number of experiments. But the fact that X-rays could ionize gases, meaning X-rays exhibit particle-like behavior, led one of the major X-ray investigators, **Sir William Bragg (1862–1942)** to suggest X-rays are not E/M waves [4], which was not accurate. This is a perfect example of someone drawing incorrect conclusions because of the lack of a key piece of information. As will be discussed later, Einstein's explanation of the photoelectric effect had already demonstrated that E/M energy can also act as a particle. Perhaps Bragg was not aware of this fact because Einstein wrote it in German. The issue was settled in 1922 by **Arthur Compton (1892–1962),** who demonstrated the scattering of electrons by light, which can only happen if light can behave as a particle [5].

Max von Laue (1879–1960), a German physicist, was the 1914 recipient of the Nobel Prize in physics for his discovery of X-ray diffraction in crystals [6]. Von Laue's discovery has had far-reaching consequences because most of our knowledge of the structure of molecules, the DNA molecule in particular, has been obtained via X-ray diffraction.

Von Laue was aided in his discovery by a serendipitous chance encounter with crystallographer Paul Peter Ewald during a Christmas recess walk. At the time, Ewald was exploring the diffraction of light by crystals. Realizing light waves were too long to have much effect, von Laue wondered if the shorter wavelengths of X-rays would be more effective and devised the apparatus demonstrated as a very general sketch in figure 3-4.

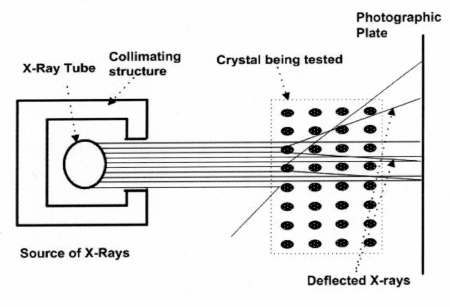

Figure 3-4. Pictorial of X-ray diffraction by crystals

Since X-rays tend to go in all directions, some means of forcing them into a beam (or collimating them) had to be provided. The regular structure of the crystal acts like a diffraction grating, and each atom in the crystal causes the X-rays to be deflected [7], [8]. If the deflected X-rays are captured on a photographic plate, it is possible to "backtrack" to find where the X-rays came from and thus deduce the shape of the crystal. So Von Laue's hunch paid off, and he officially discovered X-ray diffraction in 1912. X-ray diffraction soon became an indispensable tool for examining the atomic structure of crystals, especially molecules such as DNA, as will be explained in chapter 12.

In 1897, two years after Röntgen's serendipitous discovery of X-rays, French physicist and Nobel laureate **Antoine Henri Becquerel (1852–1908)** discovered the spontaneous emission of "rays" by uranium, a phenomenon he termed "radioactivity" [9].

Becquerel's discovery was also serendipitous. To protect photographic plates prior to an experiment, Becquerel unwittingly wrapped the plates with a paper containing uranium salts. Imagine his surprise when before he was able to perform the experiment, he found that the photographic plates had been fully exposed.

After examining all possible sources of exposure, Becquerel was forced to employ Sherlock's theorem and conclude that something in the uranium-containing substance must have somehow exposed the plates. Besides demonstrating that the atom was probably the source of these "rays" and thus not "indivisible" as had been believed for thousands of years, the ability of the rays to penetrate matter suggested that there was indeed space between them as Democritus had believed.

Further investigation by Becquerel and others demonstrated that uranium emitted three types of rays—alpha, beta, and gamma—as well as other heavy elements such as radium, polonium, and thorium [9].

Becquerel, of course, had no idea what these particles and rays were, which is why he named them by the first three letters in the Greek alphabet[1]. As we will see later, the particles are associated with the decay of unstable nuclei (the central portion of an atom).

These three rays were identified by the effect upon their motion when they passed between electrically charged plates. The alpha particle is relatively heavy and attracted to a negatively charged plate; hence it carries a positive charge. The beta ray, or actually a beta particle, is relatively light and attracted to a positively charge plate; hence it is negatively charged. On the other hand, the gamma ray is unaffected by electrically charged plates. Gamma "rays" were eventually determined to be very short wavelengths, very high-energy E/M radiation.

DISCOVERY OF THE ATOMIC NUCLEUS.

In 1911, **Ernst Rutherford (1871–1937)**, an associate of Thompson, made an exhaustive study of the effects of the particles emitted by radioactive matter, with the aid of Ernest Masden and Hans Geiger [10]. They focused in particular on the effects of the relatively heavy positively charged alpha particle [11]. The apparatus they set up is shown in figure 3-5.

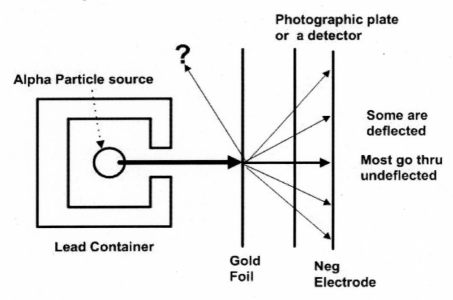

Figure 3-5. Rutherford's gold foil experiment

This apparatus allowed them to observe the interactions of the alpha particles with the gold foil and, in particular, the angle into which a particle was deflected after passing through the foil.

When Rutherford developed the films, he found that most of the alpha particles had passed through the "solid" gold foil, as expected. However, he was astonished by the fact that the alpha particles formed a pattern on the photographic film as if some of the particles had been deflected by a very heavy object in the foil; moreover, a few bounced back. It was "as if the alpha particle had struck a bowling ball," Rutherford later wrote [11].

From this experiment, he developed a mathematical formula for the deflection or "scattering" of the alpha particles. Employing the formula, Rutherford obtained a remarkable result: the alpha particles were deflected by a relatively massive, very small, positively charged object. Thus, Rutherford discovered that most of the matter in the gold foil was concentrated in small positively charge spheres. What appears to us as solid matter is mostly empty space. With this experiment, he resolved the apparently solid earth illusion.

Rutherford's deflection formula allowed him to deduce the approximate size of the central nucleus as .00000000001 inches in diameter. To appreciate this small size, imagine a one inch circle drawn around an atom. If the diameter of the nucleus is expanded to one inch, the edge of the circle would expand to over 15,000 miles away. It is no wonder that matter seems to be solid [11].

Reflecting on his findings, Rutherford concluded that Thompson's atom model was only partially true. The atom did consist of equal amounts of positive charge and negative electrons, but the positive charge was not dispersed as in a muffin, but rather it was concentrated in the center. Rutherford termed the concentrated charge in the center as the "nucleus." Note that Rutherford, like Thompson, did not know what the positive charge was, merely that it was concentrated at the center of the atom. He therefore had no reason to believe it was not a single positively charged object.

Having established that the atom has a positive nucleus surrounded by negative electrons, Rutherford was then faced with the problem of what to do with the electrons. Lacking any other information, Rutherford borrowed the solar system as a model and postulated that the electrons traveled around the central nucleus in orbits, a postulate that was later shown to be only partly true [11].

Although not a completely correct picture of the atom, Rutherford's atom confirmed Democritus's hypothesis formulated ca. 400 BCE, 2300 years before Rutherford.

SUMMARY OF CHAPTER 3.

At the close of the nineteenth century, considerable empirical data relative to the atom had accumulated:

- In 1897, Sir J.J. Thompson discovered the electron and deduced that atoms were like raisin muffins, with electrons mixed in with positive charges. It was the first new atomic concept in 2000 years.

- Thompson's discovery also resolved some of the E/M dilemmas:
  - There is only one type of electric charge.
  - Static electricity is created by adding or removing electrons.
  - Electric current is caused by flowing electrons.
- In 1895, Röntgen discovered X-rays, and Bragg demonstrated that X-rays are high energy E/M waves.
- In 1912, Von Laue discovered X-ray diffraction—the main tool for determining molecular configurations like DNA.
- In 1897, Becquerel discovered and identified three types of radioactivity:
  - "Heavy" positively charged alpha particles
  - Lightweight negative beta particles
  - High-energy gamma rays
- In 1911, Rutherford discovered the tiny positively charged atomic nucleus and proposed a "solar system" atom with negative electrons revolving around a positive nucleus.

While these were significant advances, the Rutherford-Thompson atom could not account for Dalton's law (*how* and *why* atoms form regular combinations), Mendeleev's table (*why* atoms have a periodic structure), or *why* the spectrum emitted by atoms consists of distinct spectral lines. There was also a gross problem with orbiting electrons.

In the realm of classical of E/M behavior, an electron traveling in a circle is constantly accelerating (i.e., it is constantly changing its velocity with respect to a tangent to the circle). A classical accelerating electron radiates and thus should lose energy by radiation and collapse into the nucleus. That it does not is obvious; thus the Rutherford-Thompson model was clearly incomplete. So many questions regarding electricity and magnetism remained, particularly the question, "What is magnetism?"

LOOKING AHEAD.

The next two chapters present the final resolution of the "solid" earth illusion. I have divided the "final resolution" into two chapters since, as will become clear, the final development of a proper explanation of the structure of matter actually occurred in two distinct phases: (1) the initial realization that the rules governing the atom are different than the rules governing the "real" world and (2) refinement of "quantum" concept into a complete picture.

CHAPTER 4:
FINAL RESOLUTION OF THE
"SOLID" EARTH ILLUSION, PART I

By the early 1900s it was clear that "solid" matter was anything but solid. Rutherford and Thompson had shown that "solid" matter is mostly made up of empty space populated by atoms containing a tiny nucleus surrounded by electrons. But the world of the atom is unlike anything we experience in everyday life, which is one of the reasons why the development of a proper explanation of matter was so difficult to attain.

The final resolution of the solid earth illusion occurred in two stages. This chapter will address the recognition that the atomic world is governed by new laws, the laws of quantum mechanics.

THE FIRST STEP TOWARD A CORRECT ATOMIC MODEL—MAX PLANCK'S QUANTUM LEAP.

The first step toward a true understanding of the microscopic world of the atom was a "quantum leap" taken by German mathematical physicist Max Planck (1858–1947). Plank, who is considered to be the founder of the quantum explanation of matter and one of the most important physicists of the twentieth century, was born in the German city of Kiel to Johann Julius Wilhelm Planck and his second wife, Emma Patzig. He was the sixth child in the family, though two of his siblings were from his father's first marriage.

Max Plank tackled one of the more perplexing problems of the day: the energy radiated from an object known as a blackbody. A blackbody is an object that is a perfect source of radiation. It can be closely approximated by employing a hollow ball heated to a desired temperature, with a small hole to allow radiation to escape. Of course, if the ball is heated enough, it doesn't appear to be black, but that is the term used. See [1] for more information regarding a blackbody.

The problem Planck sought to solve was the creation of a formula that described the radiation from a blackbody, a task that had frustrated the leading scientists of the day. Planck achieved success because he was apparently among the first persons to suspect that the microscopic world of the atom was not governed by classical physics (i.e., the visible world in which we live). In the visible world, kinetic energy—the energy acquired by a moving object—changes continuously and smoothly as the velocity of the object changes. However, to account for the behavior of things in the microscopic world, Plank hypothesized that energy does not change continuously but only in discrete amounts of energy termed a "quanta."

Thus, in order to express energy of blackbody radiation, Planck stated the energy must be E/M in nature and must be quantized. In other words, it must exist is small packets of energy (**E**) given by the formula: $E = hf$

where **h** is a constant (which is now known as Planck's constant) and **f** is the radiation frequency.

Planck's quantum assumption led him to deduce a formula that fit the blackbody curve perfectly. Plank published his findings in 1900 [2]. This was viewed as a monumental accomplishment, and Planck was awarded the Nobel Prize in Physics in 1918 in recognition.

EINSTEIN "BORROWS" PLANCK'S QUANTUM CONCEPT AND EXPLAINS THE PHOTOELECTRIC EFFECT

If you shine light on certain metals, such as selenium, electrons will be emitted. This phenomenon is called the photoelectric effect, and it is the process responsible for the ubiquitous photo cell. The photoelectric effect was first observed in 1887 by Heinrich Hertz (1857–1894) [3].

The electrons emitted when light strikes a metal like selenium have certain observable properties:

1. The *number* of electrons emitted by the metal depends on the *intensity* of the light beam applied on the metal. The more intense the beam, the higher the number of electrons emitted. This can be shown by measuring the current produced by a photo cell; however,

2. No electron is emitted until the light has reached a threshold frequency, independent of light intensity.

These observations baffled physicists for many decades because they cannot be explained if light is thought of only as a wave. Einstein solved this problem in 1905 by asserting another paradox of the microscopic world. As mentioned above, light sometimes behaves as a wave but other times as a particle. If one assumes light can sometimes behave as a particle, then individual particles of light (or "photons") could penetrate the metal and dislodge electrons from atoms. Einstein postulated that the energy of the photon is the same as that deduced by Max Plank, **E = hf** therefore unless the frequency **f** was high enough, the photon lacked sufficient energy to dislodge an electron.

In one stroke, Einstein showed that light behaves as a stream of particles, which provided solid evidence for the existence of quanta. His theory could satisfactorily explain all of the known properties of the photoelectric effect and was the first application of the new quantum theory [4]. The English title of Einstein's work is given in reference [5], pp. 132--148.

Today, Einstein is a household name and is known by many for his theories of relativity. However, ironically, Einstein was awarded the Nobel Prize for developing the explanation for the photo electric effect rather than for his more far-reaching contributions in relativity.

DEPICTING OBJECTS THAT CANNOT BE SEEN

In the electricity and magnetism section, we covered how detecting items that could not be "seen" was the central problem in establishing improved explanations for electricity and magnetism. This problem also encumbered investigations of solid matter. For example, the discoveries of Becquerel and the experiments of Rutherford involved indirect observations of the phenomena they were studying. Becquerel never actually "saw" the alpha and beta particles being emitted by the radioactive matter, and neither did Rutherford see the alpha particles in his experiments. Both observers merely detected the presence of the particles from the particles' ability to expose a photographic plate. From there the men were able to determine that the particles were charged or uncharged by observing the deflection caused by an electrically charged plate.

The studies of the electron were also indirect. Thompson had to create special electrode detectors and screens in order to detect the presence or absence of cathode rays. Thus, our knowledge of the atom is based fundamentally on a "model" of the atom. Beginning with the Greek model, each improvement in knowledge has involved modifications to the model to align it more closely with observations.

An interesting analogy to the study of the atom was presented in one of my college courses. The professor suggested that determining the structure of the atom was akin to sitting on a hillside outside an automobile factory and attempting to determine the process of automobile manufacturing by examining what entered and what came out of the factory. Of utmost importance is that knowing what actually happens inside the factory is irrelevant as long as our model of the factory process accurately accounts for the relationship between what goes into the factory and what comes out.

The same is true of the atom. As long as the model we employ explains all the observations, the model must be deemed correct. Of course, as we have seen, the model has continually been refined as more observations have been made. However—and this is extremely important—the revised models do not completely displace the old model; they merely refine it. For example, the positive and negative charges of Thompson's model did not disappear when Rutherford proposed a modified model. They were merely rearranged.

NIELS BOHR DEVELOPS FIRST QUANTUM ATOM MODEL

The next modification of the atom model was supplied by Danish scientist Niels Bohr (1885–1962), who applied the developing field of quantum mechanics to the atom. Also borrowing Max Planck's pioneering quantum concepts, in 1911 Bohr extended the Rutherford-Thompson model of the atom to a model that provided the first successful explanation of the myriad of observations associated with the atom, such as Dalton's law [6].

Bohr's model was based upon three ad-hoc postulates that apparently apply only in the microscopic world:

1. The electron moves in a circular orbit around the positively charged nucleus.

2. The energy of the electron (E_e) is quantized, given by formula: $E_e = nh$. In other words, it is equal to a multiple (**n**) of Planck's constant. This postulate implies that the electron cannot be at any arbitrary energy level but constrained to only discrete energy levels. The lowest energy is termed the "ground level" and is the normal level for the electrons in an atom.

3. Light is not emitted from an atom continuously. The frequency **f,** of the light emitted by an atom is related to the difference between an initial high-energy orbit of the electron with energy (**EH**) and final low-energy orbit with energy (**EL**) given by the formula: $hf = EH - EL$.

Thus, when an atom is struck by a source of energy (e.g., a gas atom in a vacuum tube through which an electric current passes) the electrons are "excited" and move into higher energy orbits. When the electrons "fall back" to a lower energy level, light is emitted in quanta of energy.

Applying these postulates to the simplest atom—the hydrogen atom which has one orbiting electron and one positive unit of charge in the nucleus—Bohr was able to deduce formula for the frequency of light emitted by hydrogen: $f = R_*(1/ n_1^2 - 1/ n_2^2)$, Where: $R = 2pm^2e^4/ h^2$, and **m** is the electron's mass, **e** is the electron's charge, and **h** is Planck's constant

Referring back to Balmer's empirically derived formula, it can be seen that Bohr's equation is the same but with one important difference. Bohr's equation contains the Rydberg constant (**R**), in terms of basic physical properties where m is the electron mass, e is the electron charge, and h is Plank's constant are basic physical constants. The symbols n_1 and n_2 are termed "quantum numbers" related to the quantized electron energy. Substituting values for **m, e,** and **h,** we get: **R = 1.097373X\x10^7**.

Referring back to Balmer's formula, we see that Bohr's value of the Rydberg constant is the same. Once again, Bohr's equation illustrates the important difference between empirically deduced results and results derived from an understanding of fundamentals.

In a stroke of genius, Niels Bohr had explained why atomic spectra has the patterns it does [6]. One can only imagine Bohr's emotional state when he realized that he had deduced the formula for the hydrogen spectrum and the Rydberg constant from an understanding of basic electron behavior.

Bohr's formula was a significant step toward a correct explanation. Not only had he explained the origin of spectra, but he was able to provide a reasonable explanation of Dalton's law and Mendeleev's table.

BOHR EXPLAINS DALTON'S LAW OF PROPORTIONS

Now we will cover Bohr's explanation of Dalton's law because it will be helpful in later sections.

In Bohr's formula for the hydrogen spectra, two numbers appear: n_1 and n_2. These numbers relate to the quantization of the position of electrons (i.e., electrons can only exist in certain orbits surrounding the nucleus and thus the electron energy is quantized, meaning the energy can only exist in certain quantities). Finally, n_1 is termed the "principal quantum number."

Bohr's formulation showed that electrons are constrained to "shells," depending upon the value of n_1, hence the reason for designating **n_1** as the principal quantum number. For **n_1** = 1, the smallest shell, only two electrons are allowed. For **n_1** = 2, the next largest shell, only eight electrons are permitted. Thus, as we add electrons to the atom, the first electron yields the element hydrogen, while the second electron yields the element helium. Helium is an inert gas, which is explained by the fact that the electron shell of helium is filled, meaning there can be no more electrons in the inner shell.

To add another electron, we must start with the second shell. Lithium is the element formed when we add the first electron to the second shell. If we continue adding electrons to fill the second shell, we form the elements shown in table 4-1 [reference 1, chapter 2, p. 79].

Number of Electrons	Element	Extra Electron Spaces Available
1	Lithium	7
2	Beryllium	6
3	Boron	5
4	Carbon	4
5	Nitrogen	3
6	Oxygen	2
7	Fluorine	1
8	Neon	0

Table 4-1. Extract from periodic table

Table 4-1 illustrates some important atom properties:

- Neon has a filled shell. There is no room for extra electrons, and it is inert, as would be predicted.

- Lithium has one electron in the shell. Its many unfilled shells imply great activity.

- Fluorine has one space available, and it is also very active.

- Oxygen has two spaces available which can be filled by hydrogen to make water.

As explained in [reference 1, chapter 2], chemical compounds are formed when atoms share electrons in such a manner that their outermost shell is filled. The outermost electrons are termed "valence electrons" and the sharing of electrons establishes a force balance that holds compounds together.

It is easy to see from table 4-1 that a compound should form between lithium, which has seven empty spaces, and fluorine, which has one. The compound lithium fluoride is well known. In like manner, a compound should form between beryllium and oxygen as well as between boron and nitrogen. Beryllium oxide and boron nitride are also well-known compounds.

Because hydrogen has one electron and one empty location, many compounds are possible with hydrogen and have been observed, particularly with carbon. Of particular interest is water, which Dalton deduced was formed by two hydrogen atoms and one oxygen atom. Since oxygen has two empty spaces and hydrogen one, two hydrogen atoms fill the two empty oxygen spaces, resulting in H_2O.

SUMMARY OF CHAPTER 4

In 1900, Max Planck deduced that the rules governing the microscopic world of the atom were different from the everyday world of human experience and postulated that, in this world, E/M energy could only exist in discreet quantities, **hf** where, h is Plank's constant and **f** is the E/M frequency. This allowed Planck to solve a long-standing problem in blackbody radiation.

In 1905, Einstein "borrowed" Planck's quantum concept and explained the photo-electric effect.

In 1911, Niels Bohr developed the first quantum atom model and solved several dilemmas:
the equations of motion for hydrogen, the solution of which explained Balmer's formula, and the fact that electrons in Bohr's quantum atom exist in layers of shells, which explained both Dalton's law and Mendeleev's table.

LOOKING AHEAD.

As powerful and useful as Bohr's formulation was, there were still many problems. Several ad hoc assumptions, though they appeared to "work," were later superseded by a better understanding of atomic fundamentals, which we will discuss in the next chapter.

CHAPTER 5:
FINAL RESOLUTION OF THE
"SOLID" EARTH ILLUSION, PART II

This chapter will explain the extension of Bohr's model of the atom to a more appropriate atom model during the 1920s and 1930s. It will also cover significant observations of atomic behavior by Davisson and Germer, Stern and Gerlach, and the inspired explanatory efforts of De Broglie, Schrödinger, and Heisenberg, among others. These observations extended the concepts pioneered by Bohr and completed the establishment of a proper explanation of the structure of matter.

DISCOVERIES REGARDING ELECTRON BEHAVIOR.

Prince Louis de Broglie (1892–1987) was member of the French aristocracy and a brilliant mathematical physicist. While working on his doctoral thesis, de Broglie examined the relationship between Einstein's treatment of the photoelectric effect in which Einstein had demonstrated that light can be both wave and a particle, and Einstein's development of the special theory of relativity, which produced the famous $E = mc^2$ formula and predicted that the electron could be a wave as well as a particle with a wavelength (**l**) given by: **l = h/mv**.

Where **h** is Plank's constant, **m** and **v** are the electron mass and velocity [1]. De Broglie's insight replaced Bohr's ad-hoc quantized electron orbit assumption and explained why electrons are constrained to certain orbits: an electron orbit must be a multiple of the electron wavelength.

This prediction was supposedly met with such derision that it threatened de Broglie's doctoral thesis. However, he was ultimately vindicated by the electron experiments performed by Davisson and Germer that confirmed de Broglie's hypothesis that electrons had wavelike properties.

Working at the Bell Telephone Laboratories in 1927, **Clinton Davisson (1881–1958)** and **Lester Germer (1896–1971)** were aware of Bragg's demonstration that X-rays were reflected from a crystalline surface. They reasoned that the wavelength of electrons predicted by de Broglie was relatively small. Therefore, they deduced that if electrons had wave-like behavior, they would also be reflected from a crystalline surface. They prepared an experimental apparatus that allowed them to "fire" slow-moving electrons at an angle to the surface of a nickel crystal. In the Davisson-Germer (D-G) experiment, electrons were indeed reflected at an angle that agreed with the X-ray diffraction pattern predicted by Bragg's X-ray diffraction, if the wavelength predicted for electrons was substituted for the wavelength of X-rays [2].

The D-G experiment clearly confirmed that the electron as well as the photon can behave as either particles or waves depending upon the experimental conditions. These results formed one of the cornerstones of modern quantum mechanics.

In addition to the rather bizarre electron behavior disclosed by the D-G experiments, in 1922, German physicists Otto Stern (1888–1969) and Walther Gerlach (1889–1979) demonstrated that electrons also spin or rotate. If an electron rotates, then it is essentially a moving electrical charge, and a moving charge generates a magnetic field, as was discovered by Michael Faraday.

The Stern-Gerlach (S-G) experiment involved the passage of electrons through a specially devised magnetic field that would deflect electrons if they had magnetic properties. The experiment was a success: the electrons were deflected as predicted [3].

The consequences of these results were far-reaching. The apparatus employed by Stern and Gerlach was later used to demonstrate that some atomic nuclei, such as iron, also spin [3], thereby explaining the magnetic properties of iron containing minerals.

THE DEVELOPMENT OF MODERN QUANTUM MECHANICS.

In one of the more fundamental physics papers, **Werner Heisenberg (1901–1976)** revealed more bizarre atomic world behavior by demonstrating in 1927 that it was not possible to accurately measure velocity and position simultaneously [4]. In other words, the more precisely one measures the location of an electron, the less one can know about the electron's velocity. This seeming quandary is known as Heisenberg's uncertainty principle and has profound implications, particularly philosophical, regarding the ability to predict the future.

Just the year before in 1926, noting that electrons can behave like waves, **Ernst Schrödinger (1887–1961)** further quantified the impossibility of simultaneously knowing the position and velocity of an object like the electron by developing a "wave" equation—another second order differential equation—to describe the behavior of electrons within the atom [5]. This equation, which has become to be known as the Schrödinger equation, is one of the most basic equations of quantum mechanics. A solution to Schrödinger equation for the hydrogen atom shows why all that can be known about an electron's position is where it "probably" is.

It is important to note that Schrödinger equation didn't disprove that Bohr's equation; it merely improved upon Bohr's work and provided a more accurate picture of the atom. For many applications such as spectroscopy, the Bohr formulation is quite adequate. Thus, when we hear someone say, "Why bother paying any attention to (a scientific idea)? It will only be refuted in a few years" or "Scientists are always changing their minds," what is usually occurring is that someone has most likely generated an improvement that does not normally displace prior knowledge.

"QUANTUM MECHANICAL TUNNELING"—ANOTHER CONSEQUENCE OF QUANTUM MECHANICS.

Yet another bizarre consequence of the quantum mechanics probabilistic location property is the ability of a subatomic particle such as the electron or alpha particle to penetrate or "tunnel" through a barrier that is higher than would be possible for a classically sized particle [6]. This ability is called quantum mechanical tunneling. While tunneling is a complex phenomenon, it is observed in many electronic items and instruments such as the tunneling electron microscope and various integrated circuits [6]. As will be seen, quantum tunneling is also responsible for the ejection of an alpha particle from an atom.

STRUCTURE OF THE ATOMIC NUCLEUS, IDENTIFYING THE POSITIVE CHARGE.

Now that we have collected some explanations for the behavior of electrons in an atom, let us examine the structure of the nucleus because it is fundamental to an understanding many things, particularly radioactive decay and nuclear fusion and fission.

Recall that Rutherford and Thompson stated that maintenance of electrical neutrality required the electrons in an atom to be balanced by a positive charge, a charge that Rutherford's alpha particle scattering experiments showed to be a tiny object in the center of the atom. When Rutherford discovered the existence of the nucleus, there was no reason to suspect that the positive nucleus was not a contiguous entity, probably spherical.

World War I interrupted Rutherford's investigations, but after the war, Rutherford and associates returned to alpha particle experiments. In an attempt to learn as much as possible about the atom, Rutherford fired alpha particles at everything he could find and examined the resultant effects.

In 1918, he observed that when nitrogen gas was bombarded by alpha particles, his scintillation counter detectors showed evidence of hydrogen nuclei. After examining all possibilities, Rutherford was forced to apply Sherlock's theorem and conclude that the hydrogen nuclei source was the nitrogen gas. Thus, nitrogen gas must contain hydrogen nuclei. Since the hydrogen nucleus was the smallest possible nucleus, Rutherford concluded the hydrogen nuclei must be an elementary particle, which Rutherford termed the "proton" from the Greek word for "first" [7]. David Parker suggests Rutherford was transmuting nitrogen into oxygen [8], but Rutherford's scintillation detectors probably could not have determined this because they could only count ionized particles.

Having discovered the proton, Rutherford was led to the logical conclusion that for an atom to remain neutral, the number of protons in the nucleus had to equal the number of electrons. For example, carbon (C) has six electrons which are balanced by six protons.

For convenience, the weight of a proton is given the value of one, which then leads to the conclusion that the "atomic weight" of carbon would be six. However when measured, it is found to be twelve. Rutherford solved this dilemma by hypothesizing that the nucleus must contain another neutral particle, one comprised of a proton and an electron. This particle would both preserve electrical neutrality and account for the weight of the atom. The neutral particle would have approximately the same weight as the proton since the weight of an electron is much less than a proton. Rutherford dubbed this particle the "neutron" [9]. Protons and neutrons are termed collectively, "nucleons."

Giving the same weight to a proton as a neutron causes problems with detailed measurements, so the atomic mass unit (AMU) is the current standard weight for a nucleon. One AMU is 1/12 the mass of carbon [10].

Through a series of brilliant experiments in 1932, for which he was awarded the Nobel Prize in 1935, one of Rutherford's assistants, **Sir James Chadwick (1891–1974)**, confirmed Rutherford's prescient prediction and proved the existence of the neutron [11]. Although not quite correct, a neutron can be envisioned as a combined proton and electron. Therefore, the nucleus consists of two particles: protons and neutrons.

It should be noted that an unattached neutron is unstable and decays into a proton, an electron, and a strange item called a neutrino with a "half-life" of approximately fourteen minutes. Half-life is defined as the time required for one-half of a collection of unstable particles to decay.

While the neutrino must be mentioned for completeness, it has no particular bearing on this discussion. Also, strictly speaking, the electron emitted during the decay of an unstable neutron is technically a beta particle as the decay of a neutron is termed "beta decay" [12].

Lawrence Wood

The Nucleus Spins—The Origin of Natural Magnetism.

In 1938, Galician-born physicist **Isidor Rabi (1898–1988)** extended the Stern-Gerlach experiment by employing beams of molecules passing through magnetic fields. Through this experiment, he demonstrated that nuclei with an odd number of either protons or neutrons have an intrinsic spin [13]. The spinning positive nucleus is a moving charge that generates a magnetic field—this is the source of natural magnetism.

Rabi's experiments were extended by Swiss-American physicist Felix Bloch (1905–1983) and American physicist Edward Mills Purcell (1912–1997) with the discovery of nuclear magnetic resonance (NMR) in 1948 for which they received the Nobel Prize in 1952 [13].

NMR is the phenomenon exploited in magnetic resonance imaging (MRI). It is perhaps apocryphal, but the reason MRI is not NMRI is the apparent concern that the word "nuclear" would alarm individuals and restrict use of the MRI technique.

The Nucleus Can Have a Varying Number of Neutrons.

While the number of protons in a nucleus must equal the number of electrons, neutrons are not so constrained [14] [15]. An atom of the same element with more than one set of neutrons is termed an isotope. In the nuclei of the stable lighter elements, the number of neutrons equals the number of protons, thus carbon has six protons and six neutrons. In classifying an element, the number of protons is termed the "atomic number" and is usually represented by the letter Z, while the weight of element (the combined weights of the protons and neutrons) is termed the "atomic weight" and is usually represented by A. Thus, the symbol for an atom can be written:

$$^{A}X_{Z}.$$

For example, we would write $^{12}C_{6}$ for the carbon atom.

But can we add as many neutrons as we desire since neutrons are not affected by the positively charged protons? It turns out that the answer is a qualified yes. However, the number of neutrons is limited because the number of neutrons in a nucleus has a dramatic effect upon nuclear stability. For example, if a neutron is added to the hydrogen nucleus, a stable isotope of hydrogen (deuterium) is formed [16]. On the other hand, if two neutrons are added, then the unstable isotope of hydrogen (tritium) is formed [17]. Tritium has a half-life of about twelve years and decays into helium via beta decay. See also [14] for a discussion of the deuterium-tritium interaction. We will discuss nuclear, or more popularly called radioactive decay below, but first let's answer the following question: what holds the nucleons, especially the protons, in the nucleus?

As we have established, just as electrical charges repel, the positively charged protons in a nucleus should repel each other. That they don't is obvious or we wouldn't be discussing the matter. Accordingly, there must be another force that holds nucleons in the nucleus, and there is—the aptly named "nuclear strong force." An exposition of the strong force, other than the elementary observation that the force must operate over relatively short distances, requires a discussion of the arcane subject of quantum chromodynamics, which is well beyond the scope of this book. For a beginning discussion see [18], but the simple diagram presented in figure 5-1 will suffice for this book. In the image, the strength of the strong and E/M forces are exhibited, and two protons are held in the nucleus because the strong force dominates as protons get close enough.

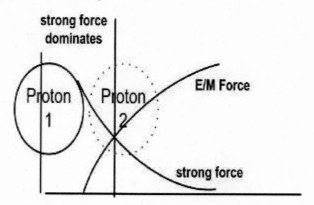

Figure 5-1. Illustration of strong force and E/M force interaction

To the left of the vertical line, the strong force exceeds the E/M force and the protons are held in the nucleus. To the right of the line, protons are repelled as the E/M force exceeds the strong force. As will be seen a little later, another term for the relationship between the strong and E/M forces is the "binding energy of the nucleus"—the amount of energy required either to pull a nucleon from a nucleus or push one into a nucleus.

In lieu of the fact that the E/M force acts farther from a nucleon than the strong force, the question then arises: How is a nucleus with two or more protons formed in the first place? Before addressing that question, we must first examine two basic nuclear processes: nuclear fusion and nuclear fission.

NUCLEAR FUSION AND NUCLEAR FISSION.

Nuclear fusion and nuclear fission release more energy than any other energy source on Earth, and they play a critical part in the story of matter.

Nuclear fusion, as the phrase implies, is the joining or fusing of two or more nuclei. Nuclear fusion was first demonstrated by **Sir Mark Oliphant (1901–2000)** in 1932 while he was working in the Cavendish Labs in Great Britain [19].

Nuclear fission, on the other hand, is the splitting of a nucleus into two "daughter nuclei" that are usually of approximately the same weight [20]. Fission was first demonstrated in 1934 by **Enrico Fermi (1901–1954)**, who bombarded uranium with neutrons [21]. Since neutrons are "neutral" and not repelled by the positive charge of the nucleus, they can strike a nucleus. However, full confirmation of uranium fission was not obtained until experiments in 1938 by German chemists **Otto Hahn (1879–1963)** and **Fritz Strassmann (1902–1980)**. Their experiments demonstrated the detection of barium after bombardment of uranium with neutrons. Barium is about half the weight of uranium, as was expected for a true splitting of uranium. Hahn's results were confirmed by **Otto Frisch (1904–1979)** in 1939. Hahn received the Nobel Prize in chemistry for his discovery [21].

Although complex in the details, the essential elements of fusion and fission are captured in figure 5-2. As the figure shows, the energy that binds a nucleus together, appropriately termed the "binding energy," is displayed for all the elements as a function number of nucleons in a nucleus (i.e., the atomic mass unit of the element) [22].

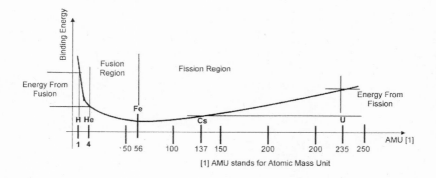

Figure 5-2. Nuclear binding energy and fusion-fission relationships

Beginning with the lightest element hydrogen (H), the binding energy per nucleon decreases until the most stable element, iron (Fe), is reached; thereafter the binding energy per nucleon increases. As noted in the diagram, the left side of iron is termed the "fusion region," and the other is the "fission region." An important aspect of this curve is that it holds the key to the life and death of stars, which will be seen to be central to the formation of the earth.

In the fusion region, atoms with higher AMU are formed from the lighter elements by the fusion of the nuclei. Of particular interest is the Helium (He) nucleus, which consists of two protons and two neutrons. It is created by the fusion of four hydrogen atoms—two of the hydrogen electrons join two of the protons to become neutrons; hence it has an atomic weight of four.

Due to the proton repulsion by the E/M force, a great deal of force must be exerted to cause protons to surmount the E/M repulsion barrier and reach the point where the strong force overcomes the E/M repulsion. In the sun and other stars, the necessary enormous forces are created by gravity, which is sufficient to cause lighter elements like hydrogen to surmount the repulsive barrier and combine to form helium.

Since the binding energy per nucleon of helium is less than that of hydrogen, when hydrogen nuclei fuse to form helium there is excess binding energy per nucleon. This excess energy is released as radiant energy. It is important to note that the source of this energy is the conversion of mass into energy, as predicted by Einstein. This is

63

illustrated by table 5-1, which compares the weight of four hydrogen atoms to the weight of one helium atom.

Element	Atomic weight (AMU)
Hydrogen	1.07825
Helium	4.002602
4 times Hydrogen	4.3176
Difference, converted to energy	0.314998

Table 5-1. Illustration of energy generated by fusing hydrogen into helium

The mass difference (0.314998 AMU) is converted to energy according to Einstein's famous equation, $e = m\, c^2$, and is the source of the sun's energy. While the sun's energy is not unlimited, it is sufficient to last for about 5 billion years, far beyond the wildest imaginings of nineteenth century scientists.

On the other end of the binding energy curve, binding energy per nucleon increases. However, as the number of neutrons in a nucleus increases, the nucleus becomes increasingly unstable until the heaviest elements (e.g., uranium) become inherently unstable and will fission via bombardment of the nucleus with neutrons that are not affected the E/M repulsive force or will spontaneously expel a particle via nuclear decay. Since the binding energy per nucleon is less in the fission particles than fusion, less energy is released in fission.

RADIOACTIVITY AND NUCLEAR DECAY.

It is now understood that what Becquerel termed "radioactivity" is actually the decay of an unstable nucleus in which one of the three objects—an alpha particle, a beta particle, or a gamma ray—is spontaneously emitted. Due to their importance in the controversy regarding radiometric dating, a short discussion of the first two is in order.

Alpha Particle Emission by Tunneling: In 1928, Russian mathematical physicist **George Gamow (1904–1968)**, who emigrated to the United States, deduced from the principals of quantum mechanics that nuclear decay is caused the emission of an alpha particle due to tunneling [23]. The alpha particle has been demonstrated to be the nucleus of a helium atom, which consists of two protons and two neutrons and is positively charged because it lacks the two electrons of the helium atom that render the helium atom neutral. Accordingly, when an alpha particle is emitted, the nucleus loses two protons and two neutrons and becomes a new element. Gamow was able to explain the wide variation of half-lives by differences in the barrier heights of different elements that the alpha particle had to tunnel through.

Beta Particle Emission by Beta Decay: Nuclear decay via the emission of a beta particle is caused by the phenomenon of beta decay. Beta decay is more complex than alpha particle emission. It was first explained in relatively simple terms by Enrico Fermi in 1931 [12] and later in more correct, albeit considerably more complex, terms by **Abdus Salam (1926–1996), Sheldon Glashow (1932–),** and **Steven Weinberg (1933–)** in 1976. They received the Nobel Prize for their work [24]. As mentioned earlier, for the purposes of this book the beta particle can be considered an electron.

Gamow, Fermi, Salam, Glashow, and Weinberg are some of the most brilliant physicists that have ever lived, each having won many awards, including the Nobel Prize. The radioactive decay processes explained by these gentlemen have been studied and employed in radiometric dating under a wide variety of conditions. Claims by pseudo-scientists representing creationism and ID that radiometric dating is unreliable are an insult to these Noble Prize winning scientists and just not supported by the facts.

Illustrations of the Effects of Nuclear Decay: Since the material provided in this chapter is background for upcoming chapters, I have provided some examples of the effects of alpha and beta particle decay for reference.

Table 5-2 illustrates the changes to a nucleus when an alpha particle is emitted beginning with protons (p) and neutrons (n).

Element	Protons	Neutrons	AMU
Original nucleus	p	n	p + n
Alpha particle	2	2	4

New nucleus	p - 2	n - 2	p + n - 4

Table 5-2. Effect of alpha particle decay on a nucleus

Table 5-2 shows the original nucleus, the constituents of the alpha particle, and the new nucleus that results from the alpha particle emission.

A typical example of nuclear decay via alpha particle emission is the decay of uranium 238 ($^{238}U_{92}$) into thorium 234 ($^{234}Th_{90}$), which is exhibited in table 5-3.

Element	Protons	Neutrons	AMU
Original $^{238}U_{92}$	92	146	238
New $^{234}Th_{90}$	92 - 2 = 90	146 - 2 =144	234

Table 5-3. Conversion of $^{238}U_{92}$ into $^{234}Th_{90}$ by alpha particle decay

Table 5-4 illustrates the changes to a nucleus when a beta particle is emitted.

Element	Protons	Neutrons	AMU
Original nucleus	P	n	p + n
New nucleus	p + 1	n - 1	Unchanged

Table 5-4. Effect of beta decay on a nucleus

In beta decay, because a beta particle is essentially an electron, a nucleus loses a neutron and gains a proton and a companion electron when a beta particle is emitted.

A typical example of beta decay is the decay of thorium 234 into protactinium 234, as shown in table-5-5.

Element	Protons	Neutrons	AMU
Original $^{234}Th_{90}$	90	144	234
New $^{234}Pro_{91}$	90 + 1 = 91	144 - 1 = 143	234

Table 5-5. Conversion of $^{234}Th_{90}$ into $^{234}Pro_{91}$ by beta decay

Note that the atomic weight does not change, but the atomic number increases due to the addition of a proton.

Another important example of beta decay is the decay of carbon 14 ($^{14}C_6$) to nitrogen 14 ($^{14}N_7$), illustrated in table 5-6.

Element	Protons	Neutrons	AMU
14C6	6	8	14
$^{14}N_7$	6 + 1 = 7	8 - 1 = 7	14

Table 5-6. Conversion of $^{14}C_6$ into $^{14}N_7$ by beta decay

It is interesting to note that in this particular process, a solid (carbon) is converted into a gas (nitrogen).

Some Typical Nuclear Half-lives: The half-lives of all of the elements that undergo either alpha particle emission or beta particle emission have been exhaustively measured and verified many times, as will be discussed in detail below. A few of the more important half lives [25] are listed in table 5-7.

Decay elements	Half life, years	Particle Emitted
$^{14}C_6 - ^{14}N_7$	5730 ± 40	beta
$^{234}Th_{90} - ^{234}Pro_{91}$	2.46×10^5	beta
$^{87}RB_{37} - ^{87}Sr_{38}$	4.88×10^{10}	beta
$^{238}U_{92} - ^{234}Th_{90}$	4.51×10^9	alpha

Table 5-7. Some typical nuclear decay half-lives

SUMMARY OF CHAPTER 5.

From 1918 to 1934, less than 20 years, the fundamental explanations of matter were finalized, and the solid earth illusion completely resolved—a stunning accomplishment. The key accomplishments of that time period are as follows:

- In 1918, Rutherford discovered the proton and proposed a nucleus consists of protons and another elementary particle, the neutron. His theory finalized the basic explanation of atomic structure. The number of protons equals the number of electrons, while the number of neutrons is determined by the number of neutral items needed to satisfy atomic weight.

- In 1924, de Broglie predicted electron waves, which explained why Bohr's electron orbits could only have certain radii. The orbit circumferences are multiples of an electron's wavelength.

- In 1927, Davisson and Germer confirmed de Broglie's hypothesis.

- In 1922, Stern and Gerlach demonstrated that electrons spin, which explained some magnetic effects.

- In 1928, Heisenberg and Schrödinger developed modern quantum mechanics, which explained that the location of an electron is probabilistic and why the electron location could not be known exactly.

- Quantum mechanical tunneling is shown to be consequence of an electrons probabilistic location.

- In 1938, Rabi extended the Stern-Gerlach experiment to show that certain nuclei spin, thereby explaining natural magnetism.

- Continued experiments disclosed that a nucleus could have differing numbers of neutrons; these nuclei are known as isotopes.

- The strong force, the force that holds the nucleus together, was confirmed and explained why protons are not ejected from the nucleus.

- In 1932, Oliphant demonstrated nuclear fusion.

- In 1934, Fermi demonstrated nuclear fission.

- The curve of binding energy was developed and explained that the fusion of four hydrogen atoms to create one helium atom releases an enormous amount of energy, which explains how the sun's energy is generated.

- Nuclear decay by either alpha particle emission, which is proved to be due to tunneling, or beta particle emission, which is proved to be due to the complex phenomenon of beta decay, explained radioactivity. This is one of the key methods for the dating of objects.

SOME SALIENT POINTS REGARDING THE DEVELOPMENT OF A CORRECT EXPLANATION OF MATTER.

The development of a proper explanation of "solid" matter was achieved by many investigators working in many locations over thousands of years. Their efforts resemble that of a group working on a giant puzzle; each investigator added a piece as he found it.

Many discoveries were serendipitous, such as Becquerel's and Roentgen's discoveries of radioactivity and X-rays.

There were many inspired conclusions, some invoking Sherlock's theorem, such as Thompson and Rutherford's conclusions regarding the structure of the atom.

There were many inspired explanations that resemble Alexander's approach to the Gordian knot[1], such as Planck's quantum hypothesis or Einstein's hypothesis that light can also be a particle.

Though some achievements were more important than others, all were important. And while there were blind alleys and mistakes, in general, the development of explanations proceeded in a rather continuous, logical flow.

Surprisingly, relatively few people—mainly those with a college education—knew of these developments until public education became more widespread.

A principal problem encountered in the attempt to understand/explain matter was the fact that the rules governing the world of the atom do not obey classical physics. A new "quantum" physics had to be invented.

LOOKING AHEAD: RESOLVING THE THREE "ASTRONOMICAL" ILLUSIONS.

Having solved the "solid" earth illusion with the establishment of a correct explanation for matter, we will next address three related astronomical illusions. They are termed astronomical because they involve objects that are not on the earth and thus are typically observed with astronomical instruments like the telescope. These illusions are:

1. The apparent motion of the sun and planets around the earth illusion;

2. The apparent same moon and sun size and distance illusions;

3. The apparent close stars around Earth illusion.

Chapter 6 will address the resolution of the first illusion and establish the correct explanation for the shape of the solar system. Chapter 7 will address the resolution of the second illusion and establish the correct explanation for the size of the solar system. Finally, chapter 8 will address the resolution of the third illusion and establish the correct explanation for the size and age of the universe, plus the formation of the stars (including the sun), the formation of planets (including the earth), and the creation of the elements (within certain stars).

At the conclusion of these three chapters, we will have located the correct place for the earth and learned how it and the sun were created. We will have also established the irrefutable fact that the universe is 13.7 billion years old.

CHAPTER 6:
SOLVING THE APPARENT MOTION OF THE SUN AND PLANETS AROUND THE EARTH ILLUSION

CREATING THE ILLUSION.

Imagine this: You are visiting Hawaii and standing on the porch of your west-facing beachfront hotel. It is early evening, and the sun is nearing the horizon, painting the low-hanging clouds brilliant shades of red and orange. A faint hint of hyacinth fills the air. It's another gorgeous sunset. As you gaze upon this scene, the sun slowly sinks toward the water and finally disappears from view.

The night is warm, so you linger. Soon it becomes dark enough to see the first stars. If you are able to return to this location and watch the sun set each night and observe the stars after the sun has set, you may notice that some of the "stars" appear to move relative to the rest of the stars. Also some of the "stars" seem much brighter than others, particularly an extremely bright star that often appears in the western sky. If you are able to view the stars over enough nights, you will see that as many as five "stars" appear to move against the background of the other stars.

But, it's just an illusion. As you contemplate the gorgeous sunset, the deep purple of the night sky, and the myriad stars, you find it difficult to remember that everything you have witnessed is one of the universe's grand illusions. The only thing that has been moving is your viewing platform—the earth. The illusion of the sun and planets around Earth is created by the earth's axial rotation, and the "stars" that appear to move against the background of the stars are not stars; they are planets, like the earth, that revolve around the sun.

However, not too many years ago, no one realized it was a very powerful illusion. Two thousand and fifty years ago, some farsighted Greek philosophers were the first to deduce that the sun around Earth was an illusion, but they were "out voted" by those who couldn't abandon the illusion and the comforting belief that we are at the center of the universe. One thousand years later, a farsighted Polish monk "rediscovered" the fact that it was an illusion, but his view was suppressed by religious authorities and others. The truth finally emerged, but today there are still those who cling to the illusion, as is easily demonstrated by entering the phrase, "The Earth Does Not Move," into a search engine like Google. You will find many Web pages devoted to this nonsensical idea.

ASTRONOMICAL OBSERVATIONS AVAILABLE TO THE UNAIDED EYE.

The following list consists of some astronomical observations that were available to early humans. These observations are also ones that anyone can make today without the aid of technology:

- The full moon rises at the same time the sun sets.

- The sun and moon occasionally change shape from full to a portion of full over a relatively short time. This phenomenon is called an "eclipse."

- The sun does not rise and set at the same point on the horizon every day. Careful observation shows that the rise and set points of the sun cycle north and south about 46 degrees over the period of one year.

- Upon several occasions when the moon is new, the entire face of the moon is dimly lit. This phenomenon is now termed "earth shine."

- Five of the "stars" that are more visible[1] move in strange patterns. The Greeks called these stars "planets" from the

Greek word for wanderer. Venus, the brightest planet, never rises more than about 40 degrees from the horizon. Therefore it never really "goes around the earth." Neither does the less easily visible planet Mercury. Other planets sometimes appear to move in the earth's direction and sometimes opposite to the earth's direction.

- Observation over considerable time shows that the planets occasionally appear at the same time as the moon, and their appearance will always "move" along a line that comes close to the moon.

Earliest Observations/Explanations of the Sun and Planets around Earth Illusion: as with the solid earth illusion, the Bible provides little insight into the astronomical illusions or into the observations listed above. Here are the lines from Genesis that discuss objects above the earth's surface:

- "And God said, Let there be light: and there was light. And God saw the light, that it was good: and God divided the light from the darkness. And God called the light Day, and the darkness he called Night. And the evening and the morning were the first day" (Gen. 1:3–4).

- "And God called the firmament Heaven. And the evening and the morning were the second day" (Gen. 1:8).

- And God said, Let there be lights in the firmament of the heaven to divide the day from the night; and let them be for signs, and for seasons, and for days, and years: And let them be for lights in the firmament of the heaven to give light upon the earth: and it was so. And God made two great lights; the greater light to rule the day, and the lesser light to rule the night: he made the stars also. And God set them in the firmament of the heaven to give light upon the earth, And to rule over the day and over the night, and to divide the light from the darkness: and God saw that it was good" (Gen. 1:14–18).

Clearly, there is little information in these passages regarding the astronomical illusions. Presumably, the greater light refers to the sun, and the lesser light refers to the moon. However, there is no discussion of the motion of these lights.

Lawrence Wood

GREEK PHILOSOPHERS SOLVE THE ILLUSION.

The solution to the illusion began with the Greeks—again. Anaxagorous, Heraclides, and Aristarchus, knowing about the above observations of the sun and planets and perhaps some others, realized that idea of the sun revolving around Earth was an illusion. They resolved it by showing that the best explanation of these observations was a heliocentric solar system (Helios was the God of the sun) in which the planets revolved about the sun, as opposed an Earth-centered or geocentric explanation. However, they were unfortunately too far ahead of their time.

Anaxagorous (ca. 500–428 BCE) was the first to make a real contribution to the resolution of this illusion. Born in Ionia around 500 BCE, Anaxagorous first explained that the moon shined by reflected light and gave the first correct explanations for eclipses: the earth, moon, and sun are essentially in the same plane, and as the moon revolves around the earth, it is occasionally between the earth and sun, which causes a solar eclipse [2, ch. 2, p. 63]. Also, the shape of the earth's shadow in lunar eclipses showed that the earth was spherical.

Heraclides of **Pontus (387–312 BCE)** deduced that Venus and Mercury revolved about the sun, probably because he observed that Venus and Mercury rise only a finite distance above the horizon, after which they sink back to the horizon. This motion is difficult to explain if one argues the planets revolve around the earth. However, as we shall see, some ingenious solutions were created that actually allow Venus and Mercury to revolve around the earth. Heraclides also adopted the view that the earth rotates on its axis every twenty-four hours [2, ch. 2, p. 214].

But **Aristarchus of Samos (310–230 BCE)** [1] was the star of the show. He deduced that the best explanation for all the observations was a heliocentric solar system with all planets, including the earth, revolving in circles around the sun [2, ch. 2, p. 214]. Aristarchus then realized that if the earth revolved about the sun, the Earth-sun distance must be greater than earth-moon distance because the moon revolved around the earth. He actually attempted to measure the Earth-sun distance. This task was more difficult than measuring distances on the earth because the measurement of the distance to an object that cannot be reached must be measured by a technique called "triangulation".

For example, to measure the width of a river, one might use approach shown in figure 6-1.

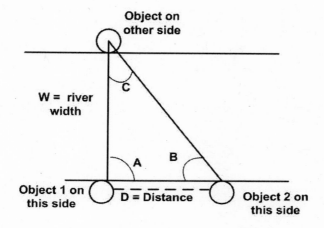

Figure 6-1. Distance measurement using triangulation

We would create a triangle by locating two objects (object 1 and object 2) on our side of the river and another object on the other side. We would then arrange the objects so that a right triangle (angle A) is formed. Then we would measure the distance (D) and angle B. We could then determine the width of the river by the following trigonometric formula: $W = D*TanB$.

TanB represents the tangent of angle B and is defined as the ratio W/D. Those who have studied some trigonometry may recognize the formula; however, if you don't, all that is necessary to know is that this formula is accurate and can be verified at this site [2]. In addition, tables of values of the tangent of an angle can easily be found on the Internet.

Aristarchus tried, with only limited success, to employ triangulation to determine the Earth-sun distance. First, he chose the earth-moon distance as the baseline. He estimated the earth-moon distance by observing how Earth's shadow moved across the moon during a lunar eclipse [3].

Next he tried to establish a triangle, similar to that above, with the sun as the object across the river and the earth and moon as the objects on his side of the river. Unfortunately, Aristarchus had rather crude angle-measuring equipment and since angle C is very small due to the great Earth-sun distance, Aristarchus's estimate of nineteen times the earth-moon distance was highly inaccurate. The actual value is approximately 400 times the distance between the earth and the moon.

Heliocentric View Rejected: While ultimately proven correct, the three Greek philosophers' heliocentric view invoked the fifth impediment discussed in the prologue: conflict with authority, especially family and religious authority. It was just "too revolutionary to be accepted by their contemporaries who debunked the theory because it conflicted with geocentric religious principles" [4]. Moreover, the heliocentric view conflicted with the third impediment—explanations often conflict with 'common sense' and 'intuition'—and with the natural desire of humans to be at the center of things, as well as with the powerful Aristotle's concept that all objects fall to the center of the earth, thus the heavy earth must occupy the lowest and central point in the cosmos. **Aristotle (384–322 BCE)** was one of the foremost of the Greek philosophers, a student of Plato and teacher of Alexander the Great [5]. He wrote on many subjects, however, in the case of sun and stars around the earth illusion, Aristotle often got it wrong. As the noted philosopher **Bertram Russell (1872–1921)** comments regarding Aristotle's influence, "Ever since the beginning of the seventeenth century, almost every serious intellectual advance has had to begin with an attack on some Aristotelian doctrine in logic" [2, ch. 2]

Aristotle also believed the stars were attached to a spherical shell, the outermost of first heaven, that enclosed the universe. This shell revolved around the earth and carried the stars with it. This view was thought reasonable due to the illusion that the stars were relatively close and appeared to be about the same distance away.

Inside the heavenly shell, the five visible planets, the sun, and the moon moved in their orbits around the central earth. The movements of planets were not as simple to explain as the stars, especially Venus and Mercury, as mentioned previously. However, the clever Aristotle is reported to have used the computations of an older astronomer, Eudoxus of Cnidus, to fabricate an elaborate scheme of fifty-five major and minor orbits known as epicycles and cycles to explain the planetary motions [6].

Ptolemy Casts Geocentric View in Concrete: Besides Aristotle, the ancient astronomer who exerted the greatest ultimate influence on Western beliefs about the solar system was **Claudius Ptolemy (83-161 CE)**. Ptolemy conducted his observations in Alexandria, Egypt around 150 CE and published his findings and explanations of the universe in a thirteen-volume treatise he entitled *Mathematical Composition* [7].

Ptolemy adopted the views of Aristotle and also rejected the idea that the earth moves. He pointed out that the earth was round and gravity was directed toward the center of the earth. He placed the motionless earth at the center of the universe, around which the sun, moon, and stars revolved at various speeds. Ptolemy believed the stars were brilliant spots of light in a concave hemisphere that arched over everything. He traced the motion of the planets against the "fixed stars" and expanded Aristotle's epicycles and cycles to further improve this method of explaining the motion of the planets.

Ptolemy's work, which eventually became known as the Almagest (Greek-Arabic for "the greatest"), was the revered and accepted view of the universe for twelve centuries [8]. Furthermore, it became incorporated into the doctrine of the Catholic Church, which gave it religious authority. Regarding religious authority, Bertram Russell suggests Aristotle's authority was equally as great. He writes, "And after his death it was two thousand years before the world produced any philosopher who could be regarded as approximately his equal. Towards the end of this long period his authority had become almost as unquestioned as that of the Church" [2, ch. 2].

EMPIRICAL DEVELOPMENT OF A PROPER EXPLANATION OF THE SUN AND STARS AROUND EARTH ILLUSION.

Nicolas Copernicus (1473–1543 CE), a Polish monk, mathematician, and astronomer, was the first person in two thousand years who dared question the geocentric view. He did this by reviving the heliocentric view proposed by Aristarchus et al. in ancient Greece.

It is not clear if Copernicus was aware of Aristarchus. Some such as Bertram Russell argue he did not [2, ch. 2], but it seems reasonable to suggest that because the monks preserved much of Greek and Roman culture and Copernicus was a monk, he may have been aware of

Aristarchus. However, whether Copernicus knew of Aristarchus is of no import. Important findings are sometimes lost and rediscovered, as will be discussed in chapter 12 regarding the case of Gregor Mendel.

Copernicus's great treatise, *De Revolutionibus Orbium Coelestium* (On the Revolutions of the Celestial Spheres), was essentially one of the starting points of modern astronomy [9]. The book was dedicated to Pope Paul III and organized in six parts [10]:

- The first part contains a general vision of the heliocentric theory and a summarized exposition of his idea on the world.

- The second part is mainly theoretical and describes the principles of spherical astronomy and a list of stars (as a basis for the arguments developed in the subsequent books).

- The third part is mainly dedicated to the apparent movements of the sun and to related phenomena.

- The fourth part contains a similar description of the moon and its orbital movements.

- The fifth and the sixth parts contain the concrete exposition of the new system.

It is important to note that although the Copernican explanation was perhaps no more accurate than the Ptolemaic, it had one significant advantage: it was much simpler. Instead of the cycles and epicycles required by Ptolemy, Copernicus could explain the visible phenomenon by simply having planets go around the sun. As will become apparent, reduction in explanation complexity is often a significant aspect of improved explanations.

Copernicus also realized that the planets moved in the same plane and in the same direction around the sun, however, *why* they did was not determined until the formation of stars and planets became understood, which will be explained in chapter 8.

Realizing that his view was very controversial, Copernicus delayed the publication of the treatise until he was at the end of his life. Rumor has it that a draft manuscript reached Copernicus on his death bed. The religious reaction following the theory's publication proved the wisdom of Copernicus's caution.

Before discussing the religious reaction to Copernicus, Galileo's contribution to the controversy via his invention of a practical telescope must be added. As noted in [11]: "The telescope was one of the central instruments of what has been called the Scientific Revolution of the seventeenth century. " The telescope, which revealed hitherto unsuspected phenomena in the heavens, was made possible by the discovery of glass and the glass lens[2].

The first useful telescope was exhibited in October 1608 in the Netherlands, the country that produced the first useful microscope, as will be discussed in chapter 12. But "it was Galileo who made the instrument famous. He constructed his first three-powered 'spyglass' in June or July 1609, presented an eight-powered instrument to the Venetian Senate in August, and turned a twenty-powered instrument to the heavens in October or November" [12].

Using his improved telescope, Galileo rubbed salt into the wounds inflicted by Copernicus's heliocentric concept when he discovered something that was never mentioned in the revealed truth of the Bible—the four moons of Jupiter. Revealed truth held that celestial objects were perfect. The four moons of Jupiter upset that celestial perfection. Galileo published his findings in *Sidereus Nuncius* in March 1610 [11].

The Church Reacts: Copernicus and Galileo's findings profoundly upset what had been the accepted dogma for thousands of years. At first the church paid little attention to Copernicus's concept and Galileo's observations, but as the "heresy" gathered adherents, the church realized it could undermine the entire church dogma. The fifth impediment—conflict with authority, especially the family and religious authority—reared its ugly head, and the church cracked down. The monk Bruno, an outspoken proponent of the Copernican concept, was burned at the stake, and the mighty Galileo was forced by the Inquisition to recant [12]. But the genie was out of the bottle so to speak, and more liberal countries such as Protestant Denmark supported continued astronomical investigation.

Tyge Ottesen Brahe (1546–1601)—whose name was Latinized to Tycho Brahe—was born in what is now modern-day Sweden. While Copernicus had to labor with the relatively poor astronomical observations available to him, Brahe was granted an estate on the Danish island of Haven by Protestant King Fredrick II of Denmark. King Frederick was no friend of the Catholic Church, and provided

Brahe with funding to build an astronomical/astrological observatory (*uraniborg* in Swedish). Brahe equipped the observatory with large astronomical instruments and became famous for making accurate astronomical observations, especially of the planets. In keeping with the transitional nature of the times, Brahe was also well known as an astrologer and alchemist [13].

From 1576, when the observatory was constructed, until the death of his patron in 1586, Brahe and his assistants amassed a vast amount of planetary position data that was more accurate by far than any previously recorded [13]. Brahe, however, was not able to use the observations to provide an improved explanation of planetary motion. It remained for Brahe's most brilliant assistant, Johannes Kepler, to analyze the data and extract the mathematical formulas that accurately describe planetary motion.

Johannes Kepler (1571–1630) was born at Wurtemberg in 1571 and studied astrology, mathematics, and astronomy. He became obsessed with the idea of finding the mathematical harmonies in the mind of the Creator, and the data collected by Tycho Brahe was ideal for this endeavor because the he believed the planetary orbits must surely demonstrate these harmonies. Kepler struggled with the mass of data Tycho Brahe had so laboriously collected for a number of years, but his interest in astrology and other mysticisms prevented him from fitting the correct curve to the data because he was trying to fit Brahe's data to the celestially perfect circle.

Finally, after nearly twenty years of frustration, he had to abandon his quest. Bowing to Sherlock's theorem, Kepler accepted the inescapable conclusion that the equation of a planetary orbit was not a circle; it was an ellipse [14]. An ellipse is a "flattened" circle that has one "radius" that is longer than the other, whereas the radii of a circle are all the same length (see figure 6-2).

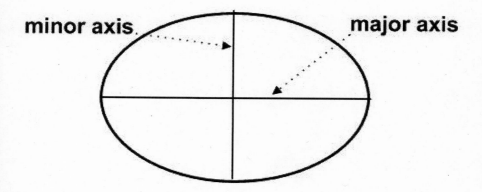

Figure 6-2. Sketch of an ellipse

Kepler's struggle was compounded by the fact that the deviation of a planetary orbit from a circle was extremely small; hence, Kepler's desire to fit the planetary equation to a circle is understandable.

Kepler set forth his three "laws" of planetary motion in *De Harmoice Mundi*, which was published in 1619. The first "law" is the most important [14]: "The orbit of a planet is elliptical, and the sun, the source of the motion, is in one of the foci."

Kepler's explanation of the planetary orbits completed the second phase of the determination of the shape of the solar system. The orbits of all the planets are ellipses[3].

It is important to note that, due to his religious leanings, Kepler incorporated religious arguments and reasoning into his work, motivated by the conviction that God had created the world according to an intelligible plan that was accessible through the natural light of reason [15]. To this end, he was perhaps one of the first proponents of intelligent design.

THEORETICAL DEVELOPMENT OF A PROPER EXPLANATION OF THE SUN AND STARS AROUND EARTH ILLUSION.

A few years after Kepler published his classic analysis of planetary orbits, solar system investigations moved to next phase with the amazing contributions of **Sir Isaac Newton (1642–1727)** regarding the derivation of planetary motion from fundamental principles.

Newton was born at Woolsthorpe, Lincolnshire, and entered Cambridge University in 1661 at age nineteen. His progress was rapid: he was elected a Fellow of Trinity College in 1667 and Lucasian Professor of Mathematics in 1669, just eight years after entering Cambridge. Newton's most productive years were 1665–1666, which he spent largely in Lincolnshire because of plague in Cambridge [16].

Newton's interest in the problem of planetary motion was aroused by correspondence with British scientist **Robert Hooke (1635–1703)**, who played an important role in the scientific revolution, making contributions in many areas [17] through both experimental and theoretical work. Hooke's experiments led him to deduce that gravity obeys an inverse square law, which Newton expanded upon during his eighteen months in Lincolnshire. Newton combined two fundamental concepts related to planetary motion to calculate the equation of a planetary orbit from basic principles.

First, Newton finalized Hooke's experimental deductions into a formula for gravitational attraction between two bodies, such as the earth and the sun. As the gravitational force between two bodies is proportional to the product of the two bodies masses and inversely proportional to the square of the distance separating them [16].

Second, Newton completed Galileo's investigation of accelerating bodies. Galileo used the clever approach of rolling cannon balls down an inclined plane so that they accelerated slowly, thereby enabling him to time the motion of the cannonballs using his pulse as a crude but effective timing device. Through this experiment, Galileo had shown that acceleration was caused by a force acting upon an object; in the case of the cannonballs, it was the force of gravity. Prior to Galileo's discovery, it was thought velocity was related directly to force [18].

Building on Galileo's observations, Newton deduced the formula for the motion of an object subject to a force. He claimed, that A force F acting on a body of mass M, produces an acceleration A proportional to the force and inversely proportional to the mass [19].

Newton combined these formulae for gravitational force with his formula for the motion of an object to create an equation for a planet orbiting the sun. Unfortunately, the equation is a differential equation, probably the first one ever devised, which led Newton to his third and perhaps most important contribution—the invention of the calculus, which he used to solve the differential equation[4]. The result was identical to Kepler's.

In a matter of weeks, Newton had arrived at the formula Kepler had labored on for almost twenty years, and more importantly, he arrived at it using fundamentals. Newton understood the inverse square law of gravitational force acting on the planets was *why* the planetary orbits were elliptical.

Newton's accomplishment ranks as one of the turning points in astronomy and science in general. One report of his success states that "He determined it to be an ellipse, so informing Edmond Halley in August 1684. Halley's interest led Newton to demonstrate the relationship afresh, to compose a brief tract on mechanics, and finally to write the *Principia* ..." [19].

The full name of the Newton's work is *Philosophiae Naturalis Principia Mathematica* (*Mathematical Principles of Natural Philosophy*), commonly known as the *Principia*. It was published in 1687 and became revered as one of greatest written scientific documents.

Pierre-Simon Marquis de la Place (1749–1827), a successor to Newton and a celebrated French mathematician and astronomer, was central to the development of mathematical astronomy. La Place extended Newton's work and published a three-volume treatise *Celestial Mechanics*, which placed celestial mechanics on a firm footing [20].

Emperor Napoleon III was ruler of France at the time and heard about La Place's treatise and that it contained no mention of the Creator. When La Place implored Napoleon to accept a copy of his work, Napoleon is reported to have teasingly asked,

"*M. Laplace, on me dit que vous avez écrit un grand livre sur le système de l'univers, sans mentionner le Créateur,*" the English translation of which is, "*M. Laplace, they tell me you have written this large book on the system of the universe, and have never even mentioned its Creator.*"

La Place, always sure of himself, answered bluntly, "*Je n'avais pas besoin de cette hypothèse-là,*" meaning, "I did not have need for that hypothesis" [21].

I have read that some believe the science-religion conflict began with this exchange. However, sources for the exchange make no mention of this, and there are many origin places for the science-religion conflict, the reaction to Copernicus in particular. I believe the reaction to

Copernicus was the true beginning of the conflict because Copernicus had the unmitigated gall to displace man from his cherished position at the center of the universe.

Einstein Demonstrates "It's all Relative" Early in the twentieth century one of the world's most famous scientists, **Albert Einstein (1879–1955)**, published two extraordinary papers introducing the concepts of special and general relativity. While relativity is complex subject, a little knowledge of the basic ideas is essential to the understanding of the earth and the universe. Einstein's fundamental paper on special relativity was published in 1905, while the equally fundamental paper on general relativity was published in 1917 [22].

The term "relative" in special relativity refers to two observers riding on two separate platforms that are moving at constant velocity *relative* to each other. Einstein refined and extended previous efforts to develop equations that showed what an observer on one platform would see on the other platform. Central to Einstein's formulation is the fact that light has a constant velocity that is independent of the motion of an observer (see chapter 7 for a discussion of the discovery that light has a finite velocity), hence there would be a time lag in receiving information from the moving platform, especially if the platform was moving at a velocity close to that of light (relativity really only applies when moving objects approach the speed of light).

In order to accommodate the time delay, Einstein showed that a proper description of the physical universe required the inclusion of time as a fourth dimension to properly locate an object. Einstein termed this "space-time,"[5] see also Stephen Hawking's celebrated book A Brief History of Time [23].

One of the most significant results of special relativity was the demonstration that mass and energy are equivalent and interchangeable as embodied in an equation everyone knows: $E = m_o c^2$.

In this equation, m_o represents the "rest mass" of an object (i.e. the mass of an object that is not moving and c is the speed of light). The equation is normally written without the qualification of mass at rest, however, the energy of a moving object increases with increasing velocity because the effective mass increases with increasing velocity, becoming infinite when the velocity reaches c.

This increase in mass as velocity approaches the speed of light renders such phrases as, "Ahead warp fact 4 (speed = 4 × the speed of light) Scotty," as famously intoned by Captain Kirk of Star Trek, impossible. The mass increase with increasing velocity has been amply demonstrated in particle accelerators where large increases in energy are required to gain relatively small increases in velocity as the velocity approaches that of light. A material object cannot exceed the speed of light.

The special theory makes many predictions that are counter to common sense, such as the slowing of time or the shrinking of objects traveling near the speed of light, and was initially rejected by many. However, all *legitimate* attempts to disprove the theory have failed.

In the special theory, Einstein derived a set of equations that showed that gravity was a consequence of the curvature of "space-time" by a massive body, thereby solving a problem that perplexed Newton, Laplace, and others. Neither Newton nor Laplace could explain what gravity was. The fact that it acted at a distance was particularly troublesome.

The general theory addresses accelerating objects. As shown earlier, Sir Isaac Newton deduced an equation that established the relationship between the motion of an object when acted upon by a force (i.e., the object accelerates). Einstein extended Newton's formulation to show that there is no experiment that can be performed that would distinguish between a person in a closed box on the earth and a person in a closed box being accelerated with a force equal to that of gravity.

Einstein's general theory solved a couple of apparently minor issues, such as the slight difference in the orbital period of Mercury when Kepler's laws are applied. No experiment to date has disproved the general theory. The general theory's crowning achievement has been in the field of cosmology, the field that investigates the universe and will be discussed in chapter 13.

SUMMARY OF CHAPTER 6.

Many observations that did not agree with sun around the earth concept led early Greeks, especially Aristarchus, to deduce the truth: a heliocentric (sun-centered) solar system existed in which the earth and planets revolved around the sun. However, this view was in conflict with the more established geocentric (sun around earth) views that were championed by such philosophical giants as Aristotle; so it was abandoned.

Aristotle, and later Ptolemy, solved the varying planetary motion problem with a series of ad hoc orbits that properly predicted the planet motion, even Venus and Mercury, to within the measurement accuracy available. This explanation remained unchallenged for almost 1500 years.

In 1543, Polish monk Nicolas Copernicus reopened the issue with new observations and calculations and reproposed the heliocentric explanation.

In 1608, Galileo perfected the telescope and with it found the four largest moons of Jupiter, causing another outburst from the church.

Despite determined opposition by the church, the Copernican explanation eventually prevailed.

Danish King Fredrick II facilitated the construction of a large observatory in 1576 where Tycho Brahe labored for twenty years, observing planetary motion. Brahe's associate, Kepler, spent the next twenty years laboriously analyzing Tycho's measurements, finally and reluctantly deducing that planets revolved around the sun in elliptical orbits.

Between 1635 and 1636, Sir Isaac Newton deduced the equation for gravitational attraction, finished Galileo's work on accelerating bodies, and deduced the equation that describes an object moving under the influence of a force. Newton combined these equations into the world's first differential equation—the equation for the motion of a planet revolving around the sun—invented the calculus to solve it, and arrived at Kepler's empirically derived equation.

Albert Einstein extended Newton's work by postulating that the proper frame of reference of motion in the presence of a large mass is four dimensions. From this, Einstein developed the special theory of relativity, whose equations also yielded elliptical orbits but were more accurate because as they resolved some small discrepancies in the orbit of Mercury due to its closeness to the sun.

Thus the shape of the solar system had been properly explained.

As a footnote regarding the religious controversy originally engendered by Copernicus and Galileo, it is interesting to note that

Pope John Paul finally pardoned Galileo hundreds of years after Galileo's heresy [24].

LOOKING AHEAD.

Having solved the sun around earth illusion by establishing the proper explanation of the basic structure of the solar system, we next address the size of the solar system, which is a critical stepping stone to determining the size of the universe. It will also reveal the age of the universe and, by implication, the age of the earth.

CHAPTER 7:
RESOLVING THE APPARENT
SAME MOON AND SUN SIZE AND
DISTANCE ILLUSIONS

CREATING THE ILLUSION.

Size and distance illusions are created by the manner in which we estimate distances with the unaided eye. Our eyes can only present a truly three-dimensional image for objects within about twenty feet. Beyond that, we determine distance by comparing the perceived size of familiar objects. If you look at two automobiles, one will appear to be more distant if it appears to be smaller. As shown in figure 7-1, with two same size objects, possibly automobiles, the farther object appears to be smaller because the "subtended angle" (the angle the object makes with the eye) is smaller, thus creating the illusion that the farther object is smaller and thus more distant[1].

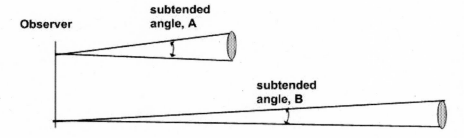

Figure 7-1. Distance estimation by relative object size

Of critical importance is our ability to recognize the two objects. If we have confidence that the two objects are the same (e.g., cars, horses, etc.) then we can conclude that B is farther away than A. Our ability to determine distance fails when we try to judge distance to unfamiliar objects like the box and oval in figure 7-2.

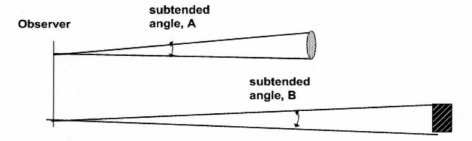

Figure 7-2. Incorrect distance estimation due to similar subtended angles

The box is larger than the oval and farther away, however it subtends the same angle; therefore it will appear to be the same distance from the observer. Illusionists often use this device.

A distance size problem occurs when we lose relationships between objects. An interesting example of this problem occurs under foggy conditions. Objects that are close and small appear large because we perceive them to be farther away.

If we consider an estimate of the size and distance to the moon and sun, we cannot really judge size and distance because we have no size by which to judge. The sun and moon appear to be the same size due to the fact that they subtend the same angle, as shown in the box-oval illusion in figure 7-2.

It is important to note that knowing the shape of an object provides no particular information regarding its size, just as knowing the size of an object may not convey any information regarding its distance from the observer. This is particularly true of the solar system; hence, we must examine both the size and distance of the various visual objects in the solar system, (e.g., the planets and the sun) to gain a complete picture of the solar system.

MEASURING DISTANCE TO THE PLANETS AND SUN,

As discussed in the previous chapter, the first person to attempt a measurement of the Earth-sun distance was Aristarchus by means of triangulation. However, due to the limited mathematical knowledge available to Aristarchus (decimals had not even been invented), he was unable to make more than a crude distance estimate. The first practical measurements of distances to objects beyond the earth also employed triangulation but of a different type. The basic technique involves not the right triangle employed by Aristarchus but the triangle shown in figure 7-3.

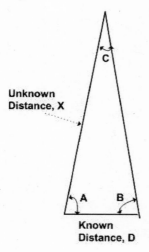

Figure 7-3. Distance measurement via triangulation with an isosceles triangle

The triangle is known as an isosceles triangle. Isosceles means equal angles; angles A and B are equal. Given a measurable known distance (c) and the ability to measure angles A and B, we can find the magnitude of the unknown distance (x) by using a trigonometric identity known as the law of sines (this trigonometry was also not available to Aristarchus). Details are provided in the trigonometry reference listed in the previous [ch 6, 2]. The desired formula that yields the valueof x, which I provide for those who have either had trigonometry or are just curious, is: x = c*SinA/SinC.

In this equation, angle C = 180 degrees - (angle A + angle B). Of course, the accuracy of the determination of x depends upon the accuracy of measuring angles A and B, as well as the known distance c. The first application of this technique was the measurement of the distances to the planets and sun, but first an accurate measurement of the circumference of the earth had to be made because portions of the circumference provide the baseline (c).

Eratosthenes of Cyrene (275–194 BCE) was the first person to make a reasonable attempt at measuring the circumference of the earth was. Eratosthenes was a Greek scholar who lived and worked in Cyrene and Alexandria [1]. He noted that on certain days the sun shined to the bottom of wells in Cyrene, which indicated the sun was directly over head. Over the course of a few years, he found that the sun was overhead on the same day each year. Having traveled back and forth from Alexandria to Cyrene on many occasions, he knew the distance between the cities. Armed with this knowledge, Eratosthenes devised a clever but simple method for measuring the earth's circumference, as shown in the diagram in figure 7-4.

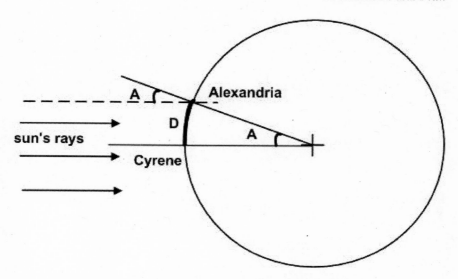

Figure 7-4. Eratosthenes' measurement of Earth's circumference

On the day the sun was overhead in Cyrene, Eratosthenes stationed an associate in Alexandria with instructions to measure the angular deviation of the sun from vertical, the angle A, in the figure. Knowing there were 360 degrees in a circle, Eratosthenes used a simple formula to determine the circumference of the earth: Circumference = D*360/A.

Eratosthenes' measurement of 25,000 miles was remarkably close to the correct value of 24,900 miles.

Giovanni Domenico Cassini (1625–1712), an Italian mathematician, astronomer, engineer, and astrologer born in what was at that time the Republic of Genoa, made the first reasonably correct Earth-sun distance measurement in 1672 via triangulation [2]. As with any triangulation measurement, Cassini first had to establish a baseline. Because the orbit of Mars is outside that of Earth, Cassini knew that at some time Mars would be in a position to provide a proper baseline. Accordingly, Cassini decided to measure the Earth-Mars distance to use as his baseline.

To accomplish this, Cassini sent a fellow astronomer Jean Richer to Cayenne, French Guiana in South America—a long distance from Genoa to therefore provide a long baseline. They observed Mars at the same time, carefully measuring its position in the sky relative to background stars. Since the two astronomers were located at different points on Earth, they saw Mars in slightly different positions and thus at slightly different angles.

So, knowing the distance between the two observation points and the angles, they were able to calculate the actual distance to Mars in miles, which they in turn were able use to determine the distance from the earth to the sun [3] [4].

It should also be noted that the measurement Cassini and Richer made was very difficult to do, and their number wasn't terribly accurate. However, it was the first big step, and the number was later refined. Today we know the distance to an accuracy of meters.

DETERMINING THE SIZE OF THE SOLAR SYSTEM.

Once the distance from Earth to the sun and Mars had been determined, it was relatively easy to employ the same triangulation techniques to determine the size of other planets' orbits. One planet in particular—Jupiter—provided a Danish astronomer an opportunity to make a startling discovery. In addition, the Earth-sun distance—also known as an "astronomical unit" (AU)—provides a convenient means for measuring distances to the nearer stars, which we will discuss in the next chapter.

A number of interesting developments followed the determination of the Earth-sun distance; one in particular was the serendipitous discovery by Cassini and **Olaf Roëmer (1644–1710)** of another illusion, the apparent infinite velocity of light. Cassini had been observing Jupiter's moons for two years beginning in 1666 and noticed some measurement discrepancies. In 1672, Danish astronomer Roëmer joined Cassini as his assistant and continued the Jupiter moon investigations.

After many measurements, Roëmer noticed that the time for the appearance of Jupiter's moons varied by about 1,000 seconds depending upon the relative position of Earth and Jupiter with respect to the sun. By carefully examining the positions of Earth and Jupiter in their orbits, Roëmer found that the 1,000 second delay occurred when Jupiter and Earth were on opposite sides of the sun, but he found there was no delay when Jupiter and Earth were on the same side [5].

After some deliberation, Roëmer was forced by Sherlock's theorem to conclude the delay existed because the velocity of light was not infinite as had been believed, but that it was indeed finite. Because the diameter of the earth's orbit is 186 million miles, Roëmer estimated the velocity of light at 186,000 miles per second, which is quite close to the currently accepted value.

This relatively simple discovery would have far-reaching consequences as we shall see in the next chapter when we employ the position of the earth at times six months apart to begin the measurement of the size of the universe, which then allows the determination its age.

SUMMARY OF CHAPTER 7.

The same moon and sun size and distance illusions were created by lack of proper reference. Measuring distance objects above the earth's surface (e.g., the planets) requires triangulation. The obvious baseline is Earth, so first the size of the earth had to be measured.

In approximately 200 BCE, Greek scholar Eratosthenes measured the earth as 25,000 miles in circumference, which is close to the presently accepted value. Then in 1672, using triangulation from two locations on Earth, Cassini made the first reasonably accurate measurement of the Earth-Mars distance.

The Earth-sun distance and observations of Jupiter's moons by Danish astronomer Olaf Roëmer showed that the appearance time of Jupiter's moons differed by approximately 1,000 seconds and in 1672 led Roëmer to realize light had a finite velocity of approximately 186,000 miles per second.

LOOKING AHEAD.

Having solved two major illusions—(1) the sun and planets around the earth illusion, which led to a proper explanation of the solar system shape, elliptical planetary orbits; and (2) the same moon and sun size illusion, which led to a proper explanation of the size of the solar system, especially the 186-million-mile major axis size of the earth's orbit—the stage was set for the resolution of the apparent closeness and motion of the stars around Earth illusion, which led to a determination of the size and age of the universe.

CHAPTER 8:
RESOLVING THE APPARENTLY
CLOSE STARS AROUND EARTH
ILLUSION

CREATING THE ILLUSION.

It's a clear night and you are gazing up at the stars. They seem so close, "you can almost touch them," I believe someone once wrote. As you stand there, you arrive at the same conclusion regarding the distance to the stars as the ancients, who believed the stars appear to be a close, fixed distance from the earth and attached to a celestial or crystal sphere. You continue to observe the stars for a while, and you notice that they seem to "drift" toward the west. Closer observation will disclose that the stars appear to be rotating around a star in the northern sky, a star now termed the "pole star."

The closeness illusion is caused by the same reference problem that causes the apparently same size and closeness of the sun and moon illusion, but there are other causes. While we can see the sun and moon, we cannot "see" any of the stars in the conventional sense. Because of the extreme distance to the stars, the human eye and even the most powerful telescope lacks the power to resolve a star. We can "see" a star because it is a source of light, but what we "see" is an optical phenomenon known as an Airy disc [1], named for its discoverer, British mathematician and astronomer **Sir George Biddell Airy (1801–1892)**. Along with

Airy's other accomplishments, he played a leading role in establishing Greenwich, England as the prime or zero meridian. Because we only see a star's Airy disc and because of the earth's atmosphere, all stars appear to be about the same distance from Earth.

The apparent rotation of the stars around the earth is caused by the earth's rotation, as noted earlier. However, the westerly drift is caused by Earth's revolution around the sun. In the winter, we are looking at the stars in the opposite direction as in the summer.

A WORD ABOUT COSMOLOGY.

The solution of the apparent close stars illusion involves the arcane subject of cosmology, which is defined as "The study of the physical universe considered as a totality of phenomena in time and space." [2]." This is, perforce, quite a broad subject, and I intend to only extract those parts that will:

- explain in general terms *how* and *why* we know the size and age of the universe;

- explain *how* and *why* the elements were created;

- explain *how* and *why* the sun and the earth were formed;

- explain *why* all planets revolve around the sun in essentially the same plane and direction;

- provide additional alternate and independent methods for determining age of the earth.

EARLIEST OBSERVATIONS AND EXPLANATION OF THE CLOSE STARS ILLUSION.

Besides the observations of the "wandering stars" (i.e., the planets that were discussed in the previous chapter), the only observations other than the ones that created the illusion were the sudden appearances of unusually bright stars that could be seen even in daylight. One of the first stars appeared in 185 CE. Chinese astronomers recorded a strange, extra bright star that could be seen during the daytime. Another "extra bright star" observed by both Chinese and Arabian astronomers occurred in 1006 CE. Yet another widely observed star appeared in 1054 CE and produced what we now term the "Crab Nebula."

The last "extra bright star" observable by the unaided human eye appeared in 1604 CE and was employed to refute Aristotle's belief that universe beyond the planets was incapable of change [3]. Other than this attack on Aristotle, no attempt to explain the close star illusion occurred until an estimate of the real size of the solar system was achieved in the seventeenth century, which was described in the previous chapter.

EMPIRICAL DEVELOPMENT OF A PROPER EXPLANATION OF THE CLOSE STARS ILLUSION.

Resolving the problem of the "extra bright stars" and the close stars illusion ultimately revealed that the universe is extremely large and very old. However, as with the resolution process of other illusions, this resolution required contributions from many persons. It began with the measurement of the distance to the nearest stars, which built upon our knowledge of the size of the solar system.

Knowing the Earth-sun distance, we can in principle measure the distance to any star using the triangulation method discussed in the previous chapter and illustrated in figure 8-1.

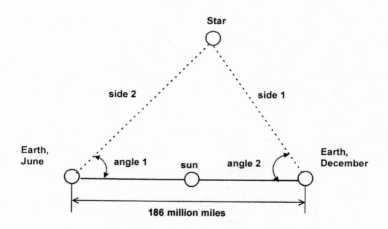

Figure 8-1. Measurement of distance to a star via triangulation

We merely measure angles 1 and 2 in a triangle formed by the Earth-sun-Earth line and the star. The difference between the two angles is known as the parallax of a star[1]. Unfortunately, stars are so far away that the difference between the two angles is very small and imperceptible to the human eye; hence the length of either side 1 or side 2 will also be the distance to the star.

The same trigonometric formula discussed previously in relation to Figure 7-3

will yield the distance of side 1, where EOD represents the earth orbit distance, which is 186 million miles.

Table 8-1 presents the eight nearest stars as measured by parallax.

Star	Distance from Earth in light-years
Proxima Centauri Alpha Centauri C	4.3
Rigil Kentaurus Alpha Centauri A	4.3
Alpha Centauri B	4.3
Barnard's Star	5.9
Wolf 359	7.6
Lalande 21185	8.1
Sirius A	8.6
Sirius B	8.6

Table 8-1. Eight of Earth's nearest stars

Due to extreme distance to even the closest star, distance to the stars is measured in light-years. A light-year is the distance light travels in one year, which is equal to 5,878,499,810,000 miles or 5 trillion 878 billion 499 million 810 thousand miles [4]. Thus our nearest star, Proxima Centauri, is 25,277,000,000,000 miles away. Clearly, astronomical distances are almost impossible to comprehend. It is also worth noting that Sirius, the brightest star, is not the closest.

There are limits to parallax measurements. As the distance to a star increases, the ability to use triangulation is limited due to angular measurement accuracy. This limit occurs at a distance of about 65 light-years [5], but serendipity solved this distance measurement limitation via the discovery of an unusual star.

John Goodricke (1764–1786), an eminent and profoundly deaf amateur astronomer [6], discovered an unusual star in 1784, whose brightness varied with a fixed period. The variable star was termed Delta Cephi as it was found in the constellation Cephus (the king). It is for this reason stars of this type are termed Cepheid variables [7]. The use of a Cepheid variable as an astronomical yardstick was developed by Henrietta Lovett.

Henrietta Swan Leavitt (1868–1921) graduated from Radcliffe College in 1893 and obtained a position at the Harvard College Observatory in the unchallenging capacity of "computer," assigned to measure and count the brightness of star images on photographic plates. Leavitt's study of star images, in particular the Cepheid variables discovered by John Goodricke[2], led Leavitt to develop a groundbreaking concept that became the basis for the extension of astronomical distance measuring beyond parallax.

As Leavitt cataloged thousands of variable stars in images taken of a group of stars known as the Magellenic Cloud stars, she was the first to realize the importance of Goodricke's discovery. She noted that a few of the variables showed a pattern: brighter stars appeared to have longer brightness variation periods. Leavitt published her results in the 1908, *Annals of the Astronomical Observatory of Harvard College* [8]. After further study, by 1912 she had confirmed that the variable stars of greater intrinsic brightness were actually Cepheid variables that did indeed have longer variation periods. The relationship of brightness to period was quite close and predictable, which permitted Leavitt to produce a precise mathematical formula for this relationship. Leavitt published her formula in a 1912 paper [9] confirming that a Cepheid variable can be used as a standard to determine the distance to the stars beyond the capability of parallax.

The reason that Cepheid variables can be used as standards relates to this simple fact: as an object of known, or intrinsic, brightness moves into the distance, the "apparent" brightness decreases as the square of the distance [10]. An object's intrinsic brightness is the amount of visible energy emitted from the object per unit time, and it is a fixed quantity known as an object's luminosity [11].

The relationship between period-luminosity and distance can be calibrated with great precision using the nearest Cepheid stars, which can be measured by parallax. Therefore, greater distances can be determined, as demonstrated with great success by Edwin Hubble, who we will discuss in more detail shortly.

Vesto Melvin Slipher (1875–1969), an American astronomer, spent his entire career at the Lowell observatory in Flagstaff, Arizona [12] that was founded in 1894 by businessman, author, mathematician, and astronomer **Precival Loevel (1855–1916)**. Slipher became observatory director in 1926.

His specialty was the use of spectroscopy to investigate astronomical phenomenon. In 1912 when examining the hydrogen spectra of various galaxies, he made a startling discovery: the spectrum of hydrogen for galaxies at great distance from the earth was shifted toward the red end of the spectrum (see figure 8-2).

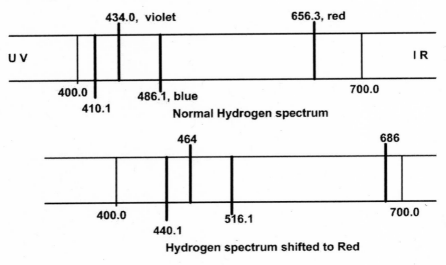

Figure 8-2. Depiction of hydrogen redshift

Slipher had made a momentous astronomical discovery called the galactic redshift [13]. As with Leavitt's discovery of Cepheid variables, the redshift discovery was another important piece of the astronomical puzzle[3].

Alexander Alexandrovich Friedman (1988–1925), a Russian cosmologist and mathematician, made another astonishing discovery in 1922: a solution to the equations of Albert Einstein's general theory of relativity that demonstrated the possibility of an expanding universe. This discovery was something entirely foreign to any existing explanation of the universe because it contradicted the prevailing view that all the stars were contained within the Milky Way galaxy [14]. Friedman's calculations were one more piece of the "what is the universe?" puzzle and set the stage for Edwin Hubble, one of the foremost astronomers who ever lived.

Edwin Powell Hubble (1889–1953) properly assembled the pieces of data that had been accumulating and profoundly changed our understanding of the universe [15]. To place his contributions into context, on the 100th anniversary of Hubble's birth, another celebrated astrophysicist named Alan Sandage paid Hubble this tribute: "From 1922 to 1936 Hubble solved four of the central problems in cosmology, any one of which would have guaranteed him a position of the first rank in history" [16]. We will discuss three of the four central problems Sandage talks about.

In 1919, Hubble accepted a position at the Mount Wilson observatory in California. He arrived just after the completion of the 100-inch Hooker Telescope, then the world's largest telescope in existence. When Hubble arrived at Mount Wilson, astronomers believed all stars were contained in the collection of stars known as the Milky Way galaxy. Astronomy had progressed beyond the initial concept of a very small universe due to the illusions discussed previously, but not nearly as much as it would during Hubble's tenure.

Hubble was aware of Harriet Leavitt's work on Cepheid variables, and using the power of the 100-inch telescope, he soon discovered a Cepheid variable in the star cluster Andromeda. When Hubble applied Leavitt's distance formula to this Cepheid variable, he arrived at the almost unimaginable distance of one million light-years. He concluded that all of the stars were not in the Milky Way and that Andromeda was a separate galaxy [15]. Since then, more precise measurements have been made, and the present value of the distance is 2.5 million light-years[4].

Hubble examined other galaxies and soon found Cepheid variables in them. As he assembled the distances, he realized the universe was vastly larger than anyone had imagined.

Next, Hubble and his assistant, Milton L. Humason, compared the distances they were measuring with the redshift data, which was originally provided by Slipher but was also being collected at Mount

Wilson. They found that a galaxy's distance was approximately proportional to its redshift. Applying Sherlock's theorem, Hubble realized that the galaxies were moving away from Earth and that the farther away the galaxies were, the faster they were moving away. This proved that the universe was expanding.

In 1929, Hubble and Humason were able to formulate the empirical redshift distance law of galaxies, nowadays simply termed "Hubble's law," which states that the greater the distance between any two galaxies, the greater their relative speed of separation.

Hubble's observations of the expanding universe corroborated Friedman's solution to general relativity [14]. However, Friedman's solutions supported two possibilities: (1) a steady state universe in which matter is created in sufficient quantity to support the expansion; and (2) a universe that began in one place with a "Big Bang." As will be discussed below, further observations support the Big Bang.

In retrospect, an expanding universe seems quite logical. The force of gravity acting on two objects will, absent any constraints, tend to pull them together. The reason the earth is not pulled into the sun is because the centripetal force created by the earth's motion balances the force of gravity. But what prevents the galaxies from being pulled together? The answer lies in the fact that they must be moving away from each other. This is exactly what one would expect if the universe began with an unimaginably large explosion that hurled matter apart. Accordingly, Hubble's discovery leads to the conclusion that the universe began with a bang [17], but one would expect that the action of gravity should slow down the galaxies and eventually cause them to stop and begin to fall together, ending with a big pileup. The fact that this does not appear to be happening is one of the stranger aspects of the universe, but that's not relevant to our current discussion.

The determination of the size and organization of the universe is one of the more important scientific accomplishments in all of history. As mentioned by Sandage, "From this work, by him and by others of his generation, it is widely believed that some glimpse of a 'creation event of the universe' became available to science by an objective method, not as in other times by metaphysics or speculation" [16].

Hubble's determination firmly established the fact of the great size of the universe and thus the great age. However, the Big Bang was still a speculation until the discovery of radiation left over from it.

George Gamow, besides explaining alpha particle emission, also suggested that the observed abundance of hydrogen and helium in the universe could be explained if the universe began with a big bang [18]. Moreover, in a paper published as "The Origin of Chemical Elements" (*Physical Review*, April 1948), [18] Gamow predicted that the Big Bang would produce extremely hot radiation. This radiation would, over the billions of years that the universe has existed, cool to a temperature close to 5 degrees above absolute zero. At this temperature, the radiation would be in the form of microwaves. Unfortunately, microwave receivers capable of testing this hypothesis were not available at the time.

But, as we have seen several times in this book, serendipity is one of the most valuable scientific assistants. It was no different with Gamow's predicted radiation.

In 1965, two young radio astronomers, Arno Penzias 1933–) and *Robert Wilson* (1936–), accidentally discovered the predicted radiation. They were developing an exceptionally sensitive microwave receiver to study radio emissions from the Milky Way. Soon after turning the receiver on, they detected a strange radiation that was soon determined to be diffuse and emanating uniformly from all directions in the sky. It had a temperature of approximately 2.7 degrees above absolute zero. Initially they could find no satisfactory explanation for their observations and considered the possibility that their signal may have been due to some undetermined systematic noise within the receiver. They even considered the possibility the signal was due to "a white dielectric substance" (i.e., pigeon droppings) in their horn [19].

Penzias and Wilson discussed the problem with others, two of whom were physicists **Robert Dicke (1916–1997)** and **Jim Peebles (1935–)** of Princeton University. Dicke and Peebles realized that Penzias and Wilson had detected the background radiation George Gamow predicted in 1948. The background of microwaves, now termed the cosmic microwave background (CMB) was in fact the cooled remnant of the primeval fireball—an echo of the Big Bang. The CMB is perhaps the most conclusive (and certainly among the most carefully examined) pieces of evidence for the Big Bang theory, as noted by this Web site:

"When the intellectual history of the 20th century is written, a few achievements will tower over all. Einstein's theory of general relativity will be one; the laws of Quantum Mechanics will be another. The so-called Big Bang Theory of the origin of the universe will be a third" [20].

At first the CMB was believed to be isotropic, a fancy word meaning that the intensity of the CMB radiation was the same regardless of arrival direction. However, it was argued that because the universe had structure, there should be small but measurable variations, technically known as anisotropic variations in the CMB.

In 2003, NASA launched the Wilkinson Microwave Anisotropy Map (WMAP) satellite to search for the predicted variations in the CMB [21]. The WMAP has been spectacularly successful. There is not room for a full discussion of the WMAP findings; these are discussed in several articles (e.g., [22] and [23]). For our purposes, the significant and irrefutable WMAP findings are [21]:

- The age of the universe is 13.7 billion years, give or take 200 thousand years.

- The width of the universe is at least 78 billion light-years.

- The universe is composed of 4 percent ordinary matter, 23 percent of an unknown type of dark matter, and 73 percent of a mysterious dark energy.

- Current theories applied to the WMAP data and studies of supernovas indicate the universe will expand forever.

FORMATION OF THE UNIVERSE.

The renowned American physicist and Nobel laureate Steven Weinberg, introduced above, summarized the amazing WMAP findings regarding the beginning of the universe [20], starting at 0.02 seconds after the Big Bang.

Time After Big Bang	Event
0.02 seconds	Universe basically electromagnetic radiation (light)
0.11 seconds	Excess of protons over neutrons appear
1.09 seconds	Light cannot escape primordial fireball
3 min. 46 seconds.	Deuterium is now stable, helium represents about 26 percent of the mass in universe
400,000 years	Hydrogen atoms have coalesced out of cloud of electrons and protons
700,000 years	Universe is cool enough for hydrogen and helium to become stable atoms
400 million years	Stars emerge and form the first galaxies

Table 8-2. Time line of events after Big Bang

Star Formation: Star formation [24] begins with thin swirling clouds consisting mainly of the most abundant element in the universe: hydrogen. The swirling motions cause atoms to occasionally come close enough to form small pockets of gas, where each atom exerts a small gravitational attraction on its neighbor. This counters the tendency of the atoms to disperse due to electrostatic repulsion. If the number of atoms in a gas pocket becomes large enough, the accumulation of all of these separate forces will hold it together indefinitely, forming an independent cloud of gas[5].

With the passage of time, gravity's constant influence causes the gas cloud to contract toward the gravitational center of the cloud. Approaching the center, the atoms' velocities increase and heats the gas to even higher temperatures[6]. This shrinking and continuously self-heating ball of gas is an "embryonic star."

The temperature of the shrinking cloud steadily increases until it reaches the critical temperature of 20 million degrees Fahrenheit. The hydrogen atoms in the gas now have sufficient velocity to penetrate the nuclear repulsive force and fuse into helium with a concomitant release of the energy of fusion (as described in chapter 1). This release of energy marks the birth of the star. At this point in the life of the star, the diameter of the ball has shrunk to about one million miles, which is the size of our sun and other typical stars.

During the hydrogen burning phase of a star's life, the energy generated by fusion passes to the surface and is radiated away in the form of light, which enables us to see the star in the sky. The energy release generates an outward pressure that halts further contraction of the star, and the star lives out the rest of its life in a balance between the outward pressures generated by the release of nuclear energy at its center and the inward pressures created by the force of gravity.

Galaxy Formation: As soon as a few stars formed, they began to form groups which we term "galaxies." The first galaxies appeared at approximately the same time as the first stars, about 400 million years after the Big Bang [25].

Today there are billions of galaxies in the universe, of which our galaxy, the Milky Way, is one. Our solar system is located in one of the "arms" of the Milky Way, approximately 26,000 light-years from galactic center [26].

Death of Stars: Just as stars are formed, they must ultimately die when their fuel has been consumed. The end of a star's life depends completely upon its size. In all stars, the fusion of hydrogen to form helium continues until all of the hydrogen has been converted to helium. Typically, this is about 99 percent of a star's lifetime.

After a star's hydrogen is exhausted, the fate of the star depends upon its size. In the case of our sun, which is cataloged as a small star, fusion of helium becomes the star's energy source as three nuclei of helium combine to form the nucleus of the carbon atom. The fusion of helium requires higher temperatures and generates more radiation pressure; therefore the star expands greatly and becomes a red giant [27]. An example of a red giant is Betelgeuse, which can be easily seen in the Orion Galaxy. In the case of our sun, the red giant stage will not be reached until about 5 billion years in the future, at which time the sun will have expanded sufficiently to engulf the earth. Obviously, all life will be obliterated (our descendants, if there are any, will need a new home).

Once the helium is consumed, a relatively small star like our sun does not have strong enough gravity to fuse the carbon produced by helium fusion, and the star stops producing energy. At that time the star slowly cools and continues to shrink, which generates enough heat and light so that it remains visible. At this stage, it becomes a white dwarf. But the star continues to cool and eventually becomes just a black ball [28].

If the star is massive enough, the temperature produced by the star's continuing collapse will reach another critical level. When it reaches 600 million degrees, the carbon nuclei at the center of the star fuse and form even heavier elements than carbon, such as oxygen [27]. Eventually the carbon is consumed, and the next heavier nuclei fuse.

This process continues down the binding energy curve until the massive star successively manufactures all elements up to iron. As shown in the binding energy curve (figure 5-2), the next element has less binding energy than iron, so iron cannot continue the fusion process. When this happens, the star's nuclear furnace shuts down rather suddenly. With no radiation pressure to support it, the star collapses under the force of its own weight with a cataclysmic explosion. The heat generated by this explosion creates temperatures so high (up to 100 billion degrees) that fusion forces great enough to create of all the elements heavier than iron are possible [27].

After the cataclysmic collapse, there is an equally cataclysmic rebound that disperses the contents of the star at extremely high velocity. The exploding star is termed a "supernova," one of the most violent events in the universe. Supernovas are, of course, the source of the "extra bright stars" observed by ancient astronomers mentioned in an earlier chapter.

Thus, without supernovas, the elements that comprise the earth would not exist and we would not be here to appreciate them.

FORMATION OF THE PLANETS, INCLUDING EARTH.

As recently as one or two generations ago, scientists were very uncertain about how planets formed. As Robert Jastrow points out, "When I was in high school [in approximately 1900], a popular explanation of planet formation involved a close encounter with another star which ripped material from the sun which then formed the planets" [28].

However, as Jastrow further observes, this explanation can be eliminated via simple statistics. The stars are separated by vast distances; hence the probability of a stellar collision is extremely remote. In fact, perhaps only one or two would have occurred in the entire lifetime of the galaxy. On the other hand, astronomers continue to find more and more extra solar planets, as noted in [29].

A much more reasonable explanation for planet formation is now being verified by computer simulation and observation of extra solar planets[7]. It is found in the formation of stars from clouds of dust and gas explained earlier. As the gas cloud that eventually gave birth to the sun contracted, it began to rotate (just like a skater spins as the he or she draws his or her arms in). This rotation produced a disc of matter surrounding the sun. Just as eddies in the original vast, tenuous cloud gave rise to small islands of concentrated material with enough gravity to begin to condense the cloud to form the sun, very small islands of concentrated material in the disc surrounding the developing sun gave rise to the planets.

The formation of planets in a disc rotating around its star is the reason *why* all of the planets revolving around the sun move in approximately the same plane and in the same direction.

An excellent explanation of planet formation can be found in the description of a computer simulation program written by the Planetary Science Institute (PSI) [30]. The program has sufficient capability to chart a course from a cloud of dust to a solid planet. While planetary formation cannot be solved exactly because it would require the ability to follow all the particles and observe their behavior, computer power has reached the point where a computer program such as the PSI program can "follow" enough particles to determine how the terrestrial planets (and others) were formed.

The PSI program models the motions of particles at different distances from the sun and tracks the results of collisions based on actual physical and mechanical properties. The model shows that adjacent particles undergo collisions at relatively low speed almost in the same way that high-speed racecars moving around a circular track might nudge into each other. The slowly colliding particles "stick together," providing strong evidence that dust particles tend to aggregate.

When the PSI program is executed with different starting conditions, the aggregation of innumerable small particles into smaller numbers of small "building bodies" called planetesimals can be observed. Once the aggregation starts, the larger planetesimals tend to sweep up the smaller ones. As this process continues, the gravity of the biggest planetesimal, which by now resembles a planetary body, tends to dominate the aggregation. The larger planetary body sweeps up most of the other bodies, producing a system with a few planet-sized worlds. Eventually the planetesimals continue aggregating until a system of a few planets is formed and few small planetesimals remain.

Additional simulations of planetary formation has refined our understanding of planetary formation, but the basic process has not changed [31]. While we have explained *how* and *why* the universe was created, we have not explained *who* created the universe because the creation occurred without any supernatural intervention, with the possible exception of the Big Bang.

WHAT OR WHO CREATED THE BIG BANG?

Entering the query, "Did God Create the Big Bang?" into a search engine like Google, one can find several Web sites devoted to this question. For example [32] states,

"Some Christians are vehemently opposed to the 'Big Bang Theory.' They view it as an attempt to explain the origin of the universe apart from God. Others ascribe to the Big Bang Theory, with the view that it was God Himself who caused the Big Bang. God, in His infinite wisdom and power, could have chosen to use a Big Bang method to create the universe, but He did not."

The Web site's author lists a number of conflicts with material in Genesis that precludes God using the Big Bang to create the universe. The author and concludes by saying, "The true purpose of the Big Bang theory is to deny His existence."

Since God exists, at least according to the Web site's author, the statement essentially argues that the Big Bang didn't happen. However, to defend the theory that the position that the Big Bang did not happen, one has to refute all the data presented in this chapter and all of the data contained in all of the references listed—a difficult, if not impossible task. Furthermore, if one were to ascribe the Big Bang to

God, as with other explanations that invoke God, nothing will have been added to our understanding. We still wouldn't know *how* or *why* the Big Bang was initiated.

It is tempting to state that, in view of this chapter and the previous chapter, the age of the earth issue is resolved. However, it is necessary to complete the picture by demonstrating how the resolution of the unchanging earth illusion developed before the advent of cosmological explanations. This explanation will be provided in the following chapters (somewhat of a bottom-up approach) and should conclude the debate regarding the earth's age because twelve independent methods for the determination of the earth's age will be provided.

As a final note, in addition to the life on Earth, the universe is also evolving. Eventually, all of the hydrogen in the universe will have been converted to stars, which will consume the hydrogen and leave an empty universe. Of course, long before this occurs our sun will have consumed its share of the universe's hydrogen.

SUMMARY OF CHAPTER 8.

Using the earth's orbit as a baseline, astronomers began to measure the distance to the stars using parallax. They were stunned. The closest star is approximately four *light-years* distant. A light-year is over 25 trillion miles, an almost impossible distance to image. Gradually, they measured stars farther and farther away until they reached the limit of parallax measurement, which is about 65 light-years. At that distance, the parallax is too small to measure.

Fortunately, John Goodricke and Henrietta Leavitt discovered an alternate "yardstick" called Cepheid variable stars. These stars permitted measurements far beyond triangulation from the earth. Then in the 1920s, Edwin Hubble applied the Cepheid variables to demonstrate that a patch of stars named Andromeda was over one million light-years distant and concluded Andromeda was a separate galaxy. Hubble continued to measure galaxies farther and farther from Earth.

Meanwhile, in 1926 Slipher discovered the "redshift." It was a shift in the spectrum of the hydrogen of some galaxies to the red end of the spectrum. This indicated that those galaxies appeared to be receding from Earth.

In 1928, Hubble and his assistant, Humason, were able to combine Slipher's redshift data with their distance data to establish the fact that the universe was expanding—the farther away a galaxy was, the larger its redshift.

Hubble and Humason's findings corroborated a prediction by Alexander Friedman that was based upon his solution of the equations of Einstein's general relativity that the universe is expanding. And it probably started with an unimaginably large explosion—a big bang

In 1948, George Gamow published a paper that predicted the existence of microwave radiation that would be a remnant of the Big Bang. Due to the long expansion time, this radiation would be approximately 5 degrees above absolute zero. Then in 1965, Arno Penzias and Robert Wilcox discovered the predicted radiation, now termed the "cosmic microwave background" (CMB), and it seemed to be coming from all directions, as would be expected.

Initially the CMB seemed uniform, but since there were galaxies, it was argued that there should be small but finite fluctuations in the CMB. To test this hypothesis, a series of satellites were launched, culminating with the Wilkinson Microwave Anisotropy Mapping (WMAP) satellite. WMAP conclusively demonstrated that the universe was approximately 13.7 billion years old.

Cosmologists eventually deduced that stars form from the gradual accumulation of dust and gas (mostly hydrogen) into a gradually shrinking sphere. As it shrinks, the sphere begins to spin, and some of the materials form a disc around the star. As the nascent star continues to shrink under gravitational attraction, it heats up until it is hot enough to fuse hydrogen into helium, and then a star is born.

Meanwhile, in the disc of material surrounding the star, matter begins to accumulate until planets are formed. Under the correct conditions, a planet like Earth results. So far, over 300 "extra-solar" planets have been found, but current observational capability, which is continually being improved, cannot "see" an Earth-sized planet.

Regarding the "six little friends of explanation," we have now explained *what* happened at the beginning of the universe—the Big Bang—which then provides the general explanation of *when, how,* and *why* the universe was created. We next explained *how* the stars were formed, and how they then formed the galaxies, including the Milky

Way. Finally, it was shown that planet formation accompanies star formation, which leads finally to *how* and *when* the earth was created: 9.1 billion years (subtract 4.5 billion-year Earth age from 13.6 billion-year universe age) after the Big Bang.

LOOKING AHEAD: TWO PARTS TO THE UNCHANGING EARTH ILLUSION.

In the next four chapters (chapters 9–11), we will examine how the apparently unchanging features of the earth illusion was resolved and how it led to a true understanding of the earth's creation, which will help complete the explanation of Earth's formation that we began in this chapter. However, as explained in the prologue, in order to place the explanation of the apparently unchanging earth illusion into proper perspective, it is necessary to recognize that there are actually two separate but related parts to this illusion:

1. Apparently unchanging physical features of the earth (essentially the field of geology)

2. An apparently unchanging biological features of the earth, animal, and plant structures (the field of biology)

Geology and biology are generally considered separate subjects, but from a historical perspective of the development of a proper explanation of the apparently unchanging nature of the earth, these two subjects are intertwined. For example, the interrelationship of fossils (biology) and rocks (geology). We will follow both paths of discovery in four connected chapters.

Explanation of the resolution of the physical features illusion is divided into two chapters. Observations and explanations up to 1850 will be addressed in chapter 9, while observations and explanations since 1850 occupies chapter 10. As you will see, there was a significant difference in the development of understanding during these two periods.

In a similar manner, the resolution of the apparently unchanging animal and plant structure illusion is divided into two chapters also. Chapter 11 will cover the period ending with Charles Darwin's publication of *The Origin of Species*, while the period after Charles

Darwin, the time during which the full explanation of the process of evolution was developed, occupies chapter 12.

Finally, chapter 13 will integrate the material presented in chapters 9 through 12 *and* explain the true creation of the earth, and how it began as a molten ball of rock—the result of the accretion of material described in this chapter—and continued through many changes, some violent, ultimately culminating in the brilliant blue planet we enjoy today.

CHAPTER 9:
RESOLVING THE APPARENTLY UNCHANGING PHYSICAL FEATURES OF THE EARTH ILLUSION UP TO 1850

CREATING THE ILLUSION.

You are visiting a famous National Park such as the Grand Canyon or Yosemite for the nth time. As you gaze on the park's amazing beauty, all seems to be just as magnificent as you remember it. Perhaps you visit one of the park's museums to view photographs taken when it was first created. If you look carefully, except for some trees that have grown since the first photos were taken, nothing appears to have changed. Some of the parks might have paintings or drawings created by the original inhabitants of the parks, the Native Americans. If they were good artists and you look closely, again nothing appears to have changed.

In fact, any place you go on Earth that has accurate depictions of the landscape, regardless of how old the depictions are, (e.g., cave or rock paintings, especially paintings in such caves as the Lascaux in France), you will find that the depictions are indistinguishable from those that might be made today. The Earth's physical features do not appear to change.

But it's all an illusion. The apparently unchanging Earth is an illusion produced by the glacial slowness of physical processes. The processes are so slow that even over thousands of years only a careful observer would detect a change. Consequently, the illusion is totally transparent. Unfortunately, it leads to the incorrect belief that the earth is not very old.

ISLAMIC OBSERVERS BEGIN TO SUSPECT THE TRUTH.

While many clues such as fossils of plants and animals imbedded in rocks far above sea level or huge boulders in places where they shouldn't be might have suggested to early humans that the earth might not be unchanging, it was apparently Islamic observers who first began to realize the truth. Unfortunately, discussion of Islamic science gets little attention in the West, except perhaps for recognizing that Al Gebra (algebra), which derives from the Arab phrase *al-jabr* for "the reunion of broken parts." It is the recognition of Arabic contributions to mathematics [1].

One of the earliest Islamic geologists **Abu al-Rayhan al-Biruni (973–1048 CE)** [2], produced the earliest writings on the geology of India, hypothesizing that the Indian subcontinent was once covered by a sea [3]. **Avicenna (981–1037)**, another Persian polymath, also made significant contributions to geology [4]. Around 1000 CE, Avicenna was already suggesting a hypothesis about the origin of mountain ranges, which in the Christian world eight hundred years later would still have been considered quite radical. In particular, one of the principles enunciated by Avicenna that underlies geologic timescales was the principle of the superposition of rock strata, a foundational principle of geology which states, "Sedimentary layers are deposited in a time sequence, with the oldest on the bottom and the youngest on the top [4]."

While some of the Arabic learning was passed to the West, the violent decline of the Islamic world beginning in 1300 with the Mongol invasions prevented much cross-fertilization [3]. Hence, at the risk of shortchanging the Islamic contributions, the fact that Islamic investigators may have realized that the earth was indeed changing while the West was still in the Dark Ages is not really relevant to this book. For this book, the issue is the time when the West finally pulled itself out of the miasma of the Dark Ages and realized evidence of change was readily available if properly viewed.

Regarding the realization of change, it is important to understand that thousands of years of dogma had resulted in an insular worldview, which is eloquently described by Simon Winchester, extant as recently as the mid-1700s:

"... among that small muddle of warm-colored stone cottages, with thatched roofs and climbing roses, the village green and the inn and the duck pond and the old steepled parish church, beliefs about such weighty matters as humankind's beginnings were unburdened by the complications of too much thought. They [humankind's beginnings] were taken on faith as the revelations of Scripture, and when and if they were recounted, they were larded with appropriate and long-remembered quotations from the Book of Genesis. The infant [William] Smith [whose contributions are discussed in detail below], whose father and mother were an essentially unremarkable country couple was thus born into a world of which at least the basis of existence had a certainty. The origins of the planet, just like the origins of mankind, were assumed to be fixed, uncomplicated and divinely directed" [5] p-16.

Thus, it required considerable courage to confront this comfortable view that was strongly supported by the religious authority[1]. But, as will be explained below, in the early 1800s, in a manner similar to the early investigators of the astronomical illusions in the 1500s and unafraid of confronting this entrenched dogma, farsighted investigators began to realize that perhaps the earth had been subject to change.

Before we examine these investigations of the early 1800s, we will examine the earliest attempts to explain the earth's physical features. Some of the physical features readily observable by early humans were:

- The basic structure and general shape of the earth—it has mountains (both craggy and rounded), valleys of varying steepness and depth, bodies of water of varying sizes (seas, lakes, and ponds), and running water of sizes varying from rivers to creeks.

- The basic structure is generally stable except for volcanoes and earthquakes. Volcanoes build mountains and spew molten rock. Earthquakes do not disturb land much but can cause considerable physical damage, especially because they are unpredictable.

- Varied surface material is easily observable. There are basically three types of rocks—igneous, sedimentary, and metamorphic. Igneous rocks are those that were formed first from molten rock, and they are generally coarse and crystalline (e.g., granite). Sedimentary rocks are formed in layers, usually underwater, either by the accumulation of eroded materials like sand, which forms sandstone, or by the accumulation of the bodies of sea life that live in shells, which forms limestone. Metamorphic rocks that are either igneous or sedimentary rocks are rocks that have been subjected to great heat and pressure.

- Rock layering and folding is observable in mountain ranges and produces dramatic scenery.

- Periodic floods are caused by rivers swollen with melting snow in spring.

- Fossils are biological entities in which the biological material has been replaced by rocky minerals (e.g., dinosaurs, whose fossilized skeletons are found in many places and have been preserved for billions of years). Fossils will be discussed in more detail below.

All of these early observations of Earth's physical features led to various explanations. One of the more important early explanations is contained in the Book of Genesis. The Bible and the book of Genesis have had considerable historical impact that still being felt today[2]. Since, as the Gallup poll demonstrates, many people still believe Genesis is the explanation for all of the observations listed above, it is appropriate to begin with this early explanation of the creation of the earth.

THE BOOK OF GENESIS—AN OVERVIEW.

The "So just what is creationism trying to say?" Web site contains the following summary statement: "There is *no reason not to believe* that *God created* [italics in the original] our universe, earth, plants, animals, and people just as described in the book of Genesis!" [6]. It is an interesting statement, especially since Genesis is silent regarding the universe and doesn't really describe the earth's creation very well.

There is an excellent and balanced overview of Genesis by J. J. M. Roberts from the Believe, Religious Information Web site [7]. The overview places Genesis in proper historical perspective. I summarize here: While Moses has generally been considered the traditional author of Genesis, modern scholars generally agree that the book is probably a composite of at least three different authors. These authors are designated J, E, and P as, "[t]he interpretation of the book has led to many controversies. One of the most difficult problems has been distinguishing historical fact from symbolic narration intended to convey a religious message." [7]

In this regard, I believe there is an interesting parallel between Genesis, the *Iliad*, and the *Odyssey*. As mentioned above, Genesis is generally attributed to Moses, while the *Iliad* and *Odyssey* are attributed to Homer [8]. However, Moses lived around 1600 BCE, and Homer lived sometime between 800 and 900 BCE, although some scholars argue Homer lived closer to the Trojan War, (ca. 1200 BCE). Specific dates aside, in neither case had writing been developed to the point where any of the three books could have been written down by their authors. Accordingly, Genesis, the *Iliad* and the *Odyssey* were passed down via memorization and oral transmission, and with it came all the attendant problems of this type of transmission, even for the most careful of oral transmitters.

Ancient Hebrew did not develop into a written form until about 1000 BCE, therefore written versions of Genesis could not have existed until 600 years after they were conceived.

J was probably written between 848 BCE when King Jehoram gained power in Judah and 722 BCE when the Assyrians destroyed the northern kingdom of Israel and took its people into exile. Some scholars date J to the tenth century BCE [7].

E probably wrote between 922 and 722 BCE. He may have been a priest from Shiloh who viewed Moses as his spiritual ancestor [7].

P rejected the concepts of angels, dreams, and talking animals that can be seen in J and E. He believed that only Levites who were descended from Aaron could be priests. He lived after J and E because he was aware of the books of the prophets, which were unknown to the others. He also lived when the country's religion reached a priestly/legal stage, before the destruction of Jerusalem in 587 BCE. He patterned his writing after the topics in J and E [7].

Accordingly, Genesis is a composite of at least three authors dating from approximately 922 BCE to 587 BCE. Thus, Genesis represents the knowledge available between 2500 and 3000 years ago, which is the age of bronze.

Regarding the *Iliad* and *Odyssey*, the Phoenician alphabet was adopted by the Greeks circa 800 BCE. We know this because "Homer's poems appear to have been recorded shortly after the alphabet's invention. For example, an inscription from the island of Ischia in the Bay of Naples, circa 740 BCE, appears to refer to a text of the *Iliad*" [8].

I have included the first thirty-one verses of Genesis in an appendix I for general reference. However, for convenience, I have listed below the first ten verses that describe the creation. The (P) indicates the probable author.

> (P) "In the beginning God created the heaven and the earth.
>
> And the earth was without form, and void; and darkness was upon the face of the deep. And the Spirit of God moved upon the face of the waters.
>
> And God said, Let there be light: and there was light.
>
> And God saw the light, that it was good: and God divided the light from the darkness.
>
> And God called the light Day, and the darkness he called Night.
>
> And the evening and the morning were the first day.
>
> And God said, Let there be a firmament in the midst of the waters, and let it divide the waters from the waters.

And God made the firmament, and divided the waters
which were under the firmament from the waters which
were above the firmament: and it was so.
And God called the firmament Heaven.
And the evening and the morning were the second
day.
And God said, Let the waters under the heaven be
gathered together unto one place, and let the dry land
appear: and it was so.
And God called the dry land Earth; and the gathering
together of the waters he calledSeas: and God saw that
it was good." (Gen. 1:1–10)

There are many possible interpretations of these verses. They
essentially describe the work of a supernatural magician, an interpretation
apparently shared by others. I found this interesting description of the
Creator written on large mirror hanging on a wall in a Paris restaurant,
of all places. The text was written in French, of course, but this is a
rough translation:

The Magician

"The grand juggler. For me, the magician that is the omnipotent God
which created our universe, it is he who has created the marvelous force
of the paradoxical world in which we live. It is the map of the active
intelligence, of the light, of the pure energy, of the creation and of the
game."

Another interpretation of these few Genesis verses is that perhaps
the authors of Genesis understood some of the basic ideas of creation—
the earth was created first and then life. As will be discussed in chapter
13, the belief in a supernatural creator was prevalent at the time
Genesis was written. However, if one compares what could be known,
such as listed above under physical features readily observable by early
humans, with what is written in Genesis, what Genesis doesn't say is as
interesting as what it does.

The great lights in the sky are not given names, and there is no
mention of the motion of the great light, which is obviously the sun.
The basic features of the earth, such as mountains and valleys, are not
mentioned, either.

In view of the material provided in Genesis, it may not have the intention of the authors to write an "encyclopedia." They merely wanted to set down some basic facts regarding the origin of the earth as they knew them, as an introduction to the history of peoples known to the writers of Genesis.

While there is much debate regarding the correctness of Genesis (e.g., are "days" of Genesis allegorical, etc.), my view is that if one considers what could have been known to the authors of Genesis, it is a reasonable document. It merely recounts in a rather limited form what was known or believed at the time. However, those who seek to find "answers in Genesis" [9] are looking in the wrong place. There are few answers in Genesis, and those that are there are demonstrably incorrect.

While there is no specific discussion of the age of the earth in Genesis, it of course contains a chronology of humanity beginning with Adam. This chronology can be used to estimate the Earth's age, as was done famously by **James Usher, Bishop of Armagh (1581–1656)**. Around 1640, Usher concluded that the Earth was created the evening before October 23, 4004 [10]. It was a remarkable achievement, as some have noted, since people's ages were given only in years. As with most developments of an explanation of phenomenon, the bishop's estimate was a bit short[3].

FLOODS—THE NEMESIS OF HUMANITY.

Besides a brief story of the creation of the earth, the book of Genesis also contains a fairly detailed description of an immense flood, the biblical flood, which has also had and continues to have great impact on the origins debate. It is often employed to explain the formation of the earth, canyons, rocks in strange places, etc. Regarding the flood, the "Just What is Creationism Trying to Say?" Web site contains the following statement:

> "The 'geologic column', which is cited as *physical evidence* of evolution occurring in the past, is *better explained* as the result of a *devastating global flood* which happened about 5,000 years ago, as described in the Bible." [9, emphasis added]

Regarding *physical evidence*, as will be demonstrated, there is zero physical evidence of a flood as recorded in the Bible. However, there is considerable physical evidence for a real "mega flood," which will be explained shortly.

Civilization developed along river banks for these simple reasons: a supply of water for drinking and farming and to provide a practical method of transportation and trade. Unfortunately, rivers have the annoying tendency of overflowing their banks every spring due to melting snows in the mountains that create the rivers. Occasionally when there is a large snow fall and an exceptionally warm spring with above-average rainfall, a super flood occurs, causing extensive devastation and loss of life. In spite of modern technology, we still experience annual devastating floods.

Due to the prevalence of floods, especially super floods, floods form a substantial part of the stories of early humans. For the purpose of this book, there were at least three super floods of interest: the biblical flood; a flood important in Greek mythology that explains why the Greeks call themselves Hellenes; and a third, well-documented flood called the Great Missoula Flood, which occurred during the last ice ages, approximately 20,000 years ago. I will discuss the biblical and Greek floods next and the Great Missoula Flood later when enough general geological information has been established enough to understand the evidence for the flood.

The Biblical Flood: I will begin with the biblical flood, which is the second part of Genesis relevant to the origins issue. Due to the extensive use of the biblical flood to "prove" the validity of creationism, we will examine it in detail.

While Genesis is silent regarding phenomenon such as earthquakes, which would have been felt by the authors or at least mentioned in the oral tradition, the biblical flood is described in great detail. Simon Winchester, in his excellent autobiography[4] of geologist William Smith makes an interesting observation regarding the biblical flood and strange "pound stones," which are round, slightly flattened stones about four inches in diameter and weighing approximately a pound, hence the name. British local farmers had been digging them up for probably hundreds of years. Winchester writes:

"People began to wonder if these stones might actually be the relics of living things, and placed where they were found by no less an agency than what they liked to call the Noachian Deluge-Noah's flood.

Perhaps somehow the flood could be implicated in shifting these objects, even to where they now existed in the rocks of high mountain ranges and on the Oxfordshire meadows. Perhaps somehow this same flood could also be implicated in the process that created the objects in the first place. Perhaps the rocks and all that lay inside them-the Chedworth Buns, the pundibs, the oyster shells, the fern leaves, and the crystal corals, fish skulls, and lizard bones-had all somehow been precipitated or had crystallized themselves from the fluid of a universal, flood-created sea. Perhaps, if such things were demonstrably true, then maybe, just maybe, the matter of intense puzzlement that had already confused untold generations of naturalists "what were fossils and why were they found where they were"?-might be solved. *The flood, in short, was to be the eighteenth-century answer to everything".* [5, emphasis added]

The account of the biblical flood in Genesis does not say anything regarding geological occurrences as a result of the flood, such as the placement of fossils in mountains or vast rearrangements of the earth that are claimed to have been caused like those mentioned above by Winchester. However, in order to employ the biblical flood to support creationism, a number of theories based upon inventive interpretations of Genesis have been advanced to "prove" the existence of the flood. They are well documented on three Web sites, [11], [12], and [13]. Reference [13] contains this statement, "The fossils, and the sedimentary deposits they were entombed within, have simply been misinterpreted by the scientific community. The fossil record is instead a recording of a devastating global scale flood."

This is an interesting statement, but it is totally unsupported by any evidence. However, ample evidence will be provided and conclusively demonstrate that the scientific community has most definitely not

126

misinterpreted the data. Moreover, it is interesting to conjecture why Genesis authors failed to mention fossils if fossils are a "recording of a devastating global scale flood."

There are two other lines from Genesis that figure prominently in establishing that the biblical flood was global: And the waters prevailed so mightily upon the earth that all the high mountains under the whole heaven were covered; the waters prevailed above the mountains, covering them fifteen cubits deep (Gen. 7:19–20).

"High mountains" are naturally interpreted to include 26,000-foot Mount Everest (e.g., see [14], which quotes a number of biblical verses that "prove" that every mountain, including Everest, was covered).

Four Questions for Any Flood. Regarding any flood, there are four simple questions that must be answered:

1. How big was the flood?

2. Where did the floodwater come from?

3. Where did the floodwater go?

4. What evidence exists for the flood?

Regarding the first three questions relative to the biblical flood, the answers are only found in the book of Genesis. Regarding the fourth question, the answers in Genesis are quite vague. However, as mentioned earlier, we have a well-documented example of a super flood that was much smaller than the biblical flood, but it is one of the largest floods ever discovered. Its discovery has shown what evidence a super flood leaves, but similar evidence has never been found regarding the biblical flood.

Summary of the Biblical Flood: Chapters 6–8 from the book of Genesis summarize the biblical flood and describe the following:

- God is displeased with his creation and plans to destroy it (Gen. 6:5–13) and reveals that one family, Noah's, is deserving of being saved. So God instructs Noah to build an ark (boat) big enough to hold his family and one of every type of living creature, and Noah is to take all into the arc (Gen 6:14–22; 7:1–9).

- It tells of the rainfall and "fountains [of water?] from the deep" and the 240 days[5] of the ark's journey until it arrived at Mount Ararat [location undefined] (Gen. 13:4), when the waters subsided enough for Noah to embark (Gen. 7:10–24; 13:1–14).

- Noah embarks from the ark (Gen. 13:15–22).

- God promises not to destroy everything again and admonishes to Noah to "Be fruitful and multiply, and fill the earth." God also says that "The fear of you and the dread of you shall be upon every beast of the earth" and that "Every moving thing that lives shall be food for you" and "Whoever sheds the blood of man, by man shall his blood be shed; for God made man in his own image [we will encounter this phrase again in chapter 17]" (Gen 14:1–10).

Size of Biblical Flood: Interpretations of the flood described below imply that it covered the entire earth because "all the mountains were covered"—a truly remarkable flood. Obviously, a vast amount of water is required to accomplish a flood of this magnitude, an amount of water that can easily be estimated.

By assuming the average height of the earth's surface, including mountains and seas is about 5000 feet (probably a conservative estimate) and since Mount Everest is 26,000 feet high, the biblical flood would have been a blanket of water averaging 21,000 feet in height (26,000–5,000 feet) over the entire earth's surface This volume of water can easily be calculated by employing the well-known formula for the volume of a sphere: $4/3 \text{ pi } r^3$, where r is the radius of the sphere. We merely calculate the volume of the earth, assuming a 4000 mile radius plus an additional 5000 feet, and then the volume of the earth, plus an additional 21,000 feet. Subtracting the smaller number from the larger yields the volume of water as 800 million cubic miles. That's a lot of water.

Where did the water come from? According to the Bible, it rained night and day for forty nights. In order for 21,000 feet of water to fall in this period of time, approximately 500 inches of rain would have fallen all over the earth every day. This would have obviously required an enormous reservoir.

The earth's oceans only hold about 332 million cubic miles of water, and, as will be seen later, extensive seismic studies of the earth preclude any large underground storage. However, the Bible tells us in Genesis 7:11, that "in the six hundredth year of Noah's life, in the second month, the seventeenth day of the month, the same day were all the fountains of the great deep broken up, and the windows of heaven were opened."[15]

Reference [15] explains the passage, saying, "The 'fountains of the great deep' are *probably* oceanic or *possibly* subterranean sources of water. In the context of the flood account, it *could* mean both." Note the words I have italicized and how they indicate the tentativeness of this sentence, as in others below.

Moreover, Reference [15] opines, "If the fountains of the great deep were the major source of the waters, then they must have been a huge source of water." Reference [15] certainly got that right. Reference [15] further states, "Some have *suggested* that when God made the dry land appear from under the waters on the third day of creation, some of the water that covered the earth became trapped underneath and within the dry land."

Finally reference [15] states, "Genesis 7:11 says that on the day the flood began, there was a 'breaking up' of the fountains, which implies a release of the water, *possibly* through large fissures in the ground or in the sea floor. The waters that had been held back [when God made the dry land?] burst forth with catastrophic consequences."

Where did the waters go? The next logical question is "Where did the waters from the biblical flood go?" Reference [16] explores that question, stating, "There are a number of Scripture passages that identify the flood waters with the present-day seas (Amos 9.6, and Job 38:8-11)."

If the waters are still here, why are the highest mountains not still covered with water as they were in Noah's day? Psalm 104:6–9 *suggests* an answer: "After the waters covered the mountains, God rebuked them [the waters?] and they fled; the mountains rose, the valleys sank down and God set a boundary so that they [the waters?] will never again cover the earth." Reference [16] adds:

> "That is why the oceans are so deep, and why there are folded mountain ranges. Indeed, if the entire earth's surface were leveled by smoothing out the topography of not only the land surface but also the rock surface on

the ocean floor, the waters of the ocean would cover the earth's surface to a depth of 1.7 miles (2.7 kilometers). We need to remember that nearly 70 percent of the earth's surface is still covered by water. Quite clearly, then, the waters of Noah's Flood are in today's ocean basins."

However, as calculated above, the amount of water required by the flood far exceeds the amount in the oceans.

What about the Animals? Although this is, strictly speaking, a biological subject, the issues regarding the animals are appropriate here. There are a number of questions: Was the ark big enough? How did the animals such as the kangaroo get to the ark? How were the animals fed? These are just a few.

The answers to these and other related questions can be found on the Christian Answers Web site [17], which states the following:

- "The ark would have been at least 450 feet long, 75 feet wide and 45 feet high [about one half the size of the Queen Mary]."

- "That God gathered the animals [method unspecified] and brought them to Noah inside the ark two by two."

- "A number of scientists [unspecified] have suggested that the animals may have gone into a type of dormancy. Perhaps these abilities were supernaturally intensified during this period."

The Web site concludes with this statement: "It is evident, when all the facts are examined that there is no scientific evidence that the biblical account of Noah's ark is a myth or fable [17]."

As with other aspects of "answers in Genesis," there are few "facts," much supposition, and appeals (e.g., to "a number of [unspecified] scientists). However, it is not true that "there is no scientific evidence that the biblical account of Noah's ark is a myth or fable." There is considerable scientific evidence that there was no flood, as was discovered by former believer Arlan Blodgett.

Arlan Blodgett's Investigation of the Biblical Flood: Arlan Blodgett is an amateur archeologist with considerable interest in biblical archeology and a former believer in the biblical flood. Arlan describes his biblical flood investigations in an interesting article that appeared in *News and Views* [18]. I will discuss Arlan's journey from creationist to non-creationist in a later chapter, but for this section, we will concentrate on Arlan's archeological investigations.

Arlan spent some time on an archeological dig in Israel and found no evidence of a large flood. Puzzled because he was brought up to believe in a literal interpretation of the Bible and at one time believed in the flood, he was determined to leave no stone unturned in a search of evidence of the flood. He surveyed a number of archeologists working in the area asking whether they had found any evidence of a flood [19]. Of almost 100 archeologists surveyed, not one reported finding any evidence other than evidence of local flooding. A typical reply to the survey was:

> Dr. A: "There is nothing in the archaeological record that supports a universal flood such as [Genesis 6–8] depicts, not within the historical period or even in the prehistorically human period. By genre, the early chapters of Genesis are patently myth, not history, similar to the Mesopotamian myths of the Atrahasis and Gilgamesh"[19].

Arlan and I have corresponded briefly via e-mail. In one of his e-mails, Arlan stated:

> "IMHO [In my humble opinion] The central issue is really the "Genesis deluge." If there never was a biblical deluge then that throws doubt on all the stories preceding the story of Noah. I find the word FLOOD is used for rivers overflowing their banks and is not adequate to describe the so called worldwide Genesis DELUGE."

Thus, the biblical explanation and "proofs" of the biblical flood are totally lacking. They are mostly equivocal statements, some exhibited above, filled with words like possibly, probably, and suggested. At the end of the day, there is not a shred of evidence, like that of the historical Missoula flood that will be discussed shortly, to support the biblical flood. But before discussing the Missoula flood, I have a few words about another mythical flood, one that almost everyone agrees was mythical.

The Greek Flood: I find it intriguing that the Greeks also had a flood myth that loosely parallels the biblical flood. It is perhaps reasonable to believe that super flood stories appear in many civilizations, but in the Greek flood, there is the question of God/Zeus's desire to destroy human kind. The reason for this is probably lost. If God/Zeus was powerful enough to create the universe, it would seem that he could have maintained discipline among his creations.

This description of the Greek flood is adapted from [20] and [21]. Deucalion was the son of Prometheus, who was the son of the Titans Iapetus and Clymene. Zeus, king of the gods, decided to destroy humankind for their lack of piety and their evil ways by sending a great flood to eliminate humanity. It seems a bit severe, but it's just a myth. Prometheus warned the righteous Deucalion and his wife, Pyrrha, daughter of Prometheus's brother, Epimetheus, and Pandora, of Zeus's plans and that they should build an ark to save themselves. Deucalion was thus the Greek equivalent to Noah.

After sailing for nine days and nights (they only had a short voyage), they landed on Mount Parnassus near Delphi. Deucalion made an offering to Zeus the Savior, and Zeus's response was to send the god Hermes with a promise to make any wish come true. Deucalion then asked for the earth to be repopulated, and he was told to throw his mother's bones behind him. Since his mother's bones were the bones of the mother of all the earth, the couple took stones and threw them over their shoulders, and the stones became the new human race. Deucalion's stones became the men, and Pyrrha's became the women. The new race was dark and short where the previous had been tall and blond.

The couple had a son, Hellen, who in turn became the father of Aeolus, Dorus, and Xythos and grandfather of Ion and Achaios. These names all refer to the different Greek tribes, with Hellen, "Greek," as the ancestor, hence the reason the Greeks call themselves Hellenes.

A REAL MEGA FLOOD—THE LAKE MISSOULA FLOOD.

The Lake Missoula flood may have been the greatest flood ever, so it therefore provides a standard for the evidence of a mega flood that would exist if one has occurred. The Missoula flood was reenacted on PBS on September 20, 2005 [22] and has been replayed several times since. As with many of the discoveries recounted in this book, the Missoula flood escaped detection until a determined geologist, **J. Harlan Bretz (1882–1981)**, decided to investigate the strange geological features of western Washington State. Bretz was then aided by two other geologists to fully explain his findings.

Across 16,000 square miles of Washington State, the landscape changes approximately 200 miles east of Seattle from undulating farmland to a wild landscape known as the Channeled Scablands. The Channeled Scablands has abrupt rips and scars, tall canyons, immense dry waterfalls higher than Niagara Falls, gigantic potholes, and a gorge called the Columbia River Gorge that is thousands of feet deep.

For more than a century, scientists have been grappling with the formation of this landscape. It is one of the most unusual on Earth. For a long time it was assumed that the Scablands' features would have taken millions of years to create. One obvious way this could have happened was by the gradual erosion caused by rivers. After all, some of the most dramatic landscapes in the world have been slowly scoured out by rivers (e.g., the Grand Canyon).

But rivers cannot create the Scablands' enormous potholes, some which are ten times the size of potholes formed by the Columbia River. Also, no known river can move the blocks of granite, some weighing 100 tons, that are scattered throughout the Scablands. On the other hand, glaciers can move large blocks of rock, usually known as glacial erratics, but the ice sheets that flowed down from Canada during the last ice age never reached the Scablands.

The answer to the mystery of the Scablands was an audacious theory that defied all scientific convention. It was formulated in the 1920s by Harlen Bretz. Bretz had studied the Scablands for years, patiently examining the rocks and other features. He became convinced, in true Sherlock theorem form, that only a huge volume of water could explain his observations. The Scablands had been formed by a gigantic flood.

Bretz would not see aerial photographs of these hills for many years, but it can be seen from the air how these shapes begin to look like ripples, a giant version of ripples left behind on the beach by the sea.

On January 12, 1927, Bretz presented his completely unconventional theory to a specially convened meeting of fellow scientists in Washington, DC. He claimed an unimaginably huge mass of water, up to 900 feet deep, had swept across the Scablands and disappeared into the Pacific

Ocean in a matter of hours. Bretz called the flood the "Spokane Flood" because he assumed the source of the water for this flood had been somewhere near Spokane, Washington [23].

Of course his fellow geologists dismissed this "biblical flood" as totally preposterous. But like others before him, Bretz remained steadfast. But Bretz had the same problem as those supporting the biblical flood: where had the water come from?

Sitting in the audience in Washington DC was another geologist, Montana Joseph Thomas Pardee, who would ultimately supply the answer to the Missoula water source [24]. Pardee had been studying markings on the walls of a valley near Missoula, Montana and knew the watermarks could only have been made by a lake because the marks were similar to marks in other valleys that were known to have been filled with water at one time. But these watermarks represented an enormous lake, which led to the obvious question of how the lake formed.

The answer was found in scratches on the bedrock of the valley, scratches that are usually associated with the movement of a glacier over the rock. Apparently a large glacier had moved into the valley. Pardee and others realized the glacier had come from Canada during the last ice age and eventually reached what is now Idaho. The ice moved down the valley and filled it from one side to the other. It encountered a mountain at the valley's end and thus blocked the river valley.

The river began to back up against the wall of ice and filled the valley with water, eventually trapping a lake that grew bigger than lakes Erie and Ontario combined. The volume of water backed up behind the ice was vast—an astounding 520 cubic miles. It became known to geologists as Glacial Lake Missoula.

Pardee also noticed one other strange feature—ripples such as those found in a stream bed. But these ripples were not inches high; they were between ten and forty feet high. Examining the ripples, Pardee hypothesized that the huge lake had somehow emptied rapidly and created the ripples by pushing up gravel, just as is done in a stream bed.

Pardee also noticed that the ripples pointed toward the Scablands—here was Bretz's water. It would have been a huge body of water traveling at fantastic speed [24].

This leads to the question of how and why the Missoula water ice dam collapsed so rapidly. The answer to this remaining piece of the puzzle was provided by glaciologist Matthew Roberts. Robertso was studying glaciers in Iceland because he was motivated to learn how an ice dam could collapse almost instantly. An ice dam had collapsed in Iceland in 1996 and caused vast devastation. After years of analysis, Roberts eventually worked out the process that causes an ice dam to fail [25].

Normally, water freezes at zero degrees centigrade, but the pressure deep at the base of an ice dam lowers the freezing point of water, resulting in what is known as "super-cooled" water. The highly pressurized, super-cooled water begins to force its way into tiny cracks that always form in ice.

Once super-cooled water has begun to trickle through the cracks, the flowing water alone is enough to trigger a very peculiar process. This moving water creates tiny amounts of friction, which releases energy in the form of heat. Therefore, as the water moves through the glacier, it melts the ice. Soon, the minute cracks become giant ones, measuring up to several feet across. A tunnel under the dam is formed and enlarges rapidly. Suddenly, the ice dam is destabilized and collapses almost instantaneously. This was the cause of Bretz's flood.

To test whether a single flood coming from Lake Missoula could really have done all of this, scientists constructed their own mini-Scablands at the University of Minnesota. The earth-surface dynamics team there has poured water over the scale model to represent the failure of Glacial Lake Missoula. The model clearly shows miniature versions of the canyons found in the Scablands. Just like the real ones, they look as if they were gradually eroded, but the fact is that they were carved out in seconds.

The model also explained the formation of the huge potholes. A water tunnel was built to demonstrate the effects of water moving at high speeds. Water moving at high speed creates a stream of bubbles. When the bubbles burst, they exert enormous force on the rocks over which the water is flowing. At the extreme speeds experienced by the wall of water from Lake Missoula, a swirling vortex of bubbles is formed that can create the potholes seen in the Scablands.

It should be noted that it was not until 1930 that Bretz considered Glacial Lake Missoula as the possible source of water for which he was searching. But the geologic evidence was elusive, and he did not fully embrace the idea until 1956. Unable to provide a clear scientific argument for the source of flood water, Bretz went on to other activities.

Eventually, a variety of geologists working on the problem established the details of the great Missoula flood. Recent work by Richard B. Waitt has identified up to 100 floods, with the earliest and largest separated by 50–100 years and the last and smallest by only a few years [23], [26] reported in [27].

As Steven Dutch points out, "It is difficult to imagine excavating something like Grand Coulee with one or a few large floods, however huge, but several dozen make the task more manageable" [27]. Regarding the Missoula flood, Steven Dutch has created an excellent overview Web site with many photos of the Scablands area that vividly illustrate the immense forces that have operated in the area.

A Different Interpretation of the Missoula Flood: A slightly different interpretation of the evidence for the Lake Missoula flood is provided by Michael J. Oard's Web article "The planned Lake Missoula flood interpretive pathway, The story that won't be told"[28]. Mr. Oard, apparently reacting to an article in the Yakima Herald Republic, states:

> "The article in the *Yakima Herald-Republic* even had the gall to say, 'By any account, the Old Testament flood was a leaky faucet by comparison [to the largest Lake Missoula flood].' 'It is only by willful blindness to geology and Scripture that they can make a statement like this (2 Peter 3:3-6)'"[28].

Regarding gall and blindness, I would say than Mr. Oard exhibits more gall by suggesting that the Lake Missoula flood was actually the biblical flood while supplying no supporting evidence other than the observation that:

> "The Lake Missoula flood provides an analog for the thousand or more rivers over the earth that now flow through mountain barriers, sometimes through gaps

much deeper than Grand Canyon. The river[s] should have gone around the barrier[s], if the slow processes over millions of years model were true, but these water gaps through transverse barriers can be [could have been?] cut rapidly during the Genesis Flood."[28]

That all the "water gaps" on Earth were cut at one time by the biblical flood is a flight of fancy beyond belief. Only a person blinded by obdurate biblical belief could accept such nonsense.

Regarding evidence for the biblical flood, while the Missoula flood was created by a lake of 520 cubic miles, 820 *million* cubic miles of water was needed for the biblical flood. Whereas, the Missoula flood produced ripples 10 to 30 feet high spread over tens of miles, the biblical flood would have produced ripples hundreds of feet high spread over hundreds of miles. In addition, potholes perhaps several thousand feet deep would have been formed. To date, nothing like this has been found, and some have very diligently tried to find them, as in the case of Arlan Blodgett.

Having resolved the biblical flood issue (though we'll have a bit more to say about the biblical flood later on), let us address another item available to early humans—fossils.

FOSSILS—THE FIRST CLUES OF THE UNCHANGING ANIMAL AND PLANT STRUCTURE ILLUSION.

The Natural Treasures, Fossils Web site states:

"Fossils are the remains or traces of animals or plants which have been preserved by natural causes in the earth's crust, excluding organisms which have been buried since the beginning of historic time. Fossils are windows which serve as insights into nature's past.

Fossils have been known to people for many centuries. Fossil shells used for the purpose of adornment have been discovered at Paleolithic sites" [29].

As usual, the Greeks (e.g., Herodotus and Xenophanes [29]) were among the first to identify fossils as the remains of long dead animals.

Unfortunately, the powerful Aristotle disagreed with this interpretation of fossils, mistakenly believing in spontaneous generation of life—the belief that complex living organisms are spontaneously generated by decaying organic substances. According to Aristotle, it was a readily observable truth that aphids arise from the dew which falls on plants, fleas from putrid matter, mice from dirty hay, crocodiles from rotting logs at the bottom of bodies of water, and so forth [30].

Amazingly, Aristotle's ideas lasted until the Middle Ages when such enlightened minds as Leonardo da Vinci, began to identify fossils as what some of the other Greeks had known but had been over ruled by Aristotle—the remains of extinct animals [29].

In general, due to the lack of a proper framework for understanding the unchanging animal and plant structure illusion, especially the true age of the earth, early explanations of fossils were rather bizarre. In one of his excellent travel shows, Rick Steves shows a painting of a giant 10- to 12-foot man on a covered boardwalk at Lake Lucerne in Switzerland. The man's proportions were based upon bone fossils of a mammoth that the locals mistakenly believed to be the remains of a giant man [31].

In her informative discussion of *The First Fossil hunters: Paleontology in Greek and Roman Times*, Adrian Mayer suggests early explanations of fossils ascribed them to prehistoric monsters. In her book, Mayer

> "... explores likely connections between the rich fossil beds around the Mediterranean and tales of griffins and giants originating in the classical world. Striking similarities exist between the *Protoceratops* skeletons of the Gobi Desert and the legends of the gold-hoarding griffin told by nomadic people of the region, and the fossilized remains of giant Miocene mammals could be taken for the heroes and monsters of earlier times.... Building a vivid picture of how the ancient Greeks, Egyptians and Romans might encounter these strange artifacts and attempt to make sense of them." [32]

Thus, while fossils were known to exist and some of the ancients apparently realized they were probably the remains of extinct animals, they were incapable of arriving at a correct explanation of their origins.

In his review of Mayer's book, Stephen Haines points out that "with no idea of the Earth's true age, it was easy to make these [assignment of fossils to prehistoric mythical monsters] judgments" [33].

Thus, the age of the earth was an important component to many investigations.

SUMMARY OF EARLY INVESTIGATIONS OF THE UNCHANGING PHYSICAL FEATURES OF THE EARTH ILLUSION.

Before we progress farther into this chapter, I would like to take a moment to review and summarize the early investigations of Earth's physical features. Due to the glacial slowness of physical processes, no one originally realized the apparently unchanging nature of physical features was an illusion.

Of importance to this book, one of the first explanations of the physical nature of the earth is contained in the book of Genesis, which was written sometime in 922 BCE and 587 BCE. All that is said about the creation of the earth and the physical nature of the earth is contained in the first ten verses of Genesis. It explains *what* happened and *who* did it, but there is no explanation of *how* or *why* it was done. Typical statements are, "And God said, Let the waters under the heaven be gathered together unto one place, and let the dry land appear: and it was so" (Gen. 1:9). Not a particularly robust explanation, but it is typical of religiously based explanations.

The biblical flood occupies many Genesis pages, but the description of the flood cannot be supported by any outside evidence. According to an interpretation of the Genesis biblical flood, Mount Everest was covered by water. This would require twice the amount of water that is in the oceans, and there is no source for the water.

Genesis contains no specific mention of the earth's age, but the genealogy of Adam is provided in sufficient detail to allow a persevering person to make a good estimate. James Usher, bishop of Armagh concluded (ca. 1640) that the earth had been created the evening before October 23, 4004 BCE.

We explored another mythical flood, the Greek flood, and a real mega flood, the Missoula Flood, in detail to serve as counter balance to the biblical flood explanation. The Missoula Flood left telltale markings that any giant flood would leave. None have been found where the biblical flood was supposed to have taken place, as supported by former believer Alan Blodgett.

And finally, fossils are found many places and must have been known to early humans, but early humans lacked a time reference for them. Because of this, it was easy for them to assume that the fossils were the remains of prehistoric mythical monsters.

RESOLVING THE APPARENTLY UNCHANGING PHYSICAL FEATURES OF THE EARTH ILLUSION UP TO 1850 (CONTINUED).

In the 900s, Islamic observers began to suspect the truth. That the Earth had indeed changed, and made a number of contributions, but the violent decline of the Islamic world due to the Mongol invasions suppressed all but a few items, such as one of the principles relative to sedimentary rock: The oldest is on the bottom.

A somewhat reluctant recognition developed in the eighteenth and nineteenth centuries that conceded that the physical properties of the earth had not been constant and had changed over vast quantities of time. With one exception, the unraveling of the earth's unchanging physical properties illusion proceeded independently of the biological investigations.

The exception to this independent development of the explanation was rather ubiquitously distributed fossils. Fossils provide an obvious connection between the physical and biological features of the earth. They are the subject of the next discussion, which begins with a most unlikely individual— The famous polymath **Leonardo da Vinci (1452–1519) [34]**. While Leonardo is renowned for his paintings (e.g., *The Last Supper* and *The Mona Lisa*) and his intriguing inventions, he had an inquisitive mind and abhorred absurd ideas such as the biblical flood. Leonardo was one of the first to resolve the mystery of fossils, if only to disprove the biblical flood.

Leonadro was one of the first persons to conduct a reasoned, systematic study of fossils [34]. Leonardo was born the illegitimate son of Messer Piero Fruosino di Antonio da Vinci, a Florentine notary, and a peasant girl named Caterina, in the Tuscan hill town of Vinci[6], located in the lower valley of the Arno River near Florence on the edge of the Apennine mountains [35]. Though Leonardo was born more than 100 years before James Usher, his secretive way of writing (mirror image Italian) resulted in his work being little known until eighteenth century with the appearance of the Codex Leicester[7].

Leonardo represents one of the transitions from the biblical-centered explanation of the earth to an explanation based upon verifiable observations. As has been seen in other areas, this transition occurred toward the end of the fifteenth century with the rediscovery of Greek and Roman learning. However, Leonardo appears to have arrived at his conclusions independently.

In the delightful and insightful book, *Leonardo's Mountain of Clams and the Diet of Worms*, Stephen Jay Gould discusses the codex which, among other things, describes Leonardo's conclusions derived from his observations of fossils in the nearby Apennines [36, pp. 22–29]. Gould points out on page 26:

> "Leonardo did not observe fossils for pure unbridled curiosity, with no aim in mind and no questions to test. He recorded all his information for a stated and definite purpose-to confute the two major interpretations of fossils current in his day:
>
> The Biblical Flood,
> Neo-Platonism, a mystical belief system that originated in the 3rd Century."

Since the biblical flood is the only one of the two of interests in the book, we will not follow Leonardo's attack on Neo-Platonism.

Leonardo's fossil investigations are illustrative in many ways. I will present some quotes from the Codex, provided by Gould [36] and based upon MacCurdy translation of Leonardo's notebooks, followed by Leonardo's conclusions and explanation regarding the biblical flood, which are based largely on his observations of strata in various locations that contained the same fossils. In this respect, Leonardo was among the first to realize that strata are connected. With this observation, Leonardo anticipated William Smith—the man who developed the first stratigraphic map of the British Isles—by about 300 year. (The following quotations, unless otherwise stated, come from the Leicester

Codex as presented in the MacCurdy translation of Leonardo's notebooks.)

"1. Leonardo recognized the temporal and historical nature of horizontal strata by correlating the same layers across the two sides of river valleys. Leonardo observed 'How the rivers have all sawn through and divided the members of the great Alps one from another; and this is revealed by the arrangement of the stratified rocks, in which from the summit of the mountain down to the river one sees the strata on the one side of the river corresponding with those on the other[8].

2. He observed that rivers deposit large, angular rocks near their sources in high mountains, and that transported blocks are progressively worn down in size, and rounded in shape, until sluggish rivers deposit gravel, and eventually fine clay, near their mouths.

3. Leonardo observed the presence of fossils in several superposed layers proving that their deposition had occurred at different and sequential times. [hence, did he realize that life had changed over time, i.e., evolved?]

4. He noted that the tracks and trails of marine organisms are often preserved on bedding planes of strata: observing 'How between the various layers of the stone are still to be found the tracks of the worms which crawled about upon them when it was not yet dry.'

5. If both valves of a clam remain together in a fossil deposit, the animal must have been buried where it lived, for any extensive transport by currents after death must disarticulate the valves, which are not cemented together in life, but only hinged by an organic ligament that quickly decays after death. (This principle of inferring transport by noting whether fossil clams

retain both valves persists as a primary rule of thumb for everyday paleoecological analysis).

At another site, on the other hand, Leonardo inferred extensive transport after death. In such a locality there was a sea beach, where the shells were all cast up broken and divided and never in pairs as they are found in the sea when alive, with two valves which form a covering the one to the other. Leonardo notd that:

> 6. No marine fossils have been found in regions or sediments not formerly covered by the sea.

> 7. When we find fossil shells broken in pieces, and heaped one upon the other, we may infer transport by waves and currents before deposition: But how could one find, in the shell of a large snail, fragments and bits of many other sorts of shells of different kinds unless they have been thrown into it by the waves of the sea as it lay dead upon the shore like the other light things which the sea casts up upon the land?" [36, p. 23]

Leonardo summarized his observations as follows:

- "Fossils are found in separate rock layers, thereby demonstrating that the fossils could not have been deposited by a single flood.
- Worm tracks are found which would have been washed away by a giant flood.
- Clam shells still in their connected position are found which would have been torn apart by a flood"[36].

The evidence of his findings essentially refutes the idea that a single cataclysmic flood could have produced his observations.

As Gould points out, Leonardo had harsh words for those who believed spontaneous generation was a reason for the existence of fossil shells in rocks above sea level:

> "And if you should say that these shells have been and still constantly are being created in such places as these

by the nature of the locality and through the potency of
the heavens in those spots, such an opinion cannot exist
in brains possessed of any extensive powers of reasoning
because the years of their growth are numbered upon
the outer coverings of their shells [observation 9 again];
and both small and large ones may be seen, and these
would not have grown without feeding or feed without
movement, and here [that is, in solid rock] they would
not be able to move.... Ignoramuses maintain that
nature or the heavens have created [fossils] in these
places through celestial influences"[36].

I have a couple of last comments regarding Leonardo. First, he was the
illegitimate son of a peasant girl and a noble [37]. This raises a question
we will examine later: how did such a union produce Leonardo da
Vinci? Also, Leonardo made many important discoveries, but he could
not really explain them (e.g., fossils—he had no idea how they got there.
It just wasn't the flood). However, based upon his limited knowledge
base, his reasoning was exemplary and a good example.

It's not clear whether Leonardo pondered the earth's age, but he
must have known it was rather old. But of course, Leonardo hadn't met
the Bishop of Armagh, who was born sixty years after Leonardo died.

As noted earlier, due to Leonardo's secretive style, his findings were
lost for many years, and therefore his findings about fossils, strata,
and erosion disappeared. But they were picked up later when modern
geology began to develop.

THE BEGINNING OF GEOLOGY.

With the exception of Leonardo, whose work was unfortunately lost
for many years, nothing of interest occurred until approximately 1750.
Then, between 1750 and 1850 a significant portion of the unchanging
physical aspects of the Earth illusion was resolved in what can best be
described as a revolution. In the mid-1700s the biblical account, as
discussed earlier by Simon Winchester, was the accepted explanation
of the earth and its origins. By the mid-1800s an army of geologists
had demonstrated that this explanation was totally incorrect and that
the earth was indeed changing, albeit imperceptibly slowly. Moreover,
they knew the earth was considerably older than biblical reckoning

predicted because the earth had changed considerably over vast time intervals. However, only a relatively few educated individuals that shared their investigations and conclusions in scholarly journals, which were generally inaccessible to the average person, participated in and were aware of the revolution. Because of this fact, the revolution took a long time to seep down to "the man on the street." In fact, when it did, many viewed the new explanation of the earth as blasphemy. Their reaction will be discussed in chapter 15, as it is still extant.

Just as the resolution of the solid earth and sun around earth illusions required drastic revisions in our understanding of the world in which we live, resolution of the apparently unchanging and therefore relatively young Earth illusions required even more drastic revisions. Thus, the efforts of a large number of investigators were once again required to gain an understanding of the unchanging Earth illusion. Therefore, designating anyone as the "Father of Geology" is unfair and unreasonable. Each investigator contributed to the development of geology in a different way.

Among the many investigators, these four are perhaps the most important. These four were also assisted by many others, whom I will weave in as appropriate. The four we will discuss in detail in this chapter are:

- William Smith (1769–1839)
- James Hutton (1726–1797)
- Jean Louis Agassiz (1807–1873)
- Charles Lyell (1797-1875)

The acknowledgement and magnitude of the effort, as well as number of people involved, in the resolution of the apparently unchanging Earth illusion precludes an in-depth discussion. My objective will be, as in previous chapters, to provide just enough information to unequivocally establish the validity of the explanations that developed in the 100 years between 1750 and 1850.

The efforts of all investigators were per force intertwined, and it is not clear who was aware of whom, although communication between investigators was greatly facilitated by the formation of the Geological Society of London in 1807 [5]. Due to the intertwined aspect of their investigations, I will present them in the order which seems most reasonable, but I would like to first discuss two early explanations that were ultimately shown to be false: neptunism and catastrophism.

Neptunism and Abraham Werner (1749–1817). Werner was a German geologist who became famous as a gifted teacher and hence was very influential in the early development of geology [38]. He became interested in the problem of explaining the existence of rock layers, And somehow determined that the earth had been initially covered with an all-encompassing ocean. The ocean gradually receded to its present boundaries and precipitated all of the rocks and minerals of the earth. Werner termed the phenomenon "Neptunism" in honor of the Roman god of the sea.

While Neptunism gained adherents, it foundered on its inability to explain one of the most common rocks, basalt. Basalt is known to be extruded in lava flows, some which cover vast areas such as the 1,500,000 km² Siberian Traps [39]. Neptunism was ultimately supplanted by a rival theory, Plutonism [40], in honor of the Roman god of the underworld. Plutonism was first proposed by James Hutton, which accorded much geologic activity and source of rocks to volcanism (see below).

Catastrophism dominates early world development beliefs. Catastrophism hypothesizes that the earth was affected by sudden short-lived violent events that were sometimes worldwide in scope [41].

Before the development of uniformitarianism by Hutton and Lyell (discussed below) the dominant belief regarding the development of the world was catastrophism. The biblical flood is a prime example of these beliefs. It was basically the only way early geologists could rationalize their observations with the young earth belief prevalent before the eighteenth and nineteenth centuries.

Interestingly, real catastrophes have been well documented; but are spread over Earth's history. As will be seen, they have had significant impact on the earth. One such catastrophe is the asteroid impact that ended the age of the dinosaurs[42]. Chapter 13 will discuss two extinctions that were "humdingers," one of which almost ended life on Earth.

William Smith's Amazing Stratiographic Map of England. As discussed above, Simon Winchester has produced an excellent biography of geologist William Smith, so I will merely provide some highlights from Winchester's book and other sources available about William Smith, in particular reference [43].

Winchester describes William Smith's map and the effort required as:

> "… a work of genius, and at the same time a lonely and potentially soul destroying project. It was the work of one man, with one idea, bent on the all encompassing mission of making a geological map of England and Wales. It was unimaginably difficult, physically as well as intellectually. It required tens of thousands of miles of solitary travel, the close study of more than fifty thousand square miles of territory that extended from the tip of Devon to the borders of Scotland, from the Welsh Marches to the coast of Kent.
>
> The task required patience, stoicism, the hide of an elephant, the strength of a thousand, and the stamina of an ox. It required a certain kind of vision, an uncanny ability to imagine a world possessed of an additional fourth dimension, a dimension that lurked beneath the purely visible surface phenomena of the length, breadth, and height of the countryside, and, because it had never been seen, was ignored by all customary cartography [43, p. 192]

Clearly, William Smith was no ordinary individual and his map, the first stratigraphic map of England that founded the extremely important geological field of stratiography[9], was a remarkable achievement.

Smith was born, in Churchill Oxfordshire[10] on March 23, 1769[45]. His father was the local blacksmith and presumably not particularly interested in much of an education for his son, so William had to take care of himself. One of the volumes he was able to obtain was *The Art of Measuring* by Daniel Fleming [5, p. 53].

The material William acquired was particularly beneficial because it aided him in obtaining a position as assistant surveyor under Edward Webb, a professional surveyor who had been hired by a group of West Oxfordshire local squires that desired to improve the measurement of their property. One of Webb's older assistants apparently had a tendency to drink too much and did a poor job on one task. The man was fired, and Webb made Smith his assistant.

Smith's surveying tasks expanded as his ability. His reputation improved, and the expanding tasks took Smith to various parts of Oxfordshire where he began to keenly observe the geology of the area.

During the eighteenth and nineteenth centuries, coal was an extremely important and much sought after material. Unfortunately, it was buried underground in a rock layer maze. Having an accurate geological map of the coal layers, or "seams" as they are customarily called, was of inestimable value. Filling the need for an accurate geological map that would greatly aid in the search for coal occupied Smith for most of his life and was his crowning achievement.

The development of Smith's map began when Webb sent him to perform a probate survey for a Lady Jones, of whom little is known, but who was a director of the High Littleton Coal Company. The company had a mine called the Mearns Pit. The Mearns Pit, as Winchester points out, "has the standing in geology that the Galapagos islands have in evolutionary theory." [43, p. 62]

The mine had been worked for many years and was 9000 feet deep in places. During his survey, Smith made extensive notes of the mine layers, from which he drew a layer map that allowed him to perceive the succession of rock layers and fossil types. Fossil types were particularly important because they were less ambiguous when determining whether rock strata were connected rather than plain rock. Hence Smith became an avid fossil collector[11].

Examining other mines, Smith began to realize that the same formations appeared in other mines and thus the layers, the strata of rocks, must be connected. Smith also began to realize that all rocks and not just coal were connected. Eventually, Smith was able to trace the various rock layers, coal seams in particular, over wide areas.

It must be noted that William Smith was not the first to realize rock strata are connected. As mentioned above, Leonardo deduced this fact over a small area, and in 1719, **John Strachey (1671–1743)** presented to the Royal Society a geological paper detailing the stratiographical cross-section of the country around Sutton Court [43]. Smith pursued the development of a stratiographic map far beyond Leonardo's or Strachey's work.

As Smith continued to accumulate data, he made a fortuitous encounter in the influential Bath Society. Two farmers, Benjamin Richardson and Thomas Davis, had drawn a map of the area surrounding Bath that showed the geographical extent of soils and vegetation. Of particular interest to Smith was Richardson and Davis's use of color to better delineate the different items on the map.

Smith realized almost immediately that this technique could be applied to a geological map where the colors would indicate the various layers under the ground. Smith's first map was based upon a geographical representation of the area around Bath. He adapted the geographical map and added the geological features under Bath. He published his map in 1799, and it is the oldest true geologic map.

Smith eventually needed the aid of a person capable of printing large maps in color. Happily, he met John Carr, Britain's eminent cartographer, probably at a London meeting in 1794. Smith and Carr's collaboration lasted many years and eventually led to publication in 1815 of the first geological map of Great Britain, covering the whole of England and Wales and parts of Scotland. Conventional symbols were used to mark canals, tunnels, collieries, and other mines. The various geological types were indicated by different colors.

Even though the maps were hand colored, they are remarkably similar to modern geological maps of England.[45] Their publication firmly mark William Smith as the "Father of Stratiography," though some believe the title belongs to John Strachey [43], perhaps due to prior publication. However, Smith's accomplishments clearly surpassed Strachey. Smith's map provided for the first time a detailed picture of the subsurface rock layers. Since each layer represented a different time, the map was a key to England's geologic history, extending back millions of years. Of particular importance was the fact that the map clearly demonstrated that England had changed considerably over time.

James Hutton (1726-1797), one of modern geologies founders. I am unaware of a James Hutton biography as eloquent as Simon Winchester's biography of William Smith, but James Hutton has his own Web site and cheering section. Hutton is a favorite son of Scotland, whose Web site states that "James Hutton (1726-1797), was truly a man of the Earth. Founder of modern geology and farmer in the Scottish Borders, he was a hero of the Scottish Enlightenment" [46].

It is a bit hyperbolic, but the Scots can be forgiven because Hutton did indeed make important contributions. However, "founder" is a bit of a stretch.

Hutton had the good fortune, as with many successful investigators, to be born into an affluent Scottish family. He inherited his father's Berwickshire farms of Slighhouse[12], a lowland farm that had been in the family since 1713, and the hill farm of *Nether Monynut* [47]. These resources freed him from the obligation to work for a living. Instead, he could spend his time improving his mind [48]. In the early 1750s, he made improvements in his farm and introduced farming practices from other parts of Britain .

His farm work developed his interest in geology Clearing and draining his farm provided ample opportunities to examine the earth, and his theoretical ideas began to come together. In 1764, he went on a geological tour of the north of Scotland with George Maxwell-Clerk.

Hutton had a number of ideas to explain the rock formations he observed around him. But according to **John Playfair (1748–1819)**, a Scottish mathematician, physicist, and geologist who presented Hutton's methods and principles in his *Illustrations of the Huttonian Theory of the Earth* (1802)."Hutton was in no haste to publish his theory; for he was one of those who are much more delighted with the contemplation of truth, than with the praise of having discovered it" [50].

After some twenty-five years of work, Hutton's *Theory of the Earth; or an Investigation of the Laws observable in the Composition, Dissolution, and Restoration of Land upon the Globe* was read to meetings of the Royal Society of Edinburgh[13] on April 4, 1785. In it, he explained:

"The solid parts of the present land appear in general, to have been composed of the productions of the sea, and of other materials similar to those now found upon the shores. Hence we find reason to conclude:

1st, That the land on which we rest is not simple and original, but that it is a composition, and had been formed by the operation of second causes.

2nd, That before the present land was made, there had subsisted a world composed of sea and land, in which were tides and currents, with such operations at the bottom of the sea as now take place. And,

Lastly, That while the present land was forming at the bottom of the ocean, the former land maintained plants and animals; at least the sea was than inhabited by animals, in a similar manner as it is at present

Hence we are led to conclude, that the greater part of our land, if not the whole had been produced by operations natural to this globe; but that in order to make this land a permanent body, resisting the operations of the waters, two things had been required;

1st, The consolidation of masses formed by collections of loose or incoherent materials;

2ndly, The elevation of those consolidated masses from the bottom of the sea, the place where they were collected, to the stations in which they now remain above the level of the ocean." [51]

In 1787, Hutton observed what is now known as the Hutton Unconformity, one of the more striking of the many formations Hutton studied. It is located in Jedburgh, a small town that lies on a tributary of the Teviot River only ten miles from the border with Great Britain [52].

In 1788, he found a similar formation at Siccar Point, a rocky promontory in the county of Berwickshire. In both cases, Hutton reasoned that the formations had been created in three steps (see figure 9-1).

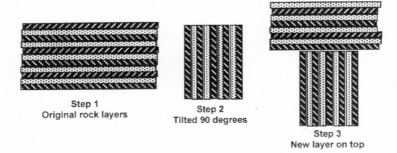

Step 1
Original rock layers

Step 2
Tilted 90 degrees

Step 3
New layer on top

Figure 9-1. Formation of Siccar Point

In the first step, many layers of sediment had been deposited on the ocean floor and over time (an enormous time) hardened into rock. In the second step, Hutton reasoned that this sedimentary rock must have be uplifted and tilted. After the rock was tilted, it was worn down by erosion. Finally new layers of sediment formed on top of the eroded rock, leaving the formations he observed at Jedburgh and Siccar Point.

Later in 1788, Hutton reported on his investigations in a paper that he presented at the Royal Society of Edinburgh. He remarked that "[t] he result, therefore, of this physical inquiry, is that we find no vestige of a beginning, no prospect of an end" [52]. While it is doubtful that Hutton had any notion regarding the source of the forces that created the Jedburgh and Siccar Point formations, he correctly reasoned that these formations were unequivocal evidence that the earth was much older than believed. Millions of years were his apparent estimates, but he had no basis with which to guess the true age of the earth.

These conclusions, which Hutton termed "gradualism," are some of the essential elements of "uniformitarianism." As will be discussed, Charles Lyell adopted and expanded Hutton's explanations and made uniformitarianism into a formal explanation of the creation of the earth.

John Louis Agassiz (1807–1873). Discovers the Ice Ages (with help). Snow typically falls in the winter months in northern hemisphere and melts in the summer. When more snow falls and then melts, snow accumulates. If enough snow accumulates, the snow will compact into ice. If snow and ice accumulation is sufficient and occurs in mountainous terrain (the usual place), the weight of the accumulated snow and ice will begin to move down the mountain and create what we call a glacier.

Glaciers are not mentioned in the Bible because they don't occur in the regions where biblical authors lived. They are, however, numerous in Switzerland. Glaciers belong to an interesting category between the apparently unchanging physical nature of the earth, those things which cannot typically be observed to change in one or more human lifetimes, and those things that can be observed within a human lifetime. However, glaciers move so slowly that careful observations are necessary to observe their motion.

On the other hand, valleys that have been traversed by glaciers are very different than valleys cut by rivers. River valleys are V-shaped, while glaciated valleys are U-shaped due to the carving action of glaciers. Thus, even though a glacier may no longer be present in a valley, a distinct U shape is irrefutable evidence of past glaciation. Besides U-shaped valleys, glaciers leave another easily visible clue for those who know how to interpret them. As a glacier moves down a mountainside, it pushes a large amount of material ahead of it. If the glacier retreats, the material (termed a "moraine") remains behind as a mound. Often lakes are impounded behind the moraines.

Besides glaciers in Switzerland, there are other objects that are quite visible such as large rocks and boulders some weighing several tons that are "strewn around the landscape" and defied explanation for a long time. Of particular interest was the fact that the rocks were often very different than the rocks they were lying on (e.g., granite boulders lying on sandstone layers).

Early humans ascribed these rocks and boulders to the work of fairies or giants. One proposal for the objects' placements was the biblical flood. As Winchester remarked, the flood was the answer to everything. However, investigators in Europe were unaware of the Missoula flood, which could move large boulders. Most dismissed the biblical flood as a myth.

Astute observers began to realize that the random positioning of large rocks and boulders indicated that an extremely large force had been active in the Swiss Alps. Three investigators in particular pursued the origin of the force: Jean de Charpentier, Karl Friedrich Schimper (1803–1867), and Louis Agassiz, finally made the connection between glaciers and strangely placed rocks. Charpentier and Schimper made the first observations and proposed tentative explanations. Agassiz built upon them and created the modern explanation of ice ages.

Jean de Charpentier's (1786–1855) father was a mining engineer, which was good profession in Switzerland [53]. Because of this, Charpentier also pursued a career as a mining engineer. He excelled in his field while working in the copper mines in the Pyrénées and salt mines in western Switzerland. However, in 1818, a lake created by a glacier ice dam created a massive flood when the dam suddenly collapsed, causing many deaths. A similar situation that had occurred in Lake Missoula. This event had a lasting impact on Charpentier.

After the disaster, he made extensive field studies in the Alps in an attempt to learn more about glaciers. Using evidence of erratic boulders, the difference between river-carved and glacier valleys, and the moraines left when a glacier melts and retreats, he hypothesized Swiss glaciers had once been much more extensive. Boulders characteristic of glaciers were strewn around as if brought there by glaciers that no longer existed, in contrast to the previous idea that ancient flooding causing the deposition. However, he wasn't sure how glaciers first formed, moved, or how they disappeared. He had only part of the answer, but his ideas were aided Agassiz, who expanded upon them

Karl Friedrich Schimper (1803–1867) was a German naturalist and poet. Born in Mannheim, he studied theology at the University of Heidelberg. His beginning research expanded the field of plant morphology [54]. He is perhaps best known as the originator of the theory of prehistoric hot and cold eras, and he was one of the initiators of the modern theories of ice ages and climatic cycles.

Bill Bryson [55] states in his celebrated book, *A Short History of Nearly Everything*, that Karl Schimper, presumably making the same observations as Charpentier, extended Charpentier's concept and, in addition to the idea of glaciation, proposed the radical idea that ice sheets had once covered much of Europe, Asia, and North America [56]. Unfortunately, Schimper was reluctant to write and never published his ideas. He did, however, discuss them with Agassiz, who went on to appropriate them as his own and, much to Schimper's dismay, undeservedly received much of the credit for the origination of the ice age concept [56].

Louis Agassiz was born in Môtier in the canton (similar to a U.S. state) of Frobourg in western Switzerland. Because it was a French-speaking area, Agassiz's natal language was most likely French. But, as with almost everyone else in Switzerland, he spoke German and probably Italian, plus a passable amount of English.

Adopting medicine as his profession, he studied at universities in Zurich, Heidelberg, and Munich. During his medical studies, Agassiz developed an interest in natural history and received a PhD in 1829 and an MD in 1830. He was a true polymath. After graduation, he moved to Paris and met George Cuvier, who ignited Agassiz's interest in both geology and zoology [57]. Cuvier will be discussed in more detail in the next chapter.

Initially, Agassiz devoted his energies to zoological interests, fish in particular. His research culminated in the 1830 issuance of a *History of the Freshwater Fish of Central Europe*. His appointment as professor of natural history at the University of Neuchatel in 1832 brought him to Monte Boloca, an area rich with fossil fish. The existence of the fossils was well known but little studied. Displaying his customary zeal, Agassiz made an exhaustive study of fish fossils, which resulted in a five volume treatise titled, *Recherches sur les poissons fossiles* (Research on Fossilized Fishes). The works were produced over the ten-year interval of 1833–1843.

During his fossil fish research, Agassiz became aware of Charpentier's and Schimper's research and conclusion that the rocks scattered over the slopes and summits of the Jura Mountains were the result of glacial action. Agassiz discussed their findings, and perhaps as Bill Bryson surmised, he may have appropriated their ideas without due credit.

Whether this is true or not, Agassiz attacked the problem with considerable more energy and in considerably greater depth than either of the other men, even going so far as to construct a hut on one of the glaciers of the Aar river, a tributary of the Rhine river, in order to study the glacier up close and personal. Agassiz made the hut his home for enough time to thoroughly investigate glacial structure and motion, which resulted in a two-volume work titled, *Etudes sur les glaciers* (Study on Glaciers). In this work, he

"... discussed the movements of the glaciers, their moraines, their influence in grooving and rounding the rocks over which they travelled, and in producing the striations and *roches moutonnees* seen in Alpine-style landscapes. He not only accepted Charpentier's and Schimper's idea that some of the alpine glaciers had extended across the wide plains and valleys drained by the Aar and the Rhone rivers, but he went still farther. He concluded that, in the relatively recent past, Switzerland had been another Greenland; that instead of a few glaciers stretching across the areas referred to, one vast sheet of ice, originating in the higher Alps, had extended over the entire valley of northwestern Switzerland until it reached the southern slopes of the Jura"[58].

In 1840, Agassiz, accompanied by the **Very Rev. Dr William Buckland (1784–1856)**. Buckland was a British geologist, paleontologist, and dean of Westminster. He wrote the first full account of a fossil dinosaur, but he was essentially a creationist attempting to reconcile Genesis by postulating that the phrase, "In the beginning" (Gen. 1:1) "meant an undefined period between the origin of the earth and the creation of its current inhabitants, during which a long series of extinctions and successive creations of new kinds of plants and animals had occurred" [59]. He was definitely an interesting companion.

The two men visited the mountainous districts of Scotland, England, and Wales, where they found extensive evidence of glaciations. Agassiz concluded that the three countries had been covered with great sheets of ice, which combined with his European investigations convinced Agassiz that a large portion of Northern Europe had been covered by a massive sheet of ice. He concluded that an ice age had occurred. It is unlikely Agassiz was able to date the ice age with any accuracy, but discovering it was a significant achievement.

Sir Roderick Impey Murchison (1792–1871) was born at Ross and Cromarty in the highlands of northern Scotland. He is known for initiating the first organized geologic time scale [60]. Murchison attended Durham military college and served eight years in the Scottish military. After leaving the military, he married Charlotte Hugonin, and after two years in Europe, they settled in Barnard Castle, County Durham, England in 1818. It was there that he made an acquaintance of Sir Humphry Davy, a noted British chemist who urged Murchison to turn his energy to science.

Murchison became fascinated by the young science of geology and joined the Geological Society of London, soon becoming one of its most active members. Among his colleagues were Charles Lyell, Charles Darwin, and noted geologist Adam Sedgwick, who devised a classification system for the Cambrian rocks. With Murchison, Sedgwick worked out the order of the carboniferous and underlying Devonian strata[14] [61].

In 1831, Murchison explored the border of Britain and Wales, attempting determine whether the greywacke rocks[15] underlying the Old Red Sandstone could be organized into a definite succession order. Murchison's efforts established the Silurian geologic "organization" system, named for a Welsh Celtic tribe, the Sillures. The new system grouped for the first time a remarkable series of formations, each replete with distinctive organic remains other than and very different from those of the other rocks of England. Together with descriptions of the coalfields and overlying formations in south Wales and the English border counties, these researches were embodied in *The Silurian System* (1839) [62].

Murchison's pioneering Silurian System was a significant contribution because geologists soon realized that most, if not all, of the various rock formations could be similarly organized into unique systems by employing their distinctive characteristics. This greatly simplified the organization of the earth's timescale and led to the modern organization of the the geologic timescale of the earth, which is presented in appendix II where the Silurian System is illustrated [63]. The geologic timescale shown in appendix II has the correct ages of the various systems, which of course were unknown in the early 1800s. Earth's age will be addressed at the end of this chapter.

Except for unusual places like the Grand Canyon, the rock layers representing the entire geologic timescale are not visible either above or below ground at any one location, and even the Grand Canyon has gaps. Sections of rock layers representing the geologic timescale at a particular location are called "stratiographic columns" [64].

Charles Lyell (1797–1875), born in Forfarshire (now Angus) in eastern Scotland, established uniformitarianism as best explanation for Earth's formation [65]. Like others such as Hutton and Agassiz, Lyell was born into a wealthy, well-educated family, which allowed him the luxury of study. Lyell attended Exeter College in Oxford, and later moved to London where he planned to become a lawyer. But his poor eyesight made that profession impossible, so Lyell turned to his real interest—science.

Geology soon became his forte, and as member of the Geological Society, he took part in the lively debates in the 1820s about how to reconcile the biblical account of the flood with geological findings[16]. Lyell rebelled against the prevailing theories of geology of the time. He thought the theories were biased and based on the interpretation of the book of Genesis. He thought it would be more practical to exclude sudden geological catastrophes to vouch for fossil remains of extinct species and believed it was necessary to create a vast time scale for Earth's history. This concept was called "uniformitarianism," a concept first identified by Hutton as "gradualism." It was a concept that Lyell and others slowly came to realize was a superior explanation for their geologic observations.

Lyell's wide-ranging geological interests included volcanoes, stratiography, paleontology, and glaciology. His best-known single contribution to geology, however, is his role in popularizing the doctrine of uniformitarianism.

Uniformitarianism: Uniformitarianism is a fundamental and axiomatic geologic principal that states that "the same processes that shape the universe occurred in the past as they do now, and that the same laws of physics apply in all parts of the knowable universe"[66]. Lyell ultimately became the leading proponent of uniformitarianism, arguing that geological processes observable at that time had not changed throughout Earth's history (e.g., mountain building and mountain eroding forces operate at the same rate today just as they did in the past).

To demonstrate that gradual processes could be responsible for great changes, Lyell used an engraving of the Temple at Serapis [67] as his frontispiece. The temple was one of the more magnificent

Roman structures produced. Later Roman Empire author Ammianus Marcellinus commented:

> "Its splendour [of the Temple] is such that mere words can only do it an injustice but its great halls of columns and its wealth of lifelike statues and other works of art make it, next to the Capitol, which is the symbol of the eternity of immemorial Rome, the most magnificent building in the whole world. It contained two priceless libraries"[68].

During the course of human history, the Temple had been above sea level, then partially submerged for a long period, and once again above sea level. This can be attested to by the dark bands of damage caused by waterborne life across the columns, thereby proving Lyell's hypothesis [69].

As the evidence for uniformitarianism continued to accumulate, other of Lyell's associates such as Roderick Murchison and George Poulett Scrope realized the superiority of this concept. They became outspoken opponents of the diluvial (biblical flood) position. To date, no informed observations have refuted the general concept of uniformitarianism. However, as noted, a number of catastrophic events have occurred in the past, therefore uniformitarianism must be leavened with a pinch of catastrophism.

Lyell's greatest overall contribution was his great geological opus: *The Principles of Geology* [70], subtitled, "An attempt to explain the former changes of the Earth's surface by reference to causes now in operation." *Principles* was initially published in three volumes from 1830–1833. In various revised editions (twelve in all through 1872), *Principles of Geology* was the most influential geological work in the middle of the nineteenth century and did much to establish modern geology.

The second edition of *Principles* introduced new ideas regarding metamorphic rocks. It described rock changes due to high temperature in sedimentary rocks adjacent to igneous rocks. His third volume dealt with paleontology and stratigraphy. Lyell stressed that the antiquity of human species was far beyond the accepted theories of that time [71].

One of the contributions that Lyell made in *Principles* was the explanation of the cause of earthquakes. It focused on recent earthquakes (less than 150 years from the time of Lyell's investigations) that were evidenced by readily observable surface irregularities like faults, fissures, stratigraphic displacements, and depressions [72].

Lyell's work on volcanoes focused largely on Vesuvius and Etna, both of which he had studied earlier. His conclusions supported gradual building of volcanoes, so-called "backed up-building" [73] as opposed to the upheaval argument supported by other geologists.

Lyell's extended the field of stratiography, which had been pioneered by William Smith. From May 1828 until February 1829, Lyell traveled to the Auvergne volcanic district in southern France and to Italy. Observations in these areas led him to conclude that the recent strata could be categorized according to the number and proportion of marine shells encased within them. Based on this, he proposed dividing the Tertiary period (see appendix II) into three parts: (1) the Pliocene (from the Greek words *pleion*, "more," and *kainos*, "new"), meaning roughly "continuation of the recent," and referring to essentially modern marine animals; (2) the Miocene, meaning "less recent"; and (3) the Eocene, which refers to the "dawn" of modern (new) mammalian animals that appeared during the epoch.

Reportedly, Lyell rejected Lamarck's view of evolution, but because he was a devout Christian, Lyell had great difficulty accepting Darwin's view of evolution, especially natural selection. Lyell was supposed to have remarked, "If Evolution is true, then Religion is a joke" [74].

SUMMARY FOR CHAPTER 9.

First, let's recap the resolution of the apparently unchanging physical features of the earth illusion up to 1850. By 1850, most of the essential elements that form the basis for modern geology had been established:

- Leonardo da Vinci initiated the proper investigation of fossils and used his observations to refute the biblical flood, at least in his eyes. The following list of his observations essentially refuted the idea that a single cataclysmic flood could have produced these observations:

- Fossils found in separate rock layers demonstrated they could not have been placed by a single flood.

- Worm tracks that would have been washed away by a giant flood were present.

  - Clam shells, still in their connected position, would have been torn apart by a flood.

- Neptunism and catastrophism were proposed as explanations of the development of the earth and had been proven incorrect. However, more recently a different form of catastrophism has been documented that has had significant impact on the earth. For example, the asteroid impact that may have ended the age of the dinosaurs. (See chapter 13 for more on this concept.)

- William Smith produced the first stratiographic map of England and laid the ground work for the important geological field of stratiography.

- James Hutton made a number of important observations (e.g., the interesting but difficult to explain sedimentary rock formation at Siccar Point, a rocky promontory in the county of Berwickshire). Hutton reasoned the formation was produced gradually over a vast amount of time, thousands or perhaps millions of years. Hutton termed his conclusion "gradualism."

- Jean de Charpentier, Karl Friedrich Schimper, and Louis Agassiz exhaustively studied the glaciers of Switzerland. Agassiz extended the other two men's studies to the rest of Europe and showed that the only rational explanation was an ice sheet that had covered most of northern Europe. The ice age had been discovered. In 1840, Agassiz et al. showed that glacial action rather than a biblical flood accounted for boulders strewn around the surface of Switzerland.

- Sir Roderick Impey Murchison showed that rock formations along the border of Britain and Wales could be organized into a system he termed the Silurian System (named for a Welsh Celtic tribe, the Sillures). This pioneering organizational approach was adapted by other geologists, who organized all of the various rock formations into unique systems, which led to the modern synthesis of the the geologic timescale of the earth.

• Sir Charles Lyell became one of the foremost geologists of the period with wide ranging geological interests including volcanoes, stratiography, paleontology, and glaciology. His single best-known contribution to geology was his role in first adapting and expanding Hutton's concept of gradualism to the more general concept of uniformitarianism and then popularizing the doctrine.

• Lyell's greatest overall contribution was his great geological opus: *The Principles of Geology* [70], subtitled, "An attempt to explain the former changes of the Earth's surface by reference to causes now in operation." *Principles* was initially published in three volumes from 1830–1833. In various revised editions (twelve in all through 1872), *Principles of Geology* was the most influential geological work in the middle of the nineteenth century and did much to establish modern geology.

LOOKING AHEAD.

By 1850, geologists had made great progress in resolving the apparently unchanging physical features of the earth illusion and had gathered sufficient evidence to demonstrate that the earth was much older than the 6,000 years calculated by the good bishop. However, for all of this progress, there were still some items regarding the earth that could not be properly explained.

For example, what is the source of the sun's energy and age of the sun? The sun had to be at least as old as the earth, but no one could explain the sun's energy source. As discussed in chapter 1, nuclear fusion provides the sun's energy, but this did not become known for almost another 100 years.

Also, what was the source of the forces that raised the sedimentary rocks at Siccar Point to a vertical position? Or that created the Alps and other mountain ranges? Geologists were completely stumped. Again, as discussed in chapter 1, the forces that raised the rocks were generated by the heat created by nuclear decay, something not determined until the twentieth century.

The next chapter will explain how these quandaries were resolved.

CHAPTER 10:
RESOLVING THE UNCHANGING PHYSICAL FEATURES OF THE EARTH ILLUSION AFTER 1850

The field of geology has seen much progress since the 1850s. We will only cover the developments of interest to this book: plate tectonics, seismology, and Earth dating methods.

Plate tectonics explains the source of the forces that raised Sicar Rock that had eluded nineteenth century geologists and provides one more indicator of the earth's age. In addition, plate tectonics is essential to the explanation of Earth's creation.

Seismology is the study of "earthquake" waves, which are created by plate motion, especially violent motion along boundaries where plates come in contact. These waves travel through the earth and reveal the details of Earth's interior structure.

Regarding Earth dating methods, this chapter will explain at least twelve independent methods for dating the earth.

PLATE TECTONICS—EARTH'S SURFACE MOVES.

The discovery that the earth's surface moves began with the inauguration of the great age of European exploration outside of local European waters, which can probably be traced to the discovery of the Azores islands by Portuguese navigators, though the exact discovery date is unknown [1] [2]. The islands are 950 miles west of Lisbon, Portugal, and obviously a significant distance west of Europe. The existence of the islands was known in the fourteenth century. Parts of them were recorded in the Atlas Catalan (Catalan is now part of Spain), the most important Catalan map of the medieval period, published in 1375 [2].

Columbus's 1492 discovery of North America set off a flurry of exploration, and in 1497–1498, Vasco de Gama rounded Cape of Good Hope enroute to India [3].

By the early 1500s, the essential coastal outlines of eastern North America, Western Europe, and Africa were well known. This information was documented in a large number of maps. When people began to examine the maps, they noticed an odd thing: if you moved the eastern edge of North America over to the western edge of Europe and Africa, it was a pretty good fit, too good to be a coincidence.

Alfred Wegner (1880–1930), a brilliant interdisciplinary scientist, advanced the preposterous idea that North America and Europe were connected at one time. Wegner was the first person to apply Sherlock's theorem to the apparent connection between the eastern and western Atlantic coasts. After extensive exploration, Wegner proposed that North America and Europe-Africa had at one time been joined in a huge continent he termed Pangaea, which is derived from the Greek word for "all the earth"[4].

Wegner's proposal was, of course, met with derision since he had no mechanism that would account for the 3000-mile movement of continents. Wegner did, however, gain some assistance from peolobiology. In 1911, he discovered a scientific paper that listed fossils of identical plants and animals found on opposite sides of the Atlantic [4]. One of the most extensively studied fossils were trilobites, extinct arthropods (spider-like creatures) [5]. Trilobites are particularly important because they were extremely diverse in type, distributed all over the globe, and relatively short-lived. Therefore, if a species was found in two places, there was certainty that they lived at the same time. The characteristics of animals such as trilobites cause them to be used as "index fossils."

Of interest to Wegner's theory, the same species of trilobite lived in Oklahoma and Morocco at the same time. Land bridges between North America and Africa were the first explanations, but Wegner's theory offered a more plausible explanation. However, one still had to explain how continents move.

Arthur Holmes (1890–1965), an exceptionally accomplished geologist [6], conceived the first realistic mechanism that could move continents. Holmes correctly hypothesized that convection currents in the earth's mantle could provide the forces that could move continents [6]. By this time, radioactive decay was well known as a potential heat source. Mantle convection currents powered by radioactively generated heat became the essential idea embodied in plate tectonics, now a major branch of geology "… that deals with the broad structural and deformational features of the outer part of the Earth, their origins, and the relationships between them [7]."

Regarding the movement of continents, the difference between the specific gravity of granite (the principal constituent continental rock) and the specific gravity of basalt (the principal constituent rock of the earth's crust) explains how continents are able to move. Granite has a specific gravity of about 2.75 [8], while basalt has a specific gravity of about 3 [9]. The ratio of 2.75 to 3 is about 0.9; hence the continents are about 10 percent lighter than the underlying crust. Interestingly, the specific gravity of ice is about 0.9 also, which is the reason ice floats. The continents "float" on the crust, just like ice floats on water. The main difference between continents floating on basalt and ice floating on water is the vast differences between the viscosities of continents on basalt vs. ice on water. It is much easier to move ice, but the strong mantle convection currents are enough to move continents.

One of the significant explanations provided by plate tectonics is the widening of the Atlantic Ocean because of seafloor spreading, a concept first proposed in the early 1960s by the American geologist and naval officer Harry H. Hess [10].

Development of highly sophisticated seismic recorders and precision depth recorders in the 1950s led to the discovery of the Mid-Atlantic Ridge in the early 1960s. It is a vast undersea mountain chain located in the middle of the Atlantic Ocean. Ultimately, it was found that the Mid-Atlantic Ridge was small segment of a globe-girdling 40,000 mile long undersea mountain system.

In many locations, this mid-ocean ridge was found to contain a gigantic gap, or rift, that was twenty to thirty miles wide and about one mile deep. At the ridge, the floor of the Atlantic Ocean is split into two sections. Lava welling up through the rift fills the gap formed as the plates move apart under the convection current forces.

In 1975, scientists of Project FAMOUS (French-American Mid-Ocean Undersea Study) used the undersea robot *Alvin* to dive on the Mid-Atlantic Ridge for the first direct observation of seafloor spreading [11].

A number of explorations have contributed to the establishment of plate tectonics as the proper explanation of the earth's surface. One of the more important was the Deep Sea Drilling Project, funded by the National Science Foundation and directed by the Joint Oceanographic Institution for Deep Earth Sampling (JOIDES), a consortium of leading U.S. oceanographic institutions [12]. The primary drilling vessel was the *Glomar Challenger*. It was the first ship of its type, displacing 10,500 tons of sediment and capable of drilling a core in 2,500 feet of sediment in 20,000 feet of water. The Deep Sea Drilling Project drilled about 600 holes into ocean floors across the world.

The project was remarkably successful, verifying that the present-day ocean basins are relatively young and confirming plate tectonics. It also discovered thick bedded salt layers from cores taken out of the Mediterranean Sea, indicating that the sea completely dried up between 5 and 12 million years ago; that Antarctica has been covered with ice for the last 20 million years; and that the northern polar ice cap was much more extensive 5 million years ago [13].

Direct observation of the effects of plate tectonics is possible at certain sites on Earth. One of the best places is the Point Reyes Peninsula on the Californian coast. The dividing line between two continental plates is termed a "fault." There are many types of faults, a principle one being the place where two plates slide by each other. Such a fault is the San Andreas Fault, one of the most famous on Earth [14]. To the west of the San Andreas Fault lies the Pacific Plate, while to the east lies the North American Plate [14]. Measurements have shown that the Pacific Plate moves northwest relative to the North American Plate at the rate of about two inches per year.

There are places on Point Reyes where the boundary between the two plates is exposed. At one spot, the west side of the fault is made of granite and the east side is a type of rock termed "chert" that is easily distinguished from granite. Approximately 350 miles south along the fault, near the city of Los Angeles on the east side, one can find granite that exactly matches the granite on the west side at Point Reyes. At two inches per year, we can calculate that approximately 11 million years have elapsed since the granite left Los Angeles.

Plate Tectonics Solves Many Other Mysteries: Uplifts of large amounts of rock like at Siccar Point or the creation of mountain ranges are also easily explained by plate tectonics.

As continents spread from a starting point, they eventually collide. When continents collide, enormous forces are created and either one of two things happen: (1) one continent slides under another, a process known as subduction [14], or (2) one continent causes the other to fold up just like a rug does when it is pushed against a wall, a process termed "orogeny" or "mountain building" [15]. The uplift Hutton observed was caused by collisions between plates, and it is how the Alps are being created [16].

Western United States is an excellent example of a subduction zone. Here the eastern edge of the pacific plate is subducting under the North American plate along the San Andreas Fault [17]. A result of this subduction was the formation of the Sierra Nevada. Approximately 250 million years ago, the pressure and friction that resulted from the grinding of the plates as the pacific plate subducted, caused the crust of the Pacific plate to melt, forming plumes of liquid rock termed "plutons" that eventually rose toward the surface.

These plutons combined into the single, massive deeply buried rock that is the Sierra Nevada. Further pressure from the collision of the two plates caused the deeply buried Sierra rock to rise, pushing up the sedimentary rock that was part of the North American plate above it. Beginning about 80 million years ago (MYA), the uplifted sedimentary rock began to erode, exposing the unique Sierra Nevada granite [18].

There is much more to the story of plate tectonics, but this should suffice for now. We now turn our attention the related subject of seismology.

SEISMOLOGY PROVIDES A "PICTURE" OF THE EARTH'S INTERIOR.

Seismology is the scientific study of earthquakes [19], which are a consequence of plate tectonics because they are usually produced by when plates rub past each other. This rubbing produces elastic "waves"[1] in the rock that propagate through the earth [20] and provide much information about the earth's interior.

John Winthrop (1714–1779), born in Boston, was one of the first American intellectuals to be taken seriously in Europe and made one of the first scientific studies of earthquakes [21]. Winthrop was noted for his attempt to explain the 1855 Lisbon earthquake as a natural as opposed to a religious phenomenon [21].

Andrija Mohorovicic (1857–1936) was an Austro-Hungarian meteorologist who discovered a boundary between the crust and an interior section of the earth known as the mantle when he was analyzing seismic data from a 1909 earthquake near Zagreb (now in Croatia). This boundary is now called the Mohorovicic discontinuity or simply the Moho [19].

U.S. seismologist **Charles F. Richter (1900–1995),** born on a farm north of Cincinnati near Hamilton, Ohio, made a number of contributions to seismological studies. One of his more significant accomplishments was the development of the seismic magnitude scale in collaboration with Beno Gutenberg, which bears Richter's name. The Richter earthquake scale describes earthquake magnitude as measured by a device known as a seismograph. Richter's scale is logarithmic, which many find confusing because each point on the Richter scale represents a ten-fold increase in amplitude of a wave launched by an earthquake [22]. Thus a magnitude 6.0 earthquake is ten times stronger that a magnitude 5.0, and a magnitude 7.0 earthquake is 100 times stronger than a magnitude 5.0.

Despite the obvious value of seismology to earthquake prediction, our interest in seismology is in its ability to determine the earth's interior structure, which is accomplished by carefully observing the two principal earthquake waves: P-waves and S-waves[2]. P-waves (principal waves) travel in the direction in which the wave is traveling, while S-waves (secondary waves) move perpendicularly to the wave direction

[23]. Figure 10-1 presents a simplified diagram of the earth's interior [24] that has been deduced by evaluating earthquake waves.

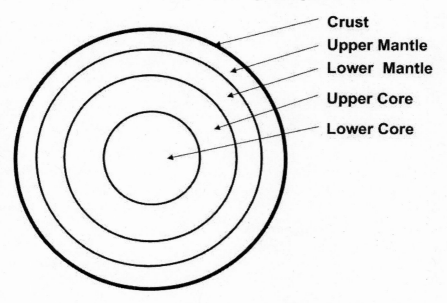

Figure 10-1. Earth's interior structure, not to scale

The application of computed tomography to earthquake waves has facilitated the creation of a complete "map" of the earth's interior to a resolution of several hundred miles. Tomography is the integration of several observations of an object taken from different observation points to obtain a three dimensional view. Computed tomography has a well-known application in medicine. The resolution it provides allows geologists to detect and identify large-scale features such as mantle plumes, which are principal sources of volcanoes.

Based upon analysis of seismic waves, the upper core has been deduced to be liquid (due presumably to radioactive element decay) because it does not transmit S-waves and because the velocity of compressional P-waves that pass through it is sharply reduced. The inner core has been deduced to be solid because the behavior of P- and S-waves passing through it is different than the upper core [24].

THE SEARCH FOR RELIABLE AND VERIFIABLE EARTH AND UNIVERSE DATING METHODS.

Much of the material that has been presented in this book either directly or indirectly supports the fact of an earth much older than 6,000 years. To firmly establish that the earth is indeed much older than 6,000 years, the next section will summarize twelve of the methods that are used to reliably and verifiably determine the age of the earth and universe. However, the development of reliable dating methods has not been an easy task.

As pointed out above, despite the obvious advantage of the data that had been collected by many, Lyell et al. had to contend with the enigma of the sun's age, which was important because the earth couldn't be older than the sun. All Earth age estimates made during the 1800s were much too low, and the true Earth's age was not determined until the twentieth century [25].

A rather bizarre attempt at age determination was undertaken 1862 by British scientist Lord Kelvin, who used Fourier's equations of heat condition to calculate the age of the earth. Knowing that the earth's temperature increased one degree Fahrenheit for each 50 feet one descends into the earth, Kelvin guessed that the earth began as molten rock at 7000° F [26]. By solving Fourier's equation, Kelvin found that it must have taken a hundred million years for the earth's temperature to level out to one degree every 50 feet. Kelvin quickly found himself caught between biblical literalists who complained it was too old and geologist who felt it was too young.

Another issue related to the dating problem was the source of the earth's inner heat. We now understand that Earth's interior heat is generated by radioactive, or more properly nuclear decay, which was not discovered until 1897 and not fully understood until the early twentieth century.

A Summary of Reliable and Verifiable Earth and Universe Dating Methods The following section will summarize several of the reliable, verifiable, and independent dating methods, beginning with one of most useful—radiometric dating.

Method 1: Radiometric Dating. As described in chapter 1, radiometric dating employs the well-known phenomenon of nuclear decay, the spontaneous emission of particles from a nucleus. Some of the more useful radioactive elements for dating are illustrated in table 10-1 [27]. See table 2-3 for the elements common names.

Parent	Daughter	Half-life, Billions of Yrs	Materials that can be dated
^{235}U	^{207}Pb	0.704	Zircon, uraninite, pitchblende
^{40}K	^{40}Ar	1.251	Muscovite, biotite, hornblende, volcanic rock, glauconite, K-feldspar
^{238}U	^{206}Pb	4.468	Zircon, uraninite, pitchblende
^{87}Rb	^{87}Sr	48.8	K-micas, K-feldspars, biotite, metamorphic rock, glauconite

Table 10-1. Some of the more useful radiometric dating materials

Arthur Holmes, introduced above, was the first person to make practical use of nuclear decay as a dating technique. Holmes published his first findings from his study of the decay of uranium (U) into lead (Pb) in which he determined that the earth was about 1.6 billion years old. This was of course much greater than other estimates and met with some skepticism. Measuring the products of nuclear decay is a tedious task, and Holmes devoted the majority of his life to the task. By 1946, he felt he had enough evidence to state that the earth was at least 3 billion years old and probably older [28].

Method 2: Isotope Ratios An informative Web article by Chris Stassen points out that ratios of isotopes such as isotopes of lead, known as isochron dating, will yield more accurate results than simple use of the nuclear decay equation [29].

As pointed out by Stassen, the most direct means for calculating the earth's age is a Pb/Pb isochron age derived from samples of the earth and meteorites. This involves measurement of three isotopes of lead: ^{206}Pb, ^{207}Pb, and either ^{208}Pb or ^{204}Pb. A plot is constructed of the ratio of ^{206}Pb to ^{204}Pb versus ^{207}Pb to ^{204}Pb.

Since most of the earliest of earth's material has been destroyed by plate tectonics, complete measurements of the earth's age must rely on extra terrestrial objects like meteorites. See for example [30].

If the solar system formed from a common pool of matter that was uniformly distributed in terms of Pb isotope ratios, then the initial plots for all objects from that pool of matter would fall on a single point. Over time, the amounts of ^{206}Pb and ^{207}Pb will change in some samples because these isotopes are decay end-products of uranium decay (^{238}U decays to ^{206}Pb and ^{235}U decays to ^{207}Pb), which gives an accurate age of the rock being dated.

Method 3: Universe Dating Techniques: Chapter 13 discussed Edwin Hubble's discovery that unique stars termed Cepheid variables could be used as a yardstick for measuring distances to the distant galaxies and determined that the closest galaxy, Andromeda, was 2.5 million light-years from Earth. Since light travels one light-year per year, the Andromeda galaxy is at least 2.5.million years old. Using this new distance measuring technique, Hubble demonstrated that the universe was expanding, which led to the hypothesis that the expansion began with a "Big Bang." The Big Bang hypothesis was shown to be correct when radiation from the Big Bang was discovered. Exhaustive study of this radiation by the Wilkinson Microwave Anisotropy Probe confirmed that the universe was 13.7 billion years old, give or take 200 thousand years.

An excellent article by Ian Plimar [31] lists six independent means for dating the earth and objects on it, one of which has already been mentioned (radiometric dating); thus we will list the other five from Plimar's list (methods 4–8).

Method 4: Electron capture in minerals: Measuring the electrons captured in minerals as a result of long periods of solar and cosmic radiation bombardment. The number of electrons is an indication of the minerals age.

Method 5: Earth magnetic field reversal: The direction of the earth's magnetic field reverses and can be used to determine when magnetic minerals formed and thus date the host material.

Method 6: Biological material decay: Biological material such as amino acids undergo decay at known rates, therefore measurement of chemicals in old biological material enables an estimate of the decay time and thus, age of the material.

Method 7: Nitrogen loss and fluorine gain in bones: As bones age, nitrogen is lost and fluorine is gained from ground water at known rates. therefore measurement of the nitrogen/fluorine ration old bones enables an estimate of the decay time and thus. Age of the bones

Method 8: seasonal sediment accumulation: This method employs the simple fact of tidal or seasonal cycles to establish the age of sediments, (e.g., more sediment from erosion occurs in summer than winter).

Method 9: Accelerator Mass Spectrometry (AMS): One can also measure ages by AMS. This technique is unique relative to other types of mass spectrometry in that it accelerates ions to extremely high energies, which gives it the ability to separate a rare isotope (e.g., ^{14}C) from an abundant neighboring mass (e.g., ^{12}C) [32]. The technique is particularly useful in dating delicate materials. For example, AMS was used to date the Shroud of Turin [33], which is so delicate that only a few fibers could be examined.

Another excellent source for dating methods is provided by UCLA Professor Edward Wright's 2005 Web article [34], which pointed out that besides the WMAP data, there are at least three ways, that the age of the universe can be estimated.

Method 10: The Age of the Chemical Elements: The age of the chemical elements can be estimated using radioactive decay to determine how old a given mixture of atoms is. The most definite ages that can be determined this way are ages since the solidification of rock samples. When a rock solidifies, the chemical elements often get separated into different crystalline grains in the rock.

Method 11: The Age of the Oldest Star Clusters: When stars are burning hydrogen to helium in their cores, they fall on a single curve in the luminosity-temperature plot known as the Hertzsprung-Russell diagram after its inventors. This track is known as the main sequence because most stars are found there. Since the luminosity of a star varies, the lifetime of a star on the main sequence varies by a constant. Thus if you measure the luminosity of the most luminous star on the main sequence, you get an upper limit for the age of the cluster.

Method 12: The Age of the Oldest White Dwarf Stars> As explained previously, a white dwarf star is an object that is about as heavy as the sun, but it has only the radius of the earth. The average density of a white dwarf is a million times denser than water. White dwarf stars form in the centers of red giant stars, but they are not visible until the envelope of the red giant is ejected into space. When this happens the ultraviolet radiation from the very hot stellar core ionizes the gas and produces a planetary nebula. The envelope of the star continues to move away from the central core, and eventually the planetary nebula fades to invisibility, leaving just the very hot core that is now a white dwarf. White dwarf stars glow just from residual heat. The oldest white dwarfs will be the coldest and thus the faintest. By searching for faint white dwarfs, one can estimate the length of time the oldest white dwarf has been cooling.

Thus there are twelve *independent methods from several independent sources* of credible, independently testable information regarding the determination of the both the age of the earth and the universe. These range from the use of various forms of nuclear decay to exotic techniques such as the age of white dwarf stars and the Wilkinson Microwave Anisotropy Probe. In order to retain the belief that the earth is only 6,000 years old, every one of these methods must be proven false, an impossible undertaking, especially because those who have attempted this feat have been uniformly unsuccessful.

Accordingly, there is no credible evidence that disputes the fact that the earth is 4.5 billion years old or that it began with a bang 13.7 billion years ago. Thus the first option in the Gallup poll, creationism, is demonstrably false. In addition, chapter 11 will demonstrate the fallaciousness of statements like the one previously quoted in chapter 9: "There is no reason not to believe that God created our universe, earth, plants, animals, and people just as described in the book of Genesis!"

In 1984, G. Brent Dalrymple of the U. S. Geological Survey presented a devastating critique of creationist dating methods in a paper titled, "How Old is the Earth, A reply to Creationism," at the 63rd Annual Meeting of the Pacific Division, AAAS. He discussed in detail scientific dating methods and demonstrated the flaws in creationist attempts to "prove" the earth is young. Dalrymple concluded: "Even a cursory reading of the literature of 'scientific' creationism, however, reveals that the creation model is not scientifically based but is, instead, a religious apologetic derived from a literal interpretation of parts of the book of Genesis" [35].

SUMMARY OF CHAPTER 10.

For this book's purpose, there were three key developments after 1850: the discovery of plate tectonics, the establishment of seismology, and refinement of Earth and universe dating techniques.

PLATE TECTONICS.

Above the mantle of the earth lies a relatively thin crust of solid rock, of which there are two kinds: basalt and granite. Granite has a slightly lower specific gravity than basalt; hence granite "floats" on top of it. Although the coefficient of friction between granite and basalt is rather large, there is sufficient strength in the mantle convection forces to slowly move the granite plates. This phenomenon is called plate tectonics. Plate tectonics explains many mysteries: the shape of the Atlantic Ocean, the building of mountain ranges, and the configuration of Siccar Point. The pace of plate tectonics is one of the many indications that the earth is very old.

SEISMOLOGY.

The study of the passage of earthquake waves through the earth revealed Earth's internal structure in considerable detail. The earth consists of a series of concentric spheres beginning with the core, which consists of a solid inner core that is mostly iron and an outer core that is liquid. Above these cores lies an upper and lower mantle. The earth's interior has been measured in sufficient detail to preclude the possibility of large pockets of water that could have produced the "fountains of the deep."

The decay of radioactive elements provides the earth's inner heat, especially the heat energy sufficient to cause convection currents in the upper mantle.

AGE OF THE EARTH.

In addition to the conclusive proof that the universe is 13.7 billion years old, there are twelve independent methods of determining Earth's age. That number does not include the controversial radioactive dating such as fluctuation in the Earth's magnetic field and the decay of biological material, which has never been convincingly proven to be incorrect. Moreover, a paper by Dalrymple that repudiates creationist objections and dubious dating methods was briefly discussed.

THE BOTTOM LINE.

Chapters 9 and 10 have explained the resolution of the apparently unchanging physical features of the Earth illusion in enough detail to establish beyond any reasonable doubt that the Earth has been changing ever since it was created from the primordial disc that surrounded the birth of the sun some 4.5 billion years ago (BYA).

These chapters compare the religious explanation of the creation of Earth's physical features that is essentially found in one book, the book of Genesis, with the scientific creation explanation found in a library full of books. We have seen how the religious explanation can answer the question of *who* created the physical features, but it is unable to explain *how* or *why* the features were created. On the other hand, science cannot answer *who* created the features, but can explain *how* and *why* the features were created in great detail.

Both religion and science can explain *when* the features were formed, but there is considerable difference between the explanations. The book of Genesis explains, with no evidence, that the features were created essentially as we see them approximately 6,000 years ago, while science explains, with considerable evidence, that the features were created with glacial slowness over a period of 4.5 billions of years. Regarding science's inability to answer *who* created the features, the reason is that no one created them.

In view of all that has been presented, Dalrymple's assessment of creationism, presented above—"Even a cursory reading of the literature of 'scientific' creationism, however, reveals that the creation model is not scientifically based but is, instead, a religious apologetic derived from a literal interpretation of parts of the book of Genesis"—should be sufficient proof for all but the most ardent believer that creationism (option 1 in the Gallup poll) is not the correct answer to which of the three mutually exclusive options posed by the Gallup poll is correct.

LOOKING AHEAD.

Having explained the resolution of the apparently unchanging physical features, in the next two chapters we will next move to the resolution of the apparently unchanging biological features illusion.

CHAPTER 11:
RESOLVING THE APPARENTLY UNCHANGING ANIMAL AND PLANT STRUCTURES ILLUSION UP TO DARWIN

CREATING THE ILLUSION.

You are visiting a famous zoo for the nth time, perhaps with your grandchild. As you look at the animals arranged in various enclosures, all seems to be just as you remember it. Perhaps you stop in at the gift shop for some animal pictures that were perhaps painted many years ago. If you look carefully, the animals look just like those in the zoo today.

In fact, any place you go on Earth where there are accurate depictions of animals and plant life, regardless of how old the depictions are (e.g., cave or rock paintings, especially paintings in such caves as the Lascaux in France), you will recognize many of the animals that you would see in a zoo or in a farmyard or just roaming free. Animal structures do not appear to change!

But It's All an Illusion: The apparently unchanging animal and plant structures are an illusion, one produced by the glacial slowness of the evolutionary process. It is a process so slow that even over thousands of years only a careful observer would detect any changes. Unfortunately, as with the illusion of an unchanging physical Earth, the illusion of unchanging animal and plant structures leads to the incorrect belief that the earth is not very old. This belief has serious consequences that will be discussed in chapter 18.

EARLIEST OBSERVATIONS/EXPLANATIONS OF ANIMAL AND PLANT STRUCTURES

Some of the biological observations available to early humans were obviously the various animals and plants. Animals are clearly of different types and so are plants. Some animals are constrained to the water, others live out their lives on the ground, and still others have the ability to rise above the ground and fly.

Then there was the annual cycle of animal life in which some animals such as birds and fish mate and produce young in the spring, while others, basically mammals, produce young in the spring but mate in the fall. Humans produce young on every day of the year. Similarly, there are plant cycles. Many plants and broad-leaved trees lose all their leaves and appear to "die" in the fall and are reborn in the spring, while other cone-bearing trees appear to remain unchanged.

Finally, in many rock formations, one finds "rocks" that resemble living animals. As discussed in the last chapter, these are fossilized remains of animals that died a long time ago. However, because the earth's age was believed to be only 6,000 years until discoveries in the late 1700s and early 1800s proved otherwise, no one knew how to interpret fossils correctly.

The Book of Genesis and the Origin of the Life: Listed below for your convenience are some of the Genesis verses from appendix I that are devoted to the origin of life. While the author(s) of Genesis devote ten verses to the creation of the earth, they devote twenty-one verses to the creation of life, perhaps a subject of more interest. It is particularly puzzling that verses 1:14–18, which are devoted to the placement of "Great lights in the firmament" are included in the discussion of life's creation rather than in the first ten verses. Perhaps this placement is due to multiple authors.

Genesis 1:11–12 reads, "And God said, Let the earth bring forth grass, the herb yielding seed, and the fruit tree yielding fruit after his kind, whose seed is in itself, upon the earth: and it was so. And the earth brought forth grass, and herb yielding seed after his kind, and the tree yielding fruit, whose seed was in itself, after his kind: and God saw that it was good."

The verses continue with Genesis 1:20–22: "And God said, Let the waters bring forth abundantly the moving creature that hath life, and fowl that may fly above the earth in the open firmament of heaven. And God created great whales, and every living creature that moveth, which the waters brought forth abundantly, after their kind, and every winged fowl after his kind: and God saw that it was good."

Finally, closing the verses dealing with the creation of life, Genesis 1:22–28 reads:

> "And God blessed them, saying, Be fruitful and multiply and fill the waters in the seas, and let fowl multiply in the earth.
>
> And God said, Let the earth bring forth the living creature after his kind, cattle, and creeping thing, and beast of the earth after his kind: and it was so.
>
> And God made the beast of the earth after his kind, and cattle after their kind, and every thing that creepeth upon the earth after his kind: and God saw that it was good.
>
> And God said, Let us make man in our image, after our likeness: and let them have dominion over the fish of the sea, and over the fowl of the air, and over the cattle, and over all the earth, and over every creeping thing that creepeth upon the earth.
>
> So God created man in his own image, in the image of God created he him; male and female created he them.
>
> And God blessed them, and God said unto them, Be fruitful, and multiply, and replenish the earth, and subdue it: and have dominion over the fish of the sea, and over the fowl of the air, and over every living thing that moveth upon the earth."

As with the section of Genesis devoted to the creation of the earth, statements such as,

> "And God said, Let the earth bring forth grass" (Gen. 1:11), only explains the existence of grass, which was somehow created. Therefore Genesis is merely a recording of what the writers observed, not an explanation of the creation of life. Moreover, there was nothing to suggest that all was not created at the same time.

Greek Explanation of Biology: While the Greeks didn't appear to have much to say about the creation of the earth, they studied life in considerable depth. Aristotle took particular interest in it. As mentioned above, he wrote on many subjects, of which biology is of primary interest in this section.

Aristotle is the earliest significant observer of biology, and his work has survived in some detail. His research included many areas, but his observation of sea life, such as those available from local fisherman, were among his best work. He separated the aquatic mammals from fish and knew that sharks and rays were part of the group he called Selache [1].

Aristotle's methods are exemplified by his work, *Generation of Animals*, where he describes breaking open fertilized chicken eggs at intervals to observe when visible organs were generated. Because of this we can reason that Aristotle had some understanding of embryogenesis but limited understanding of how reproduction functioned (i.e., he lacked the instruments to detect cells).

He gave accurate descriptions of the four-chambered ruminant stomach and of the embryological development of the hound shark [2]. Unfortunately, much of Aristotle's works were lost, and many have been superseded, but considering the knowledge base he had to work with, his findings and observations were quite remarkable. However, he was merely a recorder and partial organizer of information and provided no particularly useful explanation of his observations.

But as described earlier in chapter 9, Aristotle "stubbed his toe" on fossils [3]. Fossils, of course, have presumably been known since prehistoric times because they are found almost everywhere. Some of the early Greeks like Herodotus and Xenophanes correctly identified fossils as relics of ancient organisms in rock. However, Aristotle believed that fossils grew in rocks in response to the actions of an organic essence. Unfortunately, this incorrect belief caused fossils to be regarded as freaks and not worthy of investigation until Leonardo took up the problem.

RESOLUTION OF THE UNCHANGING BIOLOGICAL PROCESS ILLUSION LEADING TO DARWIN.

As recounted in the previous chapter, in the mid to late 1700s, a literal interpretation of the biblical account in Genesis was the accepted explanation of the existence of Earth and its inhabitants. This interpretation held that the earth was created in 4004 BCE and that nothing had changed since then.

As with the collection of information regarding the earth's physical features, considerable biological information was accumulated, beginning in the eighteenth century. The information gradually led to the recognition that the apparently unchanging animal and plant structures of the earth were an illusion. Evolution had occurred and animals had developed over a large interval of time.

Six scientists played important roles in the investigation of these new ideas: four from France, one from Germany and one from Great Britain. They had very different personalities, educations, and views. They were particularly interested in answering two key questions: had evolution occurred and if so, *how* did it work? In considering these investigators, keep in mind that none of these scientists understood very much about the engine of evolution—reproduction. A full understanding of reproduction has only been recently achieved. We begin with Georges-Louis Leclerc, Comte de Buffon.

Georges-Louis Leclerc, Comte de Buffon (1707–1788), a member of the French scientific elite, was a naturalist, mathematician, biologist, and cosmologist—truly one of the great polymaths. Buffon's views influenced the next two generations of naturalists, including Lamark and Darwin (see below). Darwin himself stated that "the first author who in modern times has treated [evolution] in a scientific spirit was Buffon" [13].

Buffon first made his mark in the field of mathematics, and in *Sur le jeu de franc-carreau* (On the game of chance), he introduced differential and integral calculus into probability theory. His most famous work was *Histoire naturelle, générale et particulière* (Natural history, general and particular), which he prepared during the years 1749–1778 and was published posthumously.

Buffon observed that different regions have distinct plants and animals despite having similar environments. These observations led him to the radical conclusion that species must have "improved"(i.e., evolved after dispersing away from a center of creation), thus following in the footsteps of such iconoclasts as Aristarchus and Copernicus who defied convention regarding the shape of solar system, and Thompson and Rutherford who defied convention regarding the shape of the atom. Buffon's observations also led him assert that climate change must have facilitated the worldwide spread of species.

In his prescient book, *Les Époques de la Nature* (*The Epochs of Nature*), published in 1778, he discussed the origins of the solar system, speculating that the planets had been created by comet collisions with the sun [4]. This was almost 200 years before the collision theory was discredited by improved explanations (as discussed in chapter 5).

Buffon continued his iconoclastic ways by making a calculation of the cooling rate of iron to estimate that the age of the earth was 75,000 years, which he also published in *The Epochs of Nature*. This obviously implied that the earth originated much earlier than Archbishop Ussher's 4004 BCE date, which made him perhaps the first person to challenge the existing position of the Catholic Church. Unsurprisingly, this invoked the fifth impediment—conflict with religious authority. The church condemned him and had his books burned.

Buffon poured salt in the wound by also denying that the biblical flood ever occurred and observed that some animals retained parts that are vestigial and no longer useful, suggesting that they had evolved rather than being spontaneously generated. Moreover, he anticipated Hutton and Lyell's uniformitarianism. Finally, his examination of the similarities between primates and humans suggested a common ancestor, which led him to believe in organic change, but he could not specify how it worked [5]. Buffon was a hard act to follow.

Jean-Baptiste Lamarck (1744–1829), like Buffon, was another member of the "French Naturalists," and he was influenced by Buffon to embrace the concept that evolution is a process governed by natural laws [6]. However, Lamarck developed the rather controversial concept that evolution occurred when animals improved by acquiring more useful capabilities, which they passed on to their offspring. This concept is known as "the inheritance of acquired characteristics."

Although the concept of passing on acquired characteristics will be shown later in this book to be impossible, Lamarck constructed what may be the first comprehensive theoretical framework of organic evolution. S. J. Gould argues that Lamarck was the "primary evolutionary theorist" [7].

Lamarck was also one of the first persons to employ the term "biology" [6] and was also one of the main contributors to the cell theory.

He was a transitional scientist, part modern and part medieval in that he opposed Lavoisier's modern chemistry in favor of alchemy and preferred the classical view of earth, wind, fire, and water as Earth's principal elements.

Étienne Geoffroy St. Hilaire (1772–1844), another French naturalist, argued in his two-volume 1818 work, *Philosophie anatomique*, that form is conserved [8]. Geoffroy asserted that vertebrate animal organization reflected one uniform structure. He saw all vertebrates as modifications of a single archetype, a single form. Currently unneeded organs (e.g., the appendix) might serve no functional purpose, but the fact that they still existed demonstrated that animals were derived from an archetype.

Georges Cuvier (1769–1832), on the other hand, a prominent French scientist who was considered one of the greatest scientists of his time [9], disagreed with Geoffroy. Based upon careful observations, he argued that form does follow function. In a famous 1830 debate with Geoffroy, Cuvier convincingly demonstrated how many apparent examples of structural unity proposed by Geoffroy to support the conservation of form concept were in fact contrived and superficial, which thus supported the argument that form follows function [10].

Regarding evolution, in his 1813 "Essay on the Theory of the Earth," Cuvier proposed that new species were created after periodic catastrophic floods [9], which eliminated many species and allowed space for new ones. Therefore, he would naturally be opposed to anyone, such a Lamarck, who proposed some type of evolution.

Among his many other accomplishments, Cuvier is credited with the foundation of vertebrate paleontology. He also convinced his contemporaries that the controversial subject of extinction is a fact. However, as was shown above, that was fairly well settled by Leonardo, though Leonardo's findings were buried under indecipherable text.

Finally, Cuvier's study of the Paris basin with Alexander Brongniart established the basic principles of biostratiography [9].

Louis Agassiz (1807–1873), a brilliant geologist, was not very effective in the field of biology. Agassiz resisted Darwin's theories on evolution and denied that species originated in single pairs, whether at a single location or many. He argued instead that multiple individuals in each species were created at the same time and then distributed throughout the continents where God meant for them to dwell, which is essentially creationism and consistent with the generally accepted views of the time. His lectures on polygenism were popular among the slaveholders in the South of the United States [11].

In opposition to Agassiz, **Ernst Haeckel (1834–1919)**, German biologist, naturalist, philosopher, and physician, shared Darwin's belief in evolution. On the other hand, he did not share Darwin's enthusiasm for natural selection as the main mechanism for generating the diversity of the biological world. Haeckel instead believed that the environment acted directly on organisms and produced new races (a version of Lamarckism). Haeckel did argue that the survival of the races depended on their interaction with the environment. His belief was a weak form of natural selection [12].

Summary of the Six Understandings of Life's Origins: If we employ these six varying understandings, can we determine if evolution has occurred? And if it has occurred, can we explain how it works? Table 11-1 summarizes the six explanations discussed above.

Investigator	Evolution occurred?	How does it work?	Life spans of investigors
Buffon	Y	Didn't know	1707–1780
Lamarck	Y	Inheritance of acquired characteristics	1744–1829
Geoffrey	Y	Not clear he had a concept	1772–1844
Cuvier	N	Improvements by extinctions and restarts	1769–1832
Agaziz	N	Believed in Creationism	1807–1873
Haeckel	Y?	Variation on Lamarckism	1834–1919

Table 11-1. Summary of six investigators' views regarding evolution.

Thus, if the votes in table 11-1 decided the issue, we could say evolution has occurred. However, the mechanism for evolution was hardly clear, except for Lamarckism, which was ultimately proven wrong. On the other hand, considering the limited state of biological understanding at this time, these findings represent remarkable prescience. They also set the stage for Darwin.

Charles Robert Darwin (1809–1882) was one of those persons fortunate enough to be in the right place at the right time [13]. While Darwin did not specifically discover the process of evolution, as shown above, he made three key contributions that provided the first significant contribution to the explanation of evolution:

1. He was the first to collect and organize enough material to allow him to conclusively demonstrate the existence of the evolutionary process.

2. In view of the unfriendly religious climate, Darwin had the courage publish his results.

3. Perhaps most importantly, Darwin provided the first useful, although incomplete, explanation of the process of evolution. [14]

Darwin was born in 1809 into a well-educated and prosperous family. He was the son of wealthy society doctor and financier Robert Darwin and the grandson of Erasmus Darwin, an English physician, natural philosopher, physiologist, inventor, and poet. Erasmus Darwin was one of the founding members of the Lunar Society, a discussion group of pioneering industrialists and natural philosophers.

Darwin was also born at a time of great intellectual ferment, particularly the controversial issue of biblical literacy. While Darwin was growing to manhood, he met a person that was to become an influential mentor, Charles Lyell, whom we encountered in chapter 9. The importance of Lyell's contribution to Darwin's efforts cannot be emphasized enough.

Darwin received a sound, fairly general education, studying first medicine at Edinburgh University and then theology at Cambridge. These studies instilled in Darwin a keen interest in natural history, which was the field he ultimately followed.

The Fateful Voyage of the Beagle: Darwin's interest in science was greatly enhanced when he had the good fortune to be selected as "a gentleman companion" by Robert Fitzroy, captain of the British survey ship *Beagle* [15]. The *Beagle's* journey was to be long, and Fitzroy dreaded a long voyage without suitable companionship, as the previous captain had committed suicide[1]. It is noteworthy that Darwin carried a copy of volume I of Lyell's treatise on geology, but Lyell argued against evolution. Lyell rejected Lamark's idea of organic evolution, proposing instead "Centers of Creation" to explain diversity and territory of species. However, many of his letters demonstrated that he was fairly open to the idea of evolution [ch 5, 65]

Originally scheduled for two years, the *Beagle* voyage lasted five (1831–1836). They circumnavigated the earth and stopping at many interesting places, including the Galapagos Islands. For Darwin, the Galapagos were one of the most ideal locations on Earth to study the development of life. Regarding the islands themselves:

"The islands lie in the Pacific Ocean about 1,000 km (approximately 600 miles) from the South American coast and straddle the Equator. There are 13 large islands, 6 smaller ones and 107 islets and rocks, with a total land area of about 8,000 km² (approximately

300 mi^2). The islands are volcanic in origin and several volcanoes in the west of the archipelago are still very active." [16]

The islands were sufficiently separated to allow independent evolution but close enough to be readily accessible. While exploring the islands, Darwin became especially interested in small birds on the islands. Darwin noted that the bird's taxonomy (physical features) varied from island to island and that beaks were particularly well adapted for the different types of seeds found on each island. Since Darwin was actually more of a geologist than ornithologist, in examining the birds, he concluded they were a mixture of blackbirds, grosbeaks, and finches. When he arrived back in England, Darwin's collection included several samples of each to exhibit.

In addition to collecting many samples of wildlife, Darwin had made many important geological discoveries, which he sent back to England on those occasions when the *Beagle* stopped at a British controlled location. These discoveries were so well received that when he arrived in England, he was celebrated as a rising naturalist of great capability.

Upon his return, Darwin wrote his first paper, demonstrating that the South American landmass was slowly rising. It was met with great enthusiasm, especially by Lyell because it supported Lyell's contention that very slow geological forces shaped the landscape [13]. With Lyell's enthusiastic backing, Darwin read his paper to the Geological Society of London on January 4, 1837.

Later the same day, Darwin presented the mammal and bird specimens he had collected during his *Beagle* voyage to the Society. Fortuitously, the bird specimens were given to ornithologist John Gould for identification.

John Gould (1804–1881) was born in Dorset, England, a poorly educated gardener's son. He adopted the gardening vocation, which gave him the opportunity to examine the birds that flocked around the garden and led to an interest in taxidermy. Gould became an expert taxidermist as well as an expert in bird identification through a bird's structure. He soon established a taxidermy business in London. His skill enabled him to become the first curator and preserver at the museum of the London Zoological Society. The position brought him into contact with the leading naturalists, which often allowed him to examine new bird collections given to the Society. This placed him in the fateful position of being the first ornithologist to view Darwin's bird specimens [17].

Gould set aside his paying work to examine Darwin's bird collection, and at the next Society meeting on January 10, 1837, he reported that the Galapagos birds Darwin had identified as blackbirds, "gros-beaks," and finches had been misidentified by Darwin and were in fact "a series of ground Finches which are so peculiar" as to form "an entirely new group, containing 12 species" [18].

This story made the newspapers. In March, Darwin again encountered Gould, who informed Darwin that his Galapagos wren "was another species of finch and the mockingbirds he had labeled by island were separate species rather than just varieties, with relatives on the South American mainland" [17].

Not realizing the importance of his bird collection, Darwin had not bothered to label his finches by island, but fortunately other expedition members had taken more care. He then sought specimens collected by Captain Fitzroy and crewmen, which allowed Darwin to establish that each species was unique to an island.

Darwin must have been stunned. The implications were clear: the finches and mockingbirds must have had a common ancestor that had somehow been carried to the islands, and that due to the separation between the islands, they had evolved over unknown thousands of years into separate species. Darwin's efforts had provided proof of the evolutionary process whereby animal structures change to match their environments.

But, why didn't John Gould arrive at the same conclusion, and since he had the information first, why didn't he "scoop" Darwin and announce his (Gould's) discovery of the process? The answer is actually rather simple. As a *News Scan Daily* article dated 28 May 1999, referenced in [18] points out:

- "The information was there, but he [Gould] didn't quite know what to make of it. He assumed that since God made one set of birds when he created the world, the specimens from different locations would be identical."

- "John Gould was a biblical literalist and couldn't understand why God would make twelve finches."

- "Gould thought the way he had been taught to think, like an expert taxonomist, and didn't see, in the finches, the textbook example of evolution unfolding right before him" "Thus he was incapable of arriving at the correct solution.".

For Darwin, there reminded two nagging questions:

How did the Evolutionary process work?

Since enormous amounts of time were necessary, where did the time come from?

Thomas Malthus and "the Survival of the Fittest" A partial answer to the first question was provided by **Thomas Malthus (1766–1834)**. Malthus's studies of population had convinced him that, all things being equal, population tends to grow geometrically while resources grow linearly. Thus, ultimately, population growth will overwhelm resources. This concept led Malthus to conclude that in such a situation, only the most fit would survive [19]. This idea had a decisive influence on Charles Darwin and his explanation of evolution.

The struggle for existence of all creatures is the catalyst by which natural selection produces the "survival of the fittest," a phrase coined by Herbert Spencer. In the book, *The Origin of Species*, Darwin gave tribute to Malthus by stating that his (Darwin's) theory was an application of the doctrines of Malthus. Darwin was a lifelong admirer of Malthus and referred to him as "that great philosopher" in a letter to J.D. Hooker on 5 June, 1860. He also wrote in his notebook that "Malthus on Man should be studied."

While Malthus's investigations had led to the correct deduction, there were three flaws in his reasoning:

1. Unless perturbed by external forces such as disease or predation, population does not grow geometrically, but rather it grows exponentially because the rate of population growth is proportional the size of the population. This sentence can be converted into a differential equation which has a solution of the form: $P(t) = P(0) + e^{C^*t}$.

Where P(t) is the population at any given time,t, P(0) is the initial population (usually zero) and $e^{C \cdot t}$ represents exponential population growth over time goverend by the constant C.. Exponential growth is more serious that geometrical growth.

2. Malthus had no way of predicting the tremendous advances in food production capability that appeared to render his predictions alarmist.

3. Malthus's studies were empirical, and as has been shown previously, empirical studies rarely reveal the underlying causes. Thus, Malthus did not know why population would expand beyond available resources. The reasons for excessive population growth that Malthus couldn't have known will be revealed in chapter 17 of this book, "Answers/Explanations in Evolution."

Illustrating the Natural Selection Process: The natural selection process can be illustrated by the simple diagram show in figure 11-1.

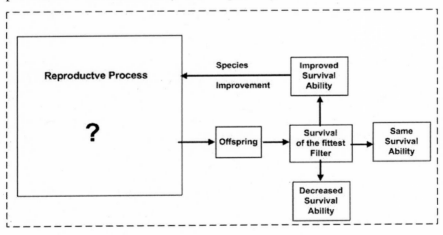

Figure 11-1. Simplified illustration of natural selection process

As will be described in detail, the reproductive process produces a child with varying survival capability. The child then passes through the natural selection filter, which compares the child's abilities with the average abilities of the general population at the time and determines if the child's ability is greater than, equal to, or less than the extant average. Species improvement and/or creation of a new species occurs when a child of superior ability appears.

However, the reason for the question mark in the reproduction process box is the lack of understanding in Darwin's time of how the process works. Such key data as cell division, which is central to the reproductive process, was unknown; therefore while Darwin understood that reproduction somehow produced the variability, neither he nor anyone else understood how the variation in children was produced.

An important qualification regarding "survival of the fittest" appeared in a recent *Wall Street Journal* article titled "Forget About Survival of the Fittest," by psychologist Gary Marcus. Marcus argues that the phrase survival of the fittest is "... perfectly ambiguous. The phrase could mean that 'of all the possible creatures that one might imagine, only the fittest possible survive.' Or ' it could mean something considerably less lavish -- that 'only the fittest creatures that happened to be around at a particular moment tend to survive'" [20].

As I discussed above, when viewed properly, the meaning of "survival of the fittest" is essentially Marcus's second option. The inputs to the natural selection filter are offspring, and in general the fittest of these offspring will survive. But these are by no means the best possible, but rather only the best of what is possible at the time. There is a significant difference between the two, but one which must be kept in mind when considering the meaning of "survival of the fittest."

As discussed above, Charles Lyell and associates deduced that the earth was much older than had been generally believed. While Lyell's geological information and age estimates were too short by modern terms, they were sufficient for Darwin. However, Lyell was equivocal regarding evolution. His book, *Geological Evidences of the Antiquity of Man* [21] brought together his views on three key themes from the geology of the most recent geologic period in Earth's history, the 1.6-million-year Quaternary period. The three key themes were glaciers, evolution, and the age of the human race. First published in 1863, the book was widely regarded as a disappointment because of Lyell's equivocal treatment of evolution. As devout Christian, Lyell had great difficulty reconciling his beliefs with natural selection, which essentially precluded a need for a creator. All will become clear in the next chapter when the details of natural selection are explained and the lack of a need for a creator will be further emphasized.

Publish or Perish: Besides the dilemma of a complete explanation of natural selection, Darwin's next dilemma was publication of his confirmation that evolution occurs and his explanation of the process. As mentioned above, the religious climate in the mid-1800s was not particularly receptive to ideas that challenged accepted dogmas:

> Darwin was well aware of the implication the theory had for the origin of humanity and the real danger to his career and reputation as an eminent geologist of being convicted of blasphemy. He worked in secret to consider all objections and prepare overwhelming evidence supporting his theory. [22]

Hence, Darwin agonized about publishing his obviously controversial findings. He was presumably mindful of the reaction to the publication of Nicolas Copernicus's groundbreaking opus, *On the Revolutions of the Celestial Spheres*, 300 years earlier.

In view of his presumed familiarity with the reaction to Copernicus's publication, Darwin's caution and desire to be as correct as possible is quite understandable. But Darwin wasn't the only person working on the problem of evolution. British naturalist, explorer, geographer, anthropologist, and biologist **Alfred Russell Wallace (1823–1913)** was also gathering facts with which to understand the origin of species [23].

In 1848, Wallace accompanied naturalist Henry Bates on a trip to the Amazon rainforest to collect specimens that he hoped would further his goal. Unfortunately, Wallace lost most of his collection during a ship fire. Undaunted, he traveled through what is now known as Malasia and Indonesia to collect additional specimens. His observed marked zoological differences across a narrow zone in the archipelago, which led him to postulate a zoological boundary, now termed the Wallace line.

Wallace's studies led him to arrive at the same conclusion as Darwin: natural selection could explain what he had been observing [23].

Interestingly, Wallace was also greatly influenced by Thomas Malthus. After finding about Darwin's work, Wallace is reputed to have called Malthus's essay "… the most important book I read …" and considered it "the most interesting coincidence" that both he and Darwin were independently led to the theory of evolution through reading Malthus [23].

Darwin's friends and associates constantly urged Darwin to publish lest he be scooped by Wallace. Finally, while Darwin was writing up his explanation of his investigations in 1858, Wallace sent him an essay which described a similar idea, prompting immediate joint publication of both of their essays in 1859. Darwin titled his book, *On the Origin of Species by Means of Natural Selection, or the Preservation of Favoured Races in the Struggle for Life*" [14].

When Wallace read Darwin's book, he realized that Darwin's was much superior and offered to withdraw his paper. Darwin, being a magnanimous person, insisted that both papers be published, and they were.

Unfortunately, Wallace's career took a strange twists and he became a spiritualist. Later Wallace maintained natural selection could not account for many human characteristics like mathematical, artistic, or musical genius. Moreover, he claimed something in "the unseen universe of Spirit" had interceded at least three times in history [24]:

1. The creation of life from inorganic matter.

2. The introduction of consciousness in the higher animals.

3. The generation of the above-mentioned faculties in mankind.

Wallace was obviously a partial believer in intelligent design. These new views greatly disturbed Darwin, who argued that spiritual appeals were not necessary and that sexual selection could easily explain such apparently non-adaptive phenomena as musical genius.

Notwithstanding Wallace's strange views, a present-day resurgence of interest in Wallace has begun. An article in the December 20, 2008, *Wall Street Journal* trumpets, "Alfred Russell Wallace's Fans Gear Up for a Darwinian Struggle – Anniversary of 'Origin of Species' Nears; Rival is Touted, Charges of Plagiarism Fly. David Hallmark, a British Lawyer having researched Wallace, including retracing Wallace's travels, wants to prove that Mr. Darwin was a cheat" [24].

These folks don't seem to realize that neither Wallace nor Darwin had an important key to the puzzle: how natural selection works.

Moreover, if Darwin or Wallace hadn't discovered evolution, someone else would have.

Darwin's Explanation of Evolution Incomplete: As mentioned above, Darwin's explanation of evolution was incomplete because Darwin had no knowledge of genetics and meiosis—the keys to sexual reproduction; otherwise his theory would have been complete. While Darwin's publication was a monumental achievement, a more complete title for his great opus would be:

On the Origin of Species by Means of Natural Selection, Based Upon_____ …

But Darwin couldn't add the ending because he had no idea how the variations, produced by reproduction, that led to improved traits and then allowed natural selection to pick the fittest were produced. Just as Lyell and others were hampered by lack of knowledge of the sun's source of energy and couldn't arrive at correct age of the sun and consequently for the earth, Darwin was also hampered by a lack of knowledge of genetics. As we will see in the next chapter, the blank above can be filled with the phrase *Reproductive Genetic Mixing*, thus completing the explanation of Evolution.

Darwin did understand that reproduction was a key and that variation was associated with inheritance, as he wrote in the conclusion to *On the Origins of Species*:

"It is interesting to contemplate a tangled bank, clothed with many plants of many kinds, with birds singing on the bushes, with various insects flitting about, and with worms crawling through the damp earth, and to reflect that these elaborately constructed forms, so different from each other, and dependent on each other in so complex a manner, have all been produced by laws acting around us. These laws, taken in the largest sense, being Growth with Reproduction; Inheritance which is almost implied by reproduction; Variability from the indirect and direct action of the conditions of life, and from use and disuse; a Ratio of Increase so high as to lead to a Struggle for Life, and as a consequence to Natural Selection, entailing Divergence of Character and the Extinction of less-improved forms. Thus, from the war of nature, from famine and death, the most exalted object which we are capable of conceiving, namely, the production of the higher animals, directly follows. There is grandeur in this view of life, with its several powers, having been originally breathed by the Creator [not sure Darwin believed in a Creator, but he had to pay homage] into a few forms or into one; and that, whilst this planet has gone cycling on according to the fixed law of gravity, from so simple a beginning endless forms most beautiful and most wonderful have been, and are being, evolved." [14]

The key phrases in this conclusion are:

- growth with reproduction;

- inheritance which is almost implied by reproduction;

- variability from the indirect and direct action of the conditions of life, and from use and disuse;

- a ratio of increase so high as to lead to a struggle for life, and as a consequence to natural selection;

- entailing divergence of character and the extinction of less-improved forms.

His statement, "Inheritance which is almost implied by reproduction," was agonizingly close, but he didn't know how reproduction worked. In the next section we will see how inheritance works and why a person can inherit their father's eyes but their mother's nose.

SCIENCE AND RELIGION DIVERGE IN RESPONSE TO EVOLUTION.

Reactions to Darwin's publication of the *Origin of Species* were very different in the scientific and religious communities.

Some scientists, especially those like Lyell who held strong religious beliefs, had their doubts and concerns about evolution and its consequences. On the other hand, progress in gaining an understanding of underlying biological processes continued, not necessarily with the objective of proving or disproving evolution. However, much of the findings such as those that led to an understanding of reproduction eventually provided unequivocal proof of evolution.

The reaction of many in the religious community, especially in the United States, was to vehemently denounce and oppose evolution. While the Copernican heliocentric concept was bad enough and logically the action that began the science-religion conflict, Darwin's concept of evolution, which displaced humans from their cherished position as a special creation of God, was the last straw. It inflamed the conflict between science and religion, which continues relatively unabated today and is the subject of chapter 14: "Religious Reaction to Evolution."

SUMMARY OF CHAPTER 11.

Early humans did not realize that the apparently unchanging animal and plant structures were an illusion. They of course observed the natural life cycle of birth and death of plants and animals, but that didn't suggest change.

The origin of life as described in the book of Genesis is merely a listing of the facts of existence—*what* can be found on the earth and *who* made them. However, there is no mention of *how* or *why* life exists as it does. In other words, Genesis does not provide much of an explanation

The Greeks provided more information, especially Aristotle, but while he cataloged the existence of many species, he was also unable to explain *how* or *why* life existed as it did. In fact, Aristotle erroneously believed in the spontaneous generation of life.

Gradually beginning in the seventeenth century, observations began to accumulate that suggested that animal and plant structures did change and that life had actually evolved over a long time period. Four French scientists, one German scientist, and one British scientist were in the forefront of this observation accumulation. Their conclusions are tabularized in table 11-1.

While none of these explanations were correct, most at least acknowledged evolution had occurred. The stage was thus set for Charles Darwin.

Darwin was the right person at the right place at the right time. While he did not specifically discover the process of evolution, he made three key contributions that provided the first significant contribution to the explanation of Evolution:

1. He was able to fortuitously tour the world, collecting geological samples and animal samples, in particular birds, which enabled him to be the first person to conclusively demonstrate the existence of the evolutionary process.

2. In view of the unfriendly religious climate, Darwin had the courage to publish his results.

3. Perhaps most importantly, Darwin provided the first useful, although incomplete, explanation of the process.

Darwin's findings initiated him into the inner circle of British science, and he was allowed to present his bird collection at a meeting of the Royal Society.

He had unfortunately misidentified his bird collection. Fortuitously, ornithologist John Gould attended the meeting and correctly identified Darwin's collection as multiple species of finch. Darwin's conclusion was that the birds had evolved from a common ancestor and thereby conclusively demonstrated that evolution had occurred.

Darwin spent several years refining his findings. With the help of Thomas Malthus, who had conceived the concept of "survival of the fittest," and Darwin's mentor Charles Lyle, who supplied the time needed for evolution, Darwin wrote his groundbreaking book, *On the Origin of Species*.

LOOKING AHEAD.

We come now to the *pièce de résistance*, the completion of the explanation of evolution and the verification of this book's title. Of course, it took almost 150 years after Darwin's publication to put all the pieces together. Chapter 12 will explain how it was done.

CHAPTER 12:
RESOLVING THE APPARENTLY UNCHANGING ANIMAL AND PLANT STRUCTURES ILLUSION AFTER DARWIN, EXPLAINING NATURAL SELECTION

Darwin's book was a triumph because it provided the first lucid argument for evolution. But there was one problem, as mentioned in the last chapter: Darwin did not fully understand how natural selection worked. It took another 150 years to work out the details of natural selection, which we will cover in this chapter.

Gregor Mendel (1822–1884), a reclusive monk, provided one of the essential elements regarding reproduction through his discovery of genes [1]. Mendel made his momentous discovery by painstakingly tracingthe characteristics of successive generations of peas. This led him to the realization that something, which he termed "genes," determined the inherited traits of peas. While Mendel had no idea what exactly a gene was, using his gene concept, he could predict the outcome of generations of peas. Mendel published his work in 1865, first in German and then in English as *Experiments in Plant Hybridization* [1].

Extending his work with peas to human children, Mendel or someone who came after him realized that the thing Mendel termed the "gene" controlled the reproduction process and influenced the children that resulted from the reproduction process. These children were subject to natural selection's survival filter, which determines the most fit to survive. The task is then straightforward: locate and identify the genes and determine the mechanism with which they control heredity and perhaps you will find the key to natural selection. This task was, of course, not simple. It took almost 150 years.

Unfortunately, Mendel's work was lost, only to be rediscovered in 1900 by three scientists: Carl Correns, Erich von Tschermak, and Hugo de Vries [2] [3] [4][1].

REPRODUCTION IS REQUIRED FOR SPECIES SURVIVAL.

It is a simple but not often articulated fact, that due to the finite life spans of multi-celled animals, reproduction is mandatory for species survival. However, there is an equally simple fact that says individuals do not need reproduction for survival. This leads to a conflict when all of the basic requirements for individual and species survival are compared (see table 12-1).

Requirement for Survival	Individual Survival	Species Survival
Oxygen	Y	N
Food	Y	N
Access to food	Y	N
Metabolism of food	Y	N
Suitable Environment	Y	N
Proper temperature		
Dry		
Safe		
Reproduction	N	Y
Territory	Y	Y

Table 12-1. Comparison of species and individuals survival requirements

The items essential for individual survival and species survival are designated with Y, while those not necessary are designated with N.

Note that N appears in many columns for species survival because a species is not interested in the survival of any particular individual, only that enough individuals—ideally, the best from the species—live long enough to reproduce and to assure their children reproduce. On the other hand, while individuals need items such as oxygen and food to survive, they do not need reproduction. In fact, reproduction gets in the way of finding food. Territory is listed as needed for individual and species because both need space in which to survive as well as reproduce.

Regarding reproduction, the only animal that can reproduce directly (i.e., make a copy of itself) is a single-celled animal; therefore all reproduction begins with a single cell.

Just as Sir J. J. Thompson's discovery of the negative electron and Rutherford's discovery of the positive nucleus largely resolved the "solid" matter illusion by explaining that "solid" matter consisted of unseen atoms as had been speculated by the ancient Greeks, the discovery of the cell ushered in a revolution in our understanding of life. Whereas the atom was the fundamental building block of matter, the cell the fundamental building block of life.

As discussed in the previous chapter, the fossil record provided some of the first clues that unchanging life is an illusion; however, a far more important clue was provided by the discovery of the basic building block of life—the cell. The cell had escaped detection until the development of the extension of the human eye—the microscope. The microscope was made possible by the invention of the optical lens.

The person generally credited with the invention of the first practical microscope was Dutch fabric merchant **Anton van Leeuwenhoek (1632–1723)** [5]. A perfectionist, Van Leeuwenhoek was not satisfied with examining his fine cloth with crude existing lenses, so he learned to grind his own. This led to the first of many microscopes, which he continued to improve.

Van Leeuwenhoek is best known for his contributions toward the establishment of microbiology, which earned him the title "Father of Microbiology." Using his handcrafted microscopes, he was the first to observe and describe single-celled organisms. He referred to these tiny organisms as "animalcules," which we now refer to as microorganisms. He was also the first to record microscopic observations of muscle fibers, bacteria, blood cells, and sperm cells.

Twenty years after he began his investigations, Van Leeuwenhoek presented his findings to the Royal Society of London. His widely circulated research revealed to the scientific community a vast array of microscopic life that had never before been observed. Lacking formal scientific training, the astounding and detailed nature of Van Leeuwenhoek's discoveries resulted in his induction as a full member of the Royal Society in 1680, where he joined the ranks of many other scientific luminaries of his day.

ESTABLISHMENT OF BASIC CELL CONCEPTS.

Encouraged by van Leeuwenhoek's findings, other investigators employed ever-improving microscopes and discovered cells were the building blocks of all tissue.

The term "cell" was coined by Robert Hooke, who while examining the cells of trees (which are rectangular), noted that a tree cell reminded him of a monk's cell [6].

By 1839, enough cellular evidence had accumulated for German biologists **Theodor Schwann (1810–1882)** [7] and **Jakob Schleiden (1804–1881)** [8] to enunciate a fundamental cell concept. Schwann had been examining the cells of animals, while Schleiden had been pursuing plant cells. As the story goes:

> "In 1838, Schwann and Schleiden were discussing their cell studies while enjoying after-dinner coffee. When Schleiden described his studies of plant cells with nuclei to Schwann, Schwann noted a similarity of Schleiden's plant cells to cells he had observed in animal tissues. The two scientists went immediately to Schwann's lab to examine his slides and quickly confirmed that both plant and animal cells had the same basic structure implying that all life was composed of cells"[9].

In 1839 they articulated their cell theory, or cell doctrine, which states all organisms are composed of similar units of organization called cells. Schleiden and Schwann's concept has remained one of the foundations of modern biology and is equivalent to the concept that all matter is composed of atoms.

Schwann published a book on animal and plant cells in 1839, which unfortunately contained no acknowledgments of anyone else's contributions, including that of Schleiden.

Interestingly, their idea predates other great biological concepts, including Darwin's theory of evolution (1859) and Mendel's laws of inheritance (1865). However, the mere awareness of cells wasn't sufficient to explain natural selection.

Schleiden made other contributions by recognizing the significance of the cell nucleus, which was discovered in 1831 by the Scottish botanist **Robert Brown (1773–1858)**, who also discovered Brownian motion. Schleiden sensed the connection of the nucleus with cell division and was one of the first German biologists to accept Darwin's evolutionary theory.

All Cells Come from Other Cells: Finally in 1855, studies by the famous German physician, anthropologist, public health activist, pathologist, prehistorian, biologist, and politician **Rudolf Ludwig Karl Virchow (1821–1902)** allowed him to state a basic axiom: "All cells come from other cells" [10]. Virchow also founded the field of social medicine and is honored as the "Father of Pathology."

Thus, nearly 200 years after the 1680 discovery of cells by Van Leeuwenhoek, the observations of Virchow, Schleiden, and Schwann established the classic, basic understanding of cells [11]:

- All organisms are made up of one or more cells.

- Cells are the fundamental functional and structural unit of life.

- All cells come from pre-existing cells.

- The cell is the unit of structure, physiology, and organization in living things.

- The cell retains a dual existence as a distinct entity and a building block in the construction of organisms.

Two Kinds of Cells: To the list of basic cell features shown above, we need to add another important piece of information: all cells contain chromosomes, which were probably discovered around 1848. Chromosomes delineate two types of cells:

1. Somatic cells, which comprise most of the body's cells, contain two sets of chromosomes, and are termed "diploid"

2. Sex cells, which contain only one set of chromosomes and are termed "haploid"

The significance of these two different cell types will become clear in the next two sections, where we will introduce one of the most important cell processes, cell division, of which there are two kinds, mitosis and meiosis.

Walther Flemming's (1843–1905) discovers Mitosis: The development of special dyes to better observe cell features probably allowed Flemming to be the first person to actually observe a chromosome, which then led to his discovery of mitosis. As [12] points out:

> ... "[Flemming] used dyes to study the structure of cells. He found a structure which strongly absorbed dye, and named it Chromatin. He observed that, during cell division, the chromatin separated into stringy objects, which became known as Chromosomes[2]. Flemming named the division of somatic cells[3] Mitosis, from a Greek work for thread."

As mentioned above, the only an animal that can replicate itself directly is a single-celled animal. Single-celled animals reproduce via mitosis.

Flemming published his work in 1882, just fifteen years after Darwin published *The Origin of Species* and three years after Darwin's death. Although Flemming did not know the purpose of chromosomes, he did observe that a complete set was inherited by each of the two cells that result from mitosis.

Some change can occur during mitosis due to mutations, as discussed in great detail in "Coming to Life" [ch 13, 4]. For example, these variations are sufficient to confer antibiotic resistance on bacteria. However, they are not sufficient by themselves to provide the variation required by natural selection. It is difficult to believe that simple mutations can produce both a Mozart and a Down syndrome child from the same couple. The needed variation is provided by a more complex form of reproduction—sexual reproduction—which involves the second type of cell division called meiosis.

Oscar Hertwig (1849–1922) discovers Meiosis: Sexual reproduction of course involves the joining of two sex cells, an egg and a sperm. Sex cells are produced by the second type of cell division, meiosis. German biologist Hertwig discovered meiosis in sea urchin eggs. Not a surprise since eggs and sperm are the only cells that undergo meiosis. Hertwig published his results in 1886 [13]. I am reasonably sure Hertwig had no idea that his was the first step along the tortuous path that led to the understanding of natural selection.

While meiosis begins with mitosis, meiosis differs from mitosis in a very significant way. Meiosis adds a second cell division. As described above, most cells are diploid, having two sets of chromosomes. In meiosis, after mitosis produces two new diploid cells, each of the two new diploid cells undergo an additional cell division. In this second cell division, the new pair of cells receives only one-half of the chromosomes from each diploid cell. These cells are haploid cells.

When a haploid egg and a haploid sperm unite in fertilization, the resultant zygote (as a fertilized egg is known), receives a set of chromosomes from male and a set from the female and thereby reestablishes the diploid state in which a cell has two sets of chromosomes. However, the zygote is genetically unique, and considerable variation can occur. This genetically unique variation is the driver of evolution. But as with others before him, Hertwig did not fully understand the significance of his discovery, although there is some indication that he suspected it was important in reproduction [13].

Meiosis was revisited by Belgian embryologist **Edouard Van Beneden (1846–1910)**. Van Beneden's studies of a horse parasite demonstrated that chromosomes were involved in meiosis [14].

In 1911, American geneticist **Thomas Hunt Morgan (1866–1945)**, worked with the fly *Drosophila melanogaster* and established the first true understanding of the role of meiosis in reproduction [15]. We will tell more about Morgan's contributions when we revisit mitosis and meiosis after we've some gathered additional background material.

ON THE TRAIL OF DNA.

While the discovery and characterizing of DNA and the determination of DNA's role in reproduction is one of science's greatest triumphs—a triumph made possible by the contribution of many investigators—it should be noted that the ***ultimate quest was the determination of how proteins are made***, since proteins are the building blocks of all animal life, which of course begins with reproduction.

One of the first clues in the mystery of reproduction was provided by **Johan Friedrich Miescher (1844–1895)** while he was working on pus cells at a hospital in Tübingen:

> ... "Miescher noted the presence of something that 'cannot belong among any of the protein substances known hitherto'[4]. A fact he was able to demonstrate by showing that it unaffected by the protein-digesting enzyme pepsin. He also showed that the new substance was derived from the nucleus of the cell alone and consequently named it 'nuclein'. Miescher was soon able to show that nuclein could be obtained from many other cells and was unusual in that it contained phosphorus in addition to the usual ingredients of organic molecules - carbon, oxygen, nitrogen, and hydrogen[16]."

Miescher published his results in 1871.

In 1881, biologist **Zacharis**, who apparently rates only an abbreviated mention in histories of microbiology, made an important discovery, when, via a series of comparison experiments, he was able to demonstrate that nuclein and chromatin are the same entity [17].

In 1889, organic chemist **Herman Emil Fischer (1852–1919)** identified two organic molecules, pyrimidines and purines. They were termed bases due their chemical reactions [18]. Pyrimidines and purines are essential elements of DNA, but, as with many other clues being discovered during the late 1800s, Fischer did not appreciate their significance.

Zacharis's discovery that nuclein and chromosomes were the same entity spurred investigation of nuclein because it was obviously involved in reproduction, being a major constituent of eggs and sperm. In 1890, while experimenting with nuclein in which extraneous material was removed, German pathologist **Richard Altmann (1852–1900)** demonstrated that nuclein was an acid. He is credited for the discovery and naming of "nucleic acid" [19].

Albrecht Kossel (1853–1927) advanced the understanding of nuclein by demonstrating that nuclein contained both protein and nonprotein (nucleic acid) parts. He further demonstrated that when broken down, nucleic acid produced nitrogen-bearing compounds: two pyrimidines (cytosine and thymine) and two purines (adenine and guanine). These four molecules are usually identified by the letters C, T, A, and G, respectively. Kossel also found that carbohydrates (i.e., sugars) were also present.

The pyrimidines C and T and the purines A and G have unique complimentary shapes so that A-C and T-G always go together, as shown figure 12-1. In the diagrams, H stands for hydrogen, which forms a rather weak bond that holds either A and T or G and C together. The A-H-T or G-H-C pairs formed are termed a "base pair" [20].

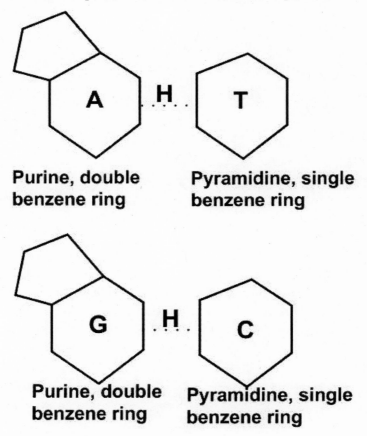

Purine, double benzene ring   Pyramidine, single benzene ring

Purine, double benzene ring   Pyramidine, single benzene ring

Figure 12-1. Diagrams of purine and pyrimidine combinations

In 1910, Kossel was awarded the Nobel Prize in Physiology or Medicine for demonstrating that two pyrimidines and two purines are present in nucleic acids [21].

In 1909, Russian-American biochemist **Phoebus Levene (1869–1940)** discovered the sugar ribose. In 1929, he discovered a similar sugar with one less oxygen atom, called deoxyribose[5] [22].

Levene's main interest was the structure and function of nucleic acids. He characterized two different forms of nucleic acid, DNA (short for deoxyribonucleic acid) and RNA (short for ribonucleic acid). Levene also extended Kossel's work by demonstrating that not only are the two pyrimidines (cytosine and thymine) and two purines (adenine and guanine) present in nucleic acids, but they are the principal constituents of DNA along with deoxyribose and a phosphate group. He further demonstrated that the DNA components were linked in the order—phosphate-sugar-base pair—to form a unit he termed a "nucleotide." RNA, on the other hand, contains ribose. It also contains the pyrimidine uracil rather than the purine thymine [22]; [23].

In 1914, German chemist **Robert Feulgen (1884–1955)** developed the fuchsin staining technique, which demonstrated that chromosomes contain DNA [24].

In 1928, British medical officer and geneticist **Fredrick Griffith (1879–1941)** demonstrated that the traits of one form of the bacteria *Pneumococcus* could be transferred to another form of the same bacteria by mixing an entity common to both that Griffith termed a "transforming principle" [25].

In 1937, British physicist and molecular biologist **William Astbury (1898–1961)** produced the first X-ray diffraction patterns of DNA, showing that DNA had a regular structure. However, Astbury was unable to produce a complete picture of the DNA structure [26].

In 1944, physician, medical researcher, and pioneer molecular biologist **Oswald Avery (1877–1955),** along with coworkers Colin Macleod and Maclyn McCarty, demonstrated that Griffith's transforming principle is DNA. Avery was also a pioneer in a branch of science known as immunochemistry, the field of chemistry concerned with chemical processes in immunology, but he is best known for his 1944 discovery that Griffith's transforming principle is actually DNA, the material where genes reside, and the carrier of inheritance [27]; [28].

In 1952, bacteriologist and geneticist **Alfred Hershey (1908–1997)** and geneticist **Martha Chase (1927–2003)** confirmed Avery's result by conducting a series of experiments, now known as the Hershey-Chase experiments. They used a tiny virus-like entity that infects bacteria, termed the T2 bacteriophage. These experiments confirmed that the T2 bacteriophage DNA contained the material of T2 bacteriophage inheritance, and therefore was where the genes were located [29]. As noted above, Miescher had discovered nuclein in 1859, which ultimately was shown to be DNA. Prior to the Avery and Hershey-Chase experiments, it was assumed by most molecular biologists that proteins carried the information for inheritance.

Knowing that DNA is the location of genes does not provide an explanation of what a gene is, but researchers were getting closer. At least they knew where genes were located. Once it became certain that genes were located on DNA, the race was on to discover the DNA shape and then perhaps the exact identity of a gene. The first attempt at a solution was made by California Technology chemist **Linus Pauling (1901–1994)**, who proposed a single-strand helix, which was later shown to be incorrect [30].

Ultimately, the race was won in 1953 when British biophysicist and X-ray crystallographer **Rosland Franklin (1920–1958)** produced X-ray diffraction images of DNA of sufficient quality for British molecular biologist, physicist, and neuroscientist **Francis Crick (1916–2004)** and American molecular biologist and mathematician, **James Watson (1928–)** to analyze the X-ray patterns and deduce the structure of DNA [31]. Considering the size of the DNA molecule and its complex double helix structure, this was an amazing accomplishment.

The significant improvement supplied by Watson and Crick was the determination that DNA was a double helix, something that we'll see later might have been obvious. Figure 12-2 shows the basic structure of a short piece of one strand of the double helix DNA.

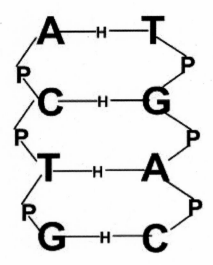

Figure 12-2. The structure of one DNA strand

The purine-pyrimidine links of A and T or C and G, held together by relatively weak hydrogen bonds, are connected into a ladder-like structure by a phosphate deoxyribose backbone, P. Each side of the "ladder" is termed a DNA strand, which are entwined to create the double helical structure. As mentioned above, Phoebus Levene termed the combination of a base pair (A and T or C and G), one phosphate molecule and a deoxyribose sugar molecule, a "nucleotide."

As suggested above, the elucidation of the DNA structure was, of course, one of the monumental achievements of science. There are many excellent books on the subject, including Watson's book *The Double Helix: A Personal Account of the Discovery of the Structure of DNA* [32]. There is also a very readable guide in Watson's book at this Web site [33].

PROTEIN MANUFACTURE, THE ULTIMATE GOAL OF DNA RESEARCH.

As mentioned previously, the ultimate objective of all this investigation was the determination of how proteins are made. However, before proceeding to answer this question, I believe it will help if we consider an analogy between protein manufacture and automobile manufacture since most peopled understand the rudiments of automobile manufacture.

An Analogy Between Protein Manufacture and Automobile Manufacture: Regarding protein manufacture we have these facts:

- Proteins are strings of amino acids [34].

- All proteins are formed from twenty amino acids.

  - Amino acids can "self construct" (i.e., the atoms and molecules that form amino acids can combine according to the laws of quantum mechanics). In fact, amino acids have been found in interstellar gas clouds [35].

- Amino acids cannot join to make protein without assistance. As will be seen, this assistance is provided by genes which will be shown to be located along the DNA strands. To place this assistance into perspective, consider automobile manufacture.

Automobile manufacture requires these essential items:

- Parts list(s) that specify the parts required to manufacture an automobile;

- Fabrication drawing(s) that specify how parts are to be manufactured;

- Assembly drawing(s) that specify how the parts are to be assembled.

Protein manufacture differs considerably from automobile manufacture based on the following facts:

- While there are thousands of automobile parts, only twenty amino acids suffice for proteins.

- As will be seen, genes provide the "fabrication drawing" for the manufacture of a protein.

- The sequence of genes on a strand of DNA provides both the list of amino acids required by a particular protein, plus the gene positions along the DNA strand specifies the order in which amino acids are assembled.

Thus, DNA is both a parts list and an assembly drawing.

So What is a Gene and What is Its Relation to DNA? Regarding genes and DNA, thus far we have established these facts:

- Genes are located in the DNA.

- DNA is a ladder-like structure with four molecules connected in pairs termed base pairs, which form the "rungs" of the ladder.

- The base pairs are formed from two purines and two pyrimidines, which are usually represented by the letters T, C and A, G.

- Due to their structure, the four bases always pair A with T and C with G.

- There are approximately three billion base pairs [36].

Intensive study of DNA has shown that a gene is a sequence of A, C, G, or T on *one side* of the DNA ladder. Thus, the positioning of the two purines and two the pyrimidines along the side of a DNA strand, form a set of codes that identifies which amino acid will be specified by a particular gene.

Presumably it was George Gamow who first realized that three of the four "letters," A, C, G, or T, would be sufficient to specify or "encode" twenty amino acids. Gamow hypothesized that the arrangement of base pairs is arbitrary; hence, all possible arrangements of the four letters from AAAA to TTTT are possible. These arrangements are known mathematically as "permutations," and the number of possible letter permutations when letter repeats (i.e., A is repeated in the combination AAAA) are permitted, can be calculated with the formula: N^r.

In the case of DNA, N represents all four bases and r represents either 1, 2, 3 or four bases. Substituting numbers into the formula, we have $4^1 = 4$; $4^2 = 16$; $4^3 = 64$; and $4^4 = 256$

Since we need enough bases to encode twenty amino acids, clearly the one or two are insufficient, but three bases that could specify up to sixty-four amino acids is more than enough. Thus a minimum of three bases is sufficient to specify twenty amino acids[6].

In 1961, **Francis Crick** and South African biologist and 2002 Nobel Laureate in Physiology or Medicine **Sydney Brenner (1927–)** experimentally demonstrated that three DNA bases do provide the code for one amino acid [37], thus confirming Gamow's conjecture.

The three base "triplets" are termed "codons" because they provide the code that identifies the amino acids associated with the gene [38].

"Extra" codons are used as start and stop codons, which specify where on the DNA sequence to begin using the codons to specify amino acids and when to stop, as will be explained below.

CONNECTING GENES TO PROTEIN MANUFACTURE.

Each cell has numerous protein "factories" known as ribosomes. Ribosomes perform the actual protein manufacture [39]. Since the DNA molecule is huge, the genetic information contained in the DNA must be carried to the ribosome. This is done via messenger ribonucleic acid (RNA), usually abbreviated as mRNA [40].

The basic steps in the manufacture of a protein are illustrated in the diagrams in figure 12-3. A short section of DNA with two sets of triplet bases, which is the same as two codons, is shown.

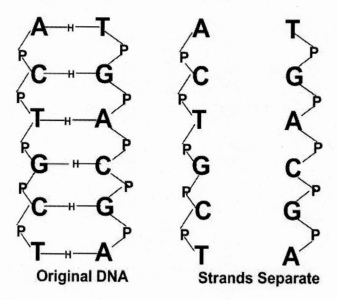

Figure 12-3. Separating DNA strands to begin protein manufacture

First, the original DNA is "unzipped" by an enzyme termed RNA polymerase, which separates the two DNA strands. Then the each letter of the DNA strand code is transferred to the mRNA. Note that the mRNA becomes a mirror image since, for example, a T transfers as an A.

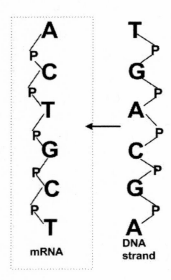

Figure 12-4. Transcribing one DNA strand to messenger RNA (mRNA)

Then the mRNA moves to the ribosome, where the letters in the mRNA are transferred to transfer RNA, usually abbreviated as tRNA, thereby restoring the original DNA code. In the diagram in figure 12-5, only one codon being transferred to the tRNA is shown.

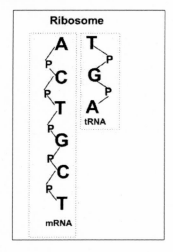

Figure 12-5. Messenger RNA (mRNA) is copied to transfer RNA (tRNA)

Next, the matching amino acid from a "pool" of amino acids is selected using the tRNA codon (figure 12-6).

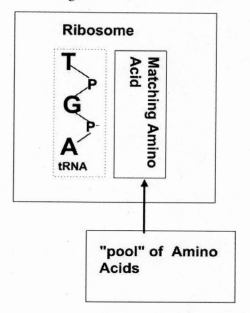

Figure 12-6. Selecting the appropriate amino acid from "pool" of amino acids

Finally, enzymes in the ribosome add the amino acid to the other amino acids (figure 12-7).

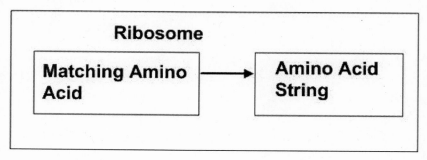

Figure 12-7. Adding the matching amino acid to the amino acid string

After all the codons have been processed, the completed protein is released from the ribosome, and the ribosome is available to make another protein [40].

American geneticist **Phillip Sharp (1944–)** and British biochemist **Sir Richard Roberts (1943–)** extended the structure of a gene when they discovered the existence of introns, which are non-coding sections of a gene [41]; accordingly, the actual gene structure is as shown in figure 12-8.

gene

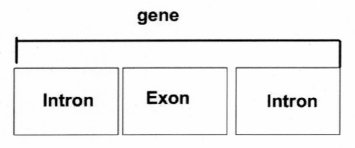

Figure 12-8. Structure of a gene

Sharp and Roberts shared the 1993 Nobel Prize in Physiology or Medicine for "the discovery that genes in eukaryotes are not contiguous strings but contain introns, and that the splicing of mRNA to delete those introns can occur in different ways, yielding different proteins from the same DNA sequence" [42]. Exons contain the triplet codons which identify amino acids [42].

So What Exactly Does a Gene Do? The gene does not actually do very much. As mentioned above, a gene provides the "blueprint" for the production of a protein, and it is protein that performs essential functions such as tissue building. So without genes, there would be no proteins and therefore no life.

As with automobile manufacture, the blueprint is essential to part manufacture. However, the part and the assembly into which the part is placed is the valuable output from the plant. Moreover, it is not uncommon to modify a blueprint slightly without destroying it to adapt a part to a special requirement in a unique assembly.

In a similar manner, epigenetics is a process whereby the functioning of a gene is modified, without modifying the gene, to produce a protein variation. Sometimes the variation is beneficial, sometimes not. There are many ways this can happen, but a detailed discussion is beyond the scope of this book. For more detail see [43]. For the purpose of this book, protein production is not simply transferring the amino acid manufactured from a gene's code to a ribosome for protein manufacture; the process can be modulated by epigenetic factors. The importance of this will become clear in chapter 15.

Once the basic genetic structure had been identified and the basic gene function determined, the next step was mapping the location of all the genes in the human genome, a mammoth undertaking considering there are approximately 3 billion base pairs in human DNA.

The Human Genome Project was initiated in 1990 by the U.S. National Institutes of Health (NIH). James Watson was the initial head, but he was forced to resign over disagreements with the NIH Director Bernadine Healy and was replaced by Francis Collins.A parallel project was undertaken by Celera Genomics Inc., lead by Craig Ventner. Ventner was aided by DNA sequencers developed by Applied Biosystems Inc.

The story of establishing the first copy of the human genome is an exciting one but beyond the scope of this book. Ventner's group at Celera published the first copy in the Feb 16, 2001 issue of *Science* [44]. Celera's genome contained approximately 20,000–25,000 genes.

RETURNING TO MITOSIS, MEIOSIS AND REPRODUCTIVE GENETIC MIXING.

I have presented a great deal of information regarding DNA and genetics because some knowledge of DNA and genetics is necessary to understand enough of the details of mitosis and meiosis. This leads to reproduction genetic mixing, which is the key to natural selection and evolution

The Mitosis Process: Because it is the simpler of the two, we will begin with explaining the process of mitosis. As noted above, mitosis was discovered by Flemming, who published his findings in 1882 in which he reported that during cell division, the chromatin separated into "stringy objects," which became known as chromosomes. We now know that the "stringy objects" are DNA molecules. So when a cell divides, the following things occur:

- Enzymes cause the DNA molecule to unwind.

- Enzymes next cause the DNA molecule to separate into two separate strands.

  - One strand goes to each new daughter cell—this is, of course, why DNA has to be a double helix.

- Enzymes in the new cell rebuild the DNA molecule, adding the appropriate bases and thereby recovering the double helix form.

However, during the rebuilding, despite elaborate error-correcting functions, occasionally the wrong base is added to a gene, resulting in a mutation. Sometime these are benign, but other times the mutated gene will not produce its proper protein. Many medical problems result from these mutations. Also, mutations in bacteria DNA can lead to antibiotic-resistant bacteria.

The Meiosis Process: Regarding meiosis, as mentioned above, most cells have two sets of chromosomes and are termed "diploid." In meiosis, each of the two new diploid cells undergo an additional cell division and thus receive only one-half of the chromosomes from either the male or female. These cells are termed "haploid."

The meiosis process begins with a normal diploid cell, shown in figure 12-9A. The normal diploid cell has just two sets of DNA for simplicity (1 and 2), one of which was contributed by the male parent (M1 and M2), and the other of which was contributed by the female parent (F1 and F2). The DNA pairs from the male and female are the same type, but they may have some slight differences due to mutations, hence the different shading.

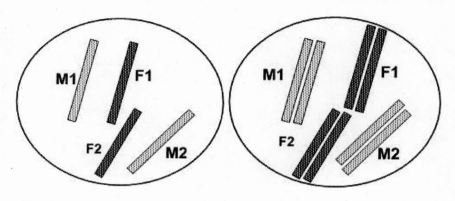

Figure 12-9A. Original diploid cell with two sets of male and female chromosomes.

Figure 12-9B. Chromosomes duplicate

Next, the DNA molecules undergo mitosis, yielding two new sets of DNA molecules (see figure 12-9B). Note that mutations can occur at this step [45].

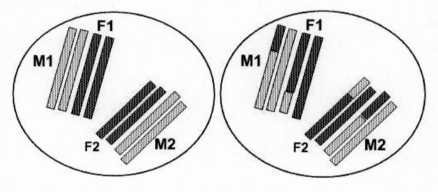

Figure 12-10A. Like chromosomes pair up

Figure 12-10B. Chromosomes swap sections of DNA in genetic recombination

Following the creation of new sets of DNA via mitosis. The DNA molecules pair up as shown in figure 12-10A and get close enough such that genetic recombination can occur in which the DNA molecules can exchange genes [46] as shown in figure 12-10B. (This could be termed genetic mixing.) Following genetic recombination, the cell depicted in figure 12-10B divides by mitosis into two daughter cells shown in figures 12-11A and B.

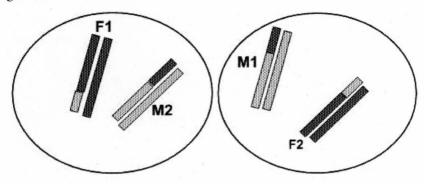

Figure 12-11A and 12-11B. Cell in figure 12-10B divides into two daughter cells via mitosis

It should be emphasized that the assignment of pairs is completely random (e.g, the DNA strands represented by F1 in 12-11A and M1 in 12-11B could have been in 11B and 11A respectively). Finally, the two daughter cells undergo a second, non-mitotic cell division, which creates the haploid gametes, shown in figures 12-12A–D. The DNA molecules from figure 12-11A have been distributed to haploid cells in figure 12-12A and B [45], while the DNA molecules from Figure 12-11B have been distributed to haploid cells Figure 12-12C and D. Again, the assignment is strictly random.

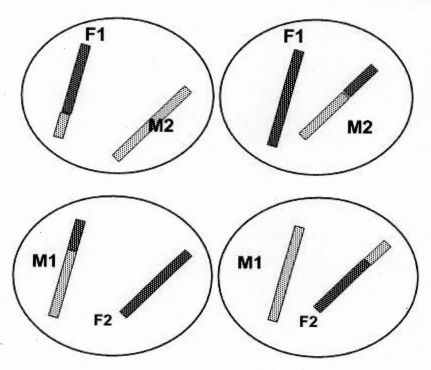

Figures 12-12 A–D. Diploid daughter cells in figures 12-11A and B divide into haploid cells

Next, one of the newly formed gametes—say Figure 12-12A, assuming it's male—designated by *M,* will fuse with another gamete from a female, designated by *F.* (Please note the DNA markings are different in figure 12-13 because this DNA pair comes from a female meiosis not previously shown.)

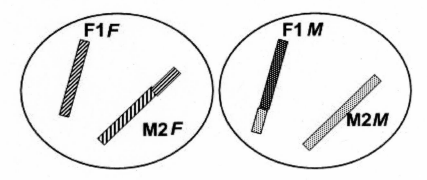

Figure 12-13. Male gamete and female gamete

Finally, the two cells fuse (see figure 12-14), which will result in a genetically new individual, thereby completing the cycle of sexual reproduction.

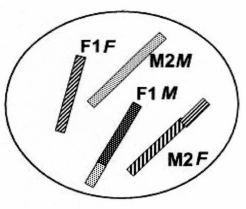

Figure 12-14. Fused gametes

NATURAL SELECTION BY REPRODUCTIVE GENETIC MIXING.

We can now modify the previous simplified natural selection diagram to show enough detail of the reproductive process to establish that natural selection proceeds via reproductive genetic mixing.

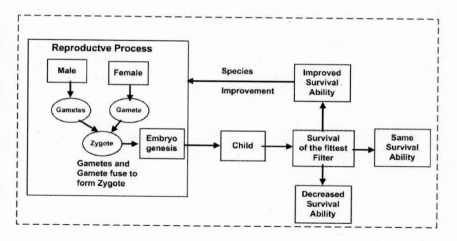

Figure 12-15. Natural selection by reproductive genetic mixing

In figure 12-15, we see the male and female gametes created by meiosis and genetic mixing joining to create a genetically unique zygote. It should be noted that, while the mitosis part of meiosis is "responsible" for mutations, meiosis is responsible for "spreading the mutations around." ***This is what drives evolution.***

The zygote then begins to divide by mitosis, initiating the process of embryogenesis whereby an offspring is created—an offspring that is genetically unique from either the father or the mother. This offspring and other offspring then pass through the survival of the fittest filter, and those offspring with improved survival ability will, in general, pass on half of their genes to their offspring. Over time the natural selection process will result in a general improvement, including new species, which fully explains the natural selection process.

Completing the explanation of natural selection, we are now able to complete the title to Darwin's magnificent opus as promised in the previous chapter. The complete title should be, *On the Origin of Species by Means of Natural Selection*, **Based Upon Reproductive Genetic Mixing**, *or the Preservation of Favoured Races in the Struggle for Life.*

LAWS OF PROBABILITY APPLY TO REPRODUCTION.

As mentioned above, the assignment of genes is random; hence the laws of probability determine the characteristics of the offspring. We will discuss the variation caused by probability in more detail in chapter

14, however, the essential effect of random gene assignment is depicted in figures 12-16 and 12-17.

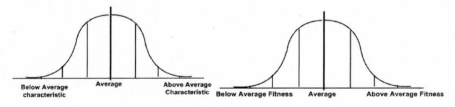

| Figure 12-16. Distribution of any characteristic | Figure 12-17. Distribution of fitness |

The left diagram illustrates how any genetically determined characteristics, such as height, hair color, intelligence, etc. will be distributed. Fitness will also be distributed in a similar manner. The vertical bars indicate the amount of the particular characteristic.

Note that, in general, most of the population will be close to average. However, a limited number will have above-average fitness, while some will have below-average fitness, as would be expected. It is, of course, those fortunate to have above average fitness that are the ones most likely to survive and create the inevitable improvements that accompany evolution.

AN INTELLIGENT DESIGNER IS NOT INVOLVED IN EVOLUTION.

We have now explained *how* reproduction functions plus *how* natural selection functions, which leads to an explanation regarding *why* there is variation such that some will have greater survival fitness than others. Those with greater survival fitness have a higher probability of passing along their genes, and as a result, there will be a gradual improvement in life. On the other hand, we have not encountered anything remotely resembling an intelligent designer or anything that can be ascribed to causation by a supernatural being. At the end of the day, there just isn't any intelligent designer. I will address this subject in more detail in chapter 14, which asks the question, "why three explanations of creation?"

SUMMARY OF CHAPTER 12.

In 1865, an obscure monk named Gregor Mendel, after thousands of experiments with peas, announced the existence of something he termed a "gene," which he claimed held the key to heredity. It was not until the late twentieth century that Mendel's genes were isolated, identified, and the mechanism by which they controlled heredity explained.

Perhaps it's self-evident, but the simple fact that reproduction is required for species survival is rarely articulated. Moreover, reproduction is the driver of evolution. Since the only animal that can reproduce itself directly is a single-celled animal, multi-celled reproduction must begin with a single cell, which of course is what we observe.

An understanding of cells began with the invention of the microscope in the late 1600s, but it required almost 200 years to establish the basic cell properties, which are:

- All organisms are made up of one or more cells.

- Cells are the fundamental functional and structural unit of life.

- All cells come from pre-existing cells.

- The cell is the unit of structure, physiology, and organization in living things.

- The cell retains a dual existence as a distinct entity and a building block in the construction of organisms.

In approximately 1848, cell structure had been refined sufficiently to understand that all cells contain a nucleus and that each nucleus contains filament-like entities termed "chromosomes." Most cells contain two sets of chromosomes and are termed "diploid," while some cells, the sex cells, contain only one set of chromosomes and are termed "haploid."

In order for reproduction and growth to occur, cells must divide. There are two kinds of cell division:

1. Mitosis was discovered in 1882, just fifteen years after Charles Darwin's death. The division is associated with most cells in which the two new "daughter" cells contain a complete set of chromosomes

2. Meiosis was discovered four years later. It is the division of sex cells in which the two new "daughter" cells contain only one set of chromosomes.

Following the discovery of cell division, especially meiosis, a series of discoveries followed that solved the puzzles of reproduction, natural selection and evolution's mechanism:

1871: Miescher discovered a nonprotein molecule in the nucleus, which he named "nuclein."

1881: Zacharis discovered that nuclein is the same as the material of chromosomes, "chromatin."

1889: Fisher discovered two molecules that would prove important to DNA structure: pyrimidine and purine.

1890: Altman discovered that nuclein is an acid and named it "nucleic acid."

1910: Kossel demonstrated pyrimidine and purine are present in nucleic acid.

1909–1929: Levene determined there are two different forms of nucleic acid (DNA and RNA). He also discovered that there are two pyrimidines (cytosine and thymine) and two purines (adenine and guanine) present in nucleic acids and that they are the principal constituents of DNA along with deoxyribose and a phosphate group. Finally, he found that the DNA components were linked in the order, phosphate-sugar-base pair, to form a unit he termed a "nucleotide."

1928: Griffith demonstrated that DNA carries the inheritance information.

1952: Hershey and Chase demonstrated that genes are located in DNA.

1953: Crick, Franklin and Watson used X-ray diffraction to determine the shape of DNA, which revealed the following important details of cell division:

- At the beginning of cell division, the DNA helix unwinds and is split into two strands by enzymes.

- One strand goes to each of the new daughter cells. The creation of a strand for each daughter cell is the reason for the double helix.

- Enzymes then add the appropriate bases to the single strand, thereby recovering the double helix form.

- Mutations occur when the wrong bases, termed "copying errors," are added to the single strand.

1955: Gamow showed that any three of the four bases (C, T, A, or G) should be sufficient to code for an amino acid.

1961: Crick and Brenner experimentally confirmed Gamow's hypothesis.

1993: Sharp and Roberts shared the Nobel Prize for the determination of gene structure—intron-exon-intron, with introns being noncoding and exons carrying the code.

2001: Ventner et al. published first sequence of the human genome.

Since 2001 the basic function of genes has been determined. Messenger RNA (mRNA) makes a copy of a code, viz. the "codon," for an amino acid contained in a gene located on DNA. MRNA is then copied onto transfer RNA, which carries the genetic information to a ribosome "protein factory." In the ribosome, the genetic information is used to select an amino acid, which is then added to other amino acids to create a protein.

At the end of this sequence of discovery, it was clear that genetic mixing occurred during sperm and egg cell creation and that the combination of male sperm and female egg cells leads to reproductive genetic uniqueness. This uniqueness provides the input to the natural selection filter, thus completing the explanation of how evolution works. This also allows us to complete the title to Darwin's magnificent opus to read: On the Origin of Species by Means of Natural Selection, *Based Upon Reproductive Genetic Mixing*, or the Preservation of Favoured Races in the Struggle for Life.

We are now one step closer to establishing the second half of objective one: answering the questions "Why are their three explanations of our origins?" and "Which explanation is correct?" We now have a complete explanation of the process of evolution, in particular how it works. However, there are a few more points to be made regarding the elimination of creationism and ID as correct explanations of or origins. These will be concluded in chapter 14.

Moreover, we still need to demonstrate that evolution is the inevitable consequence of the need to ensure species survival. This will be accomplished in the following chapters, particularly chapter 17, "Answers in Evolution," where we explain why evolution is the inevitable consequence of the need to ensure species survival.

LOOKING AHEAD.

We have now resolved the apparently unchanging physical and biological aspects of the earth by explaining how the physical processes and the biological process of evolution actually function. In the next chapter, we will explain how these processes created the earth and all the life on it.

CHAPTER 13:
RESOLVING THE APPARENTLY UNCHANGING FEATURES OF THE EARTH ILLUSION—HOW THE EARTH WAS REALLY CREATED

EARLY EXPLANATIONS OF EARTH'S FORMATION.

As discussed in chapters 9–12, the early explanation of Earth's formation most relevant to this book is first few lines from the book of Genesis. Relative to this chapter, the most closely related verses are:

- "In the beginning God created the heaven and the earth. And the earth was without form, and void; and darkness was upon the face of the deep. And the Spirit of God moved upon the face of the waters" (Gen 1:1–2).

- "And God said, Let the waters under the heaven be gathered together unto one place, and let the dry land appear: and it was so. And God called the dry land Earth; and the gathering together of the waters he called Seas: and God saw that it was good" (Gen 1:9–10).

Thus, according to Genesis, the Earth originally had no form but was apparently covered by water because as illustrated in verse 2, "the Spirit of God moved upon the face of the waters." Then later, in verse 9, "and [God] let the dry land appear." So first there was water, and then later land appeared. As we will see, this is opposite to what really happened.

INITIAL FORMATION OF EARTH.

As discussed in chapter 8, available evidence to-date demonstrates that the earth was formed by the accretion of material from the primordial solar disc. The immense energy imparted by impacts of material, striking the earth at high velocity, created a molten rocky ball. As discussed in chapter 10, the first person to propose a molten earth was Lord Kelvin, a thermodynamics expert who believed the earth was cooling down from a molten state. Presumably Kelvin got his insight from the fact that volcanoes emit molten rock. But what we now know that Kelvin couldn't have known is that the earth's internal heat is maintained by radioactive decay. Therefore, the cooling rate Kelvin used was much too fast.

A new analysis of ancient minerals called zircons, which are exceptionally resistant to chemical changes, are the oldest known materials on Earth [1] and offer a window in time back as far as 4.4 billion years ago when the planet was a mere 150 million years old. Because of these properties, "zircon crystals have become the gold standard for determining the age of ancient rocks", says University of Wisconsin-Madison geologist John Valley, who used these tiny minerals to show that rocky continents and liquid water formed on the earth about 4.2 billion years ago, much earlier than previously thought [2].

The early earth's atmosphere is believed to have contained extremely high levels of carbon dioxide, maybe 10,000 times as much as today. "At [those levels], you would have had vicious acid rain and intense greenhouse [effects]. That is a condition that will dissolve rocks" [3].

Conditions like this bring up a fundamental question: "How did this barren rock become the blue planet we know and love?"

The answer to that question is an intertwined combination of actions, both biological and physical, that performed the transformation over a period of 4.5 billion years. This is obviously a complicated story, and there is only room in this book to sketch the bare essentials. However, we will see how the resolution of the unchanging earth illusion—which led to the correct explanation of the earth's biological and physical processes—will allow us to explain how the earth was really created.

This will further confirm the findings of the previous two chapters, especially evolution, and provide additional information for the answer to the question, "Why are there three explanations for the origins of humans and the universe?"

Where did Earth's Water Come From and When did it First Appear? Pillow lavas (round structures, hence the name) provide one possible earliest date for the appearance of water. Pillow lava is formed when molten magma is excreted into water, usually fairly deep water [4]. Therefore, the presence of pillow lava is proof that water existed at the time the pillow lava was formed.

Some of the oldest pillow lava formations are found in South Africa and have been dated to over 3.4 BYA [5]. So we can safely say water has been on Earth at least since the time the pillow lava formed and probably much longer because time was required for water to accumulate.

As for where the water comes from, this site [6] points out, "Water is a common chemical compound in our solar system. In addition to Earth, it has been identified in asteroids, comets, meteorites, Mars, in the atmospheres, rings and moons of giant planets, and there is evidence for it in the poles of our Moon and of Mercury." However, the site also goes on to say, "There is no agreement on the origin of water on Earth and Mars. A number of sources have been proposed but the pieces of this puzzle do not currently fit into a coherent picture" [6].

On the other hand, there are some strong candidates. The initial molten Earth cooled sufficiently such that volatile components being out gassed, which initially escaped, were held to the earth. This allowed the atmosphere to acquire sufficient pressure for the stabilization and retention of water [7].

Meteorites known as carbonaceous chondrites, which are generally agreed to have formed in the outer reaches of the asteroid belt, have a water content of sometimes more than 10 percent of their weight [7]. One researcher, A. Morbidelli, proposed that the largest part of today's water comes from these objects when they plunged toward the earth [8].

Hidenori Genda from the Tokyo Institute of Technology and his colleagues suggest another theory. They claim there is evidence that the earth could have had a thick hydrogen atmosphere, which reacted with oxides in the earth's mantle to produce copious water [9].

There is also accumulating evidence that water existed as early as 4.4 billion years ago [6]; [3]. Reference [6] provides one of the more thorough summaries of the state of knowledge in this area, which is continually being refined. However, there is one undisputable fact—water exists on Earth, and it has been here for a long time.

When did Life First Appear? The South African pillow lava also provides information regarding the appearance of life:

> Pillow lava rims from a formation known as the Mesoarchean Barberton Greenstone Belt in South Africa contain micrometer-scale mineralized tubes that provide evidence of submarine microbial activity during the early history of Earth. The tubes formed during microbial etching of glass along fractures. [Similar tubes are] seen in pillow lavas from recent oceanic crust. The margins of the tubes contain organic carbon, and many of the pillow rims exhibit isotopically light bulk-rock carbonate ^{13}C [difference in amount of ^{13}C] values, supporting their biogenic [biologically generated] origin. Overlapping metamorphic and magmatic dates from the pillow lavas suggest that microbial life colonized these subaqueous volcanic rocks soon after their eruption almost 3.5 billion years ago. [5]

In other words, the minute tubes in pillow lava could only have been made by microbial life. In addition to providing evidence of life, the existence of pillow lava suggests plate tectonics were active at least 3.4 BYA, which will be discussed in more detail shortly.

Stromatolites: Another piece of evidence regarding the appearance of life is found in strange formations known as stromatolites [10], which are also among the oldest objects on Earth. They are one of the intertwinings between biology and geology and were formed by the growth, layer upon layer, of one of the simplest forms of carbon-based life, cyanobacteria. Cyanobacteria are known as prokaryotes, meaning they have no nucleus. Of particular interest is the fact that, like all bacteria, cyanobacteria reproduce by binary fission rather than mitosis [11]. Mitosis only occurs in cells with a nucleus. This more primitive form of reproduction is also evidence to the great age of cyanobacteria.

Stromatolites have been dated to 3.5 BYA, and along with the micrometer-scale mineralized tubes in pillow lava mentioned above, they demonstrate that primitive life has existed on Earth for billions of years. Most stromatolites became fossils billions of years ago; hence living stromatolites are perforce quite rare. But we can still study their development. The peculiar conditions in the Hamelin Pool in Western Australia's Sharks Bay contains the most diverse and abundant examples of stromatolite forms in the world today [10].

Stromatolites use light energy to create food and excrete rock. This simple organism remained the only life on Earth for the next 2 billion years. For a simplified time line of life's development on Earth as recorded in the fossil record, see appendix III.

Formation of Granite—Earth's Outer Crust Though it was not among the first rocks formed, granite is the principal constituent of continents. The first rocks formed were mainly basalt. Granite is usually formed under the earth's surface from molten magma. It is has a medium- to coarse-grained texture, in contrast to sedimentary rock which has a layered structure. Granites occasionally display some relatively large crystals and can be pink to dark gray or even black in color, depending on its chemistry and mineralogy.

In contrast, basalt is formed when magma flows onto the earth's surface as lava. It is usually gray to black and fine grained due to rapid cooling of the lava on the earth's surface. Unweathered basalt is black or gray [12]. And, as discussed in chapter 10, granite is less dense than basalt and therefore "floats" on basalt.

Due to the constant changes that occur on Earth because of to erosion and plate tectonics, most of the earliest granites have long disappeared. However, as pointed out by Dr. Lorence G. Collins, Department of Geological Sciences, California State University Northridge, "Granites in the bottom of the Grand Canyon give Precambrian ages of 1.58 and 1.65 billion years, younger than the 1.7-1.85 billion-year-old Vishnu schist" [13]. Dr. Collins's paper also contains another interesting answer to creationist arguments regarding the age of the earth's rocks.

So exactly when enough granite was formed to create the continents is difficult ascertain with great detail, but we know granite definitely had to be formed before the Grand Canyon rocks in order for the rock layers to exist.

Creation of Earth's Oxygen: Regarding the creation of Earth's free oxygen, there are two points of agreement:

1. There was no free oxygen when the earth formed, though there was oxygen locked in minerals such as silicates (SiO_2).

2. The current atmosphere contains approximately 20 percent oxygen.

Other than these rather obvious facts, there are many explanations for the formation of oxygen. One of the more prominent explanations involves cyanobacteria photosynthesis in stromatolites as the source of the earth's O_2. The stromatolite O_2 generation also played a prominent role in the formation of the iron ore deposits found throughout the earth [14].

During the formation of earth's oceans, iron compounds (probably sulphates) dissolved in the oceans. When stromatolites began generating O_2, the O_2 was initially dissolved into the oceans where it "rusted" the dissolved iron, forming layers of iron oxides now termed the banded "iron oxide layers" [15]. These iron oxide layers are one of the main sources of commercial iron ore, such as the Mesabi Range in Minnesota.

Once all the iron in the oceans had been "rusted," the O_2 began to accumulate in the atmosphere, changing the atmosphere from one dominated by CO_2 to one dominated by O_2 [11]. As a result of the oxygen generation, the oceans and sky turned blue. However, it is estimated that as much as twenty times more oxygen is sequestered in the banded iron oxides than in the atmosphere [15].

PLATE TECTONICS.

As discussed in chapter 10, plate tectonics has explained a great many mysteries regarding the earth, such as demonstrating that granitic continent plates ride across a basalt under layer and have been in constant motion since the formation of the crust. In addition to the pillow lava evidence discussed above, recent studies have also demonstrated that tectonic activity began early in Earth's history. A report summarized in the January 3, 2009 issue of *Science News* [16] states:

> Rocks in the Jack Hills of Western Australia hosted zircon crystals that contain tiny mineral inclusions.... The zircons and inclusions are more than 4 billion years old and contain evidence suggesting an early start for tectonic activity on Earth.

> The chemical composition of ancient crystals now found in Australian rocks bolsters the notion that tectonic plates may have jostled across Earth's surface more than 4 billion years ago. Scientists call the first 600 million years of Earth's history the "Hadean eon" because of the presumably hellish temperatures on the freshly coalesced and largely molten planet. Also, radioactive isotopes, which generate heat inside Earth as they decay, were much more common then than they are now, says Mark Harrison, a geologist at the University of California, Los Angeles. (p. 10)

One of the more spectacular consequences of plate motion, due to plate tectonics, that have had significant impact upon the Earth's creation has been the occasional uniting of most, if not all, of the earth's crust into a single "supercontinent," two of which are named Rodinia (from the Russian word for "motherland" or "homeland") and Pangaea (named by Wegner, see chapter 10).

The existence of supercontinents such as Rodinia and Pangaea have been disclosed by the observational technique known as paleomagnetism. Paleomagnetism is an examination of the variation of the earth's magnetic field over long periods of time, which is preserved in various magnetic materials [17] such as nickel, iron, and cobalt [18]. The study of paleomagnetism has demonstrated that, over time, the earth's magnetic field has varied substantially in both orientation and intensity [17]. Unfortunately, the record of Earth's magnetic field only provides latitude information. To obtain longitude information, strata must be compared. However, if the magnetic orientation, strata, and age of two crustal pieces match, then even if they are presently miles apart they must have been connected at some time in the past.

Formation and Ultimate Separation of Rodinia—Prelude to Complex Life Rodinia began forming about 1.3 BYA from three or four pre-existing continents, one of which is called the Grenville Province, a rock structure running from Labrador to Texas, consolidated perhaps 1,100 to 1,000 MYA [19].

Rodinia formed during a period that geologists term the Neoproterozoic Era [20], which extends from 1000–542 MYA, see appendix II. Rodinia caused severe climate change, significantly cooling the earth and leading to the Cryogenian period.

The Cryogenian period began approximately 850 MYA and ended approximately 635 MYA This period was marked by large-scale glaciation that may have covered the entire planet, the so-called "snowball" earth. However, while glaciation did occur, it uncertain whether the entire planet was covered or the glaciations were local. Regardless, it was a time when the earth was indeed very cold. Also, during the Cryogenian period, the supercontinent Rodinia began to break up [21].

EDIACARAN BACTERIA, THE NEXT STAGE IN LIFE'S DEVELOPMENT.

The last period of the Neoproterozoic was the the Ediacaran, named for the Ediacara Hills in South Australia, where some the first Ediacaran fossils were discovered [22]. The Ediacaran period lasted from about 650–540 MYA.

Evidence of extensive lava flows and volcanic eruptions during the Ediacaran period, such as those that covered an area of about 350,000 km^2 in eastern Europe [23], suggests that Rodinia began to drift apart no later than 750 MYA [24] and probably sooner, as noted above.

In the mid-1800s geologists and paleobiologists were certain that the next stage in the development of life occurred *during* a period termed the Cambrian (see next section), but some scientists, such as Charles Darwin, suspected that more primitive forms must have preexisted the Cambrian.

Improved observational techniques have now demonstrated that macroscopic fossils of soft-bodied, multi-celled organisms existed in the Ediacaran period, confirming Darwin's expectations. These organisms appeared soon after the Cryogenic period thawed. Although extremely old, Ediacaran fossils can be found in a few localities around the world [24].

The development of simple multi-celled organisms in the Ediacaran set the stage for the rapid increase in number and diversity of species known as the Cambrian Explosion. However, it must be noted that for multi-celled organisms to form, two important developments were required:

1. Cellular specialization or differentiation

2. Sexual reproduction

A multi-celled organism cannot survive unless the tasks required for life are distributed among the organism's cells. Thus, cells that can provide a protective layer must be formed as well as cells that aid in the ingestion and digestion of oxygen and food.

As explained in chapters 11 and 12, another requirement for multi-celled species' survival is reproduction. Indeed, multi-celled animals can only reproduce sexually; so the development of meiosis, which is essential for sexual reproduction, can be traced to the Ediacaran period. Also demonstrated in chapters 11 and 12, sexual reproduction involves reproductive genetic mixing—the key to evolution. Therefore it can be said that evolution can be traced to at least the Ediacaran period, 540 MYA.

It is interesting to note that binary fission reproduction, used by cyanobacteria, is over 3 BYA. Mitosis appeared sometime between binary fission and meiosis, perhaps about 1 MYA when oxygen levels became high enough to support eukaryotes [25]. But meiotic reproduction is approximately 650–540 million years old. Sexual reproduction is a relative newcomer, but with the advent of sexual reproduction, the stage was set for the Cambrian Explosion.

THE CAMBRIAN EXPLOSION—MULTIFACETED LIFE BEGINS AND IS CAPTURED IN SHALE,

Shale is the most common sedimentary rock. It consists of very fine particles that collect in very slow moving water such as off the shore of sandy beaches. The slow accumulation of fine particles accompanying the formation of shale is an ideal environment for the preservation of small animals and soft body parts.

About 530 million years ago, shale formations preserved an unprecedented increase in the number and complexity of life forms—an increase usually termed the Cambrian Explosion. Cambria, the Roman name for modern Wales, is one of the first places this fossil record was found [26]. It wasn't exactly an explosion; the process occupied over 10 million years. But most known animal phyla appeared in the fossil record as marine species during this time [27].

The seemingly rapid appearance of fossils in this "Primordial Strata" was noted as early as the mid-nineteenth century and generated extensive scientific debate [28]. Charles Darwin feared it could be used as an objection to his concept of evolution by natural selection [29] pp. 315–316. However, because Darwin didn't fully understand how natural selection worked, he didn't realize that natural selection was probably one of the principal causes of the Cambrian Explosion.

The causes of the Cambrian Explosion have been the subject of considerable debate. There is evidence that genetic factors played a significant role, and recent research suggests that during the Ediacaran period, a "genetic tool kit" of genes that govern developmental processes gradually evolved [26]. As noted above, one of the more significant aspects of this "tool kit" was the development of sexual reproduction. Reproductive genetic mixing permitted a wide diversity of life-forms to be created, and natural selection's survival of the fittest filter selected the best of them.

As would be expected, many forms seen in the fossil record of the Cambrian disappeared without trace, victims of the need for the most fit to survive in accordance with the species survival imperative. Once the body plans natural selection found most successful came to dominate the biosphere, significant evolutionary modification ceased, and evolutionary change was limited to adjustment of the body plans that already existed [26].

This "adjustment" led to considerable improvement in the development of body plans over the 500 million years since the Cambrian, which is an important aspect of evolution. For that reason, a few words regarding this important topic are in order. An extract from the survival requirements table, as shown in figure 13-1, will aid this discussion (recall that while reproduction is not necessary for individual survival, it must be provided to ensure species survival).

Requirement for Survival	Individual Survival
Oxygen	Y
Water	Y
Food	Y
Access to food	Y
Metabolism of food	Y
Suitable Environment	Y
Proper temperature	
Dry	
Safe	
Reproduction	N

Table 13-1. Basic survival needs of an individual

As discussed above, the advent of multi-cellular life required: (1) cell specialization or differentiation, and (2) sexual reproduction.

Thus the cells in a multi-celled animal must be arranged to provide the items indicated in the above table to assure its survival. Accordingly, means must also be provided for the following:

- Ingestion of O_2 and transfer to all cells

- Ingestion of water and transfer to all cells

- Access to food, which in most cases implies some form ability to locate and move to food

- Ingestion of and metabolism of food to extract the useful nutrients and transfer them to cells

- Suitable environment, which in most cases also implies some ability to locate and move to a suitable environment (e.g., "get out of the rain")

- Combining sex cells to assure reproduction occurs

In addition to these items, which are directly related to the table, the following additional capabilities will be needed:

- The ability to locate water, food, and a suitable environment, which implies the need to sense the environment. There are five items in the environment that can be sensed: light, sound, odors, taste, and objects (sense of touch); hence organs able to sense these would logically evolve.

- The ability to access water, food, and a suitable environment, which implies the need for some form of locomotion.

- The ability to transfer O_2, water, etc., to all the cells, which implies some form of a circulation system.

- Some form of control system was needed as the multi-celled animal became more complex.

Finally, as multi-celled animals became more complex, cells grouped together into organs or systems under the "prodding" of natural selection to provide specific functions such as the digestive, circulatory, and reproductive systems.

As we review the development of various animals, we will examine how each species has met these survival needs and how the ability to meet these needs improved over time. We will study the first large animal, the fish, in more detail than the others to establish the basic pattern that is followed with improved capability throughout the evolution of the all animals.

Origin of Fish in the Cambrian: One of the animal species common today that originated in the Cambrian is the fish, though the original fish barely resembles any alive today. Fish are members of the Vertebrates, animals with backbones. The first fish were jawless and toothless soft-bodied creatures that wriggled through the water and sucked up microscopic food particles [30].

Modern fish with skeletons appeared in the late Silurian period, 443–416 MYA or the early Devonian period (named for Devon Great Britain where some of the early Devonian formations were found), approximately 395 MYA[1] [30]. Fish became the dominant animal on Earth during the Devonian, which causes the Devonian to sometimes be called the Age of Fish [31].

The fish body plan is well known, but it is worth a few words to review how a fish meets the needs discussed above:

- Ingestion of O_2 is performed by gills and a circulation system transfers O_2 to all cells.

- Ingestion of water (i.e., do fish drink water?) depends on the type of fish. Freshwater fish absorb water through their skin cells, and saltwater fish drink water with the gills removing the salt [32].

- Food is ingested by mouth and metabolized in a digestive system that extracts the useful nutrients and transfers it to cells via a circulation system. The waste products of metabolism are passed to the circulation system and excreted through an anal pore at the lower rear of the fish.

- Food is generally found in the water through which a fish swims.

- Fish have eyes the front of their heads to locate the food and olfactory sensors to assess food.

- Fish have fins to provide locomotive power to propel them through water to the food (fins also allow fish to evade predators).

- In general, the water in which a fish lives provides a suitable environment.

- Female fish lay eggs, which are fertilized by sperm released by the male fish. Typically, the female deposits many eggs to improve the probability that enough will survive to reproduce.

- The fish has a rudimentary brain, primarily devoted to the eye, nose, and balance maintenance.

- The basic fish body plan has bilateral asymmetry. The vertebrate fish spine occupies the center of the fish and provides structural stability

The salient features of a fish are summarized in tabular form in appendix IV. Other species have also been added to this table, which illustrates the improvements provided by evolution over time.

Origin of Amphibians: The generation of O_2 eventually led to the development of an ozone layer, which provided protection from the sun's lethal UV rays and therefore allowed life to move from the sea to land.

While the Devonian was dominated by fish, some adventurous fish (perhaps to escape overcrowding) began to venture onto land. One of the first fish to make the critical step was the Tiktaalik, which appears in the fossil record 382 MYA. Well-preserved Tiktaalik fossils were found in 2004 on Ellesmere Island in Canada [35]. It had many of the features of tetrapods[2] [33]. It is an example from several lines of ancient fish that had lobed fins that eventually evolved into legs as an adaptation to the oxygen-poor shallow-water habitats of the time [34].

The Tiktaalik's lobed front fins featured arm-like skeletal structures that resembled a crocodile, including a shoulder, elbow, and wrist. The rear fins and tail have not yet been found. A significant difference between Tiktaalik and a fish was the Tiktaalik's ability to move its neck independently of its body, something a fish cannot do. Moreover, its neck and ribs are similar to those of tetrapods. The ribs supported its body and aided in breathing via lungs, something else fish don't do, but is a *sine qua non* for a land animal [35]. Tiktaalik is truly a transition animal[3]. Its mixture of fish and tetrapod characteristics led one of its discoverers, American paleontologist and evolutionary biologist Neil Shubin, to characterize it as a "fishapod" [36]. Shubin gained considerable fame for his discovery of Tiktaalik [37]. Lobed-finned fishapods such as Tiktaalik led to the evolution of amphibians.

Amphibians obviously represent the transition between life in the sea and life on land because amphibians lay eggs in water and fertilize the eggs in water, as do fish. When the amphibian eggs hatch, the resultant "tadpoles" spend their first few days in water, slowly metamorphosing into a frog. Anyone who has lived near a pond has witnessed this amazing transformation.

While in the water, the tadpole satisfies its needs basically, just as a fish does. Once the transformation to amphibian is complete, the amphibian loses its gills, which means it must breathe air. This is accomplished by breathing in through the mouth to lungs, which then transfers the oxygen to a circulatory system. An amphibian has four legs that provide it mobility to search for food and a mate.

Although able to live on dry land, amphibians usually live more in water [38]. Amphibian characteristics are summarized in appendix IV, where the ability to live partially on land can be seen as an improvement.

THE CARBONIFEROUS PERIOD AND THE ORIGIN OF AMNIOTES.

The geologic period following the Devonian was the Carboniferous, which began 359 MYA and ended 299 MYA. Carboniferous means "coal-bearing," and the period was named after it because many beds of coal were laid down all over the world during this period [39].

One of the greatest evolutionary innovations of the Carboniferous was the amniotic membrane (protection for an embryo). The survival advantage provided by the amniotic membrane is extremely important, and those animals whose embryos are protected by an amniotic membrane are termed "amniotes" [40].

It is difficult to over emphasize the importance of the amniotic membrane. As has been discussed in detail, reproduction is the most important function performed by any species because it is necessary for the species' survival. All new animals begin as fertilized eggs that become embryos that grow and eventually become mature animals ready to begin the reproductive cycle. All embryos require a fluid "bath" in which to develop. In the case of fish, which are constrained to live their lives entirely in water, and amphibians, which must begin their lives in water and remain near water most of their lives, the fluid bath provided for the eggs is the body of water in which the eggs are laid. The location places the eggs in considerable danger, hence the large number of eggs laid.

With the advent of the amniotic sac, fertilized eggs did not have to contend with open water. Moreover, animals could live their entire lives on land, which opened up significant possibilities: access to new territories, more diverse living conditions, and ultimately the exploitation of the earth's surface resources.

There are two groups of amniotes: (1) the sauropsida, which includes reptiles, dinosaurs, and birds, and (2) the synapsida, which include mammals and mammal-like reptiles [40].

The terms "Sauropsida" (lizard faces) and "Theropsida" or Synapsida (beast faces) were coined in 1916 by British zoologist **Edwin S. Goodrich (1868–1946)** [41] to distinguish between lizards, birds, and their relatives on one hand (Sauropsida) and mammals (Synapsida) on the other [42].

The oldest known synapsid is Archaeothyris, found in Nova Scotia. It lived in the mid-carboniferous period about 320 MYA. Archaeothyris is the precursor of all synapsids, including mammals [43] and is therefore one of our most distant ancestors.

Reptiles—the Next Evolutionary Step.

As mentioned above, reptiles are amniotes and members of the Sauropsida. Reptiles first appeared about 320 MYA during what is known as the Pennsylvanian portion of the Carboniferous period, named because the first fossils were found in Pennsylvania.

Reptiles are bilaterally symmetric with a head at the front of the body that contains eyes, ears, and a mouth. In the back of the body, it has an excretory anal pore, followed by a tail. The circulation system of most reptiles is driven by a three-chamber heart consisting of two atria and one ventricle [42]. Reptiles are air-breathing and therefore have lungs that transfer O_2 to the circulatory system. They have a digestive system whereby food is ingested through the mouth, digested in a series of organs spread along the body, and excreted as waste at the back. The products of digestion are passed to the circulatory system.

The reptile's body is supported by a skeletal system. Reptiles are cold-blooded; hence they have only rudimentary thermal control and become lethargic in the cold. They have scaly skin and four legs attached to the skeletal system for locomotion.

Reptiles also reproduce via eggs, as do fish and amphibians, but with one important improvement. Their eggs are fertilized within the protective female environment. The resultant embryo is enclosed in the all-important protective amniotic membrane within a hard shell that provides structural strength, and when mature enough, it is deposited in a protected place for further maturation until the egg "hatches." While most reptiles lay eggs, some are capable of live birth, and some have a placenta similar to mammals, which indicate a transitional variation [42].

Reptiles have a rudimentary nervous system, including a spinal cord that is terminated with an enlarged mass of tissue in the head. This mass of tissue is the first indication of a brain, and is therefore obviously termed the reptilian brain[4]. Various clumps of cells in the reptilian brain determine the brain's general level of alertness and control basic life functions such as the autonomic nervous system, breathing, heart rate, and the fight or flight mechanism. Lacking language, a reptile's impulses are instinctual and ritualistic. The reptile is concerned with fundamental needs like survival, physical maintenance, hoarding, dominance, preening, and mating [44]. The essential reptilian features are summarized in appendix IV.

While reptiles no longer dominate to the extent they did, a few such as crocodiles, alligators, turtles, and snakes have survived the various extinctions and are a part of life on Earth today.

One important member of the reptilian family, the dinosaur, will be discussed below, but first a few words about catastrophic extinction events that almost ended life's experiment.

MAJOR CATASTROPHIC EXTINCTION EVENTS.

As mentioned in chapter 9, one of the early explanations for both the physical and animal and plant structures of Earth was catastrophism. While these explanations have been shown to be incorrect because they occurred too recently, catastrophes have occurred in Earth's history. The five main catastrophic extinction events are listed in table 13-2.

Extinction Event	Time MYA	Percent Extinction
Ordovician-Silurian	450	40
Late Devonian	375	40
Permian-Triassic	250	95
Triassic-Jurassic	200	30
Cretaceous-Ternary	65	35

Table 13-2. Summary of the five most severe extinction events

Clearly, the Permian-Triassic (P-T) extinction was the most severe, and the candle of life was almost blown out. To place the P-T extinction into perspective, let's begin with a brief description of the Permian period.

Permian: The Permian period began 299 MYA and ended with a bang 251 MYA [45]. Sir Roderick Murchison, who was introduced in chapter 9 as a person interested in the Silurian system, named the Permian period after the Russian kingdom of Permia [46].

During the Permian, the supercontinent Pangaea completed its formation. Because of its large size, it caused the sea levels to remain generally low. Permian terrestrial life included diverse plants, fungi, arthropods, and a variety of tetrapods, which were dominated by amphibians. Toward the end of the Permian, reptiles grew to dominance among vertebrates because their special adaptations enabled them to flourish in the drier climate [47].

Of particular interest near the end of the Permian was when the first archosaurs (Greek for "running lizards") appeared. Archosaurs are a reptilian group from which dinosaurs evolved and which are represented by modern birds [48].

During the prolific Permian, the first cynodonts appeared. Cynodonts belonged to the synapsid class and had nearly all the characteristics of mammals. They were probably warm-blooded and had hair. Also, they survived the P-T extinction and continued their evolution into mammals [49].

Permian-Triassic Extinction Details: The P-T extinction event, informally known as the Great Dying, occurred 251 MYA, thus forming the boundary between the Permian and Triassic periods [50].

The extent of the devastation is summarized in table 13-3 from information gathered from references [51]; [52]; and [53].

Species	Percent extinction
Marine	96
Terrestrial vertebrates	70
Insect families	57
Insect genera	83

Table 13-3. Summary percent extinction during P-T extinction event

The event has been described as the "mother of all mass extinctions." Due the extreme loss of life-forms, the recovery of life on Earth was much slower than after other extinction events. A detailed listing and discussion of the P-T extinction can be found in [54] and is recommended for those who wonder how close to the edge we came.

Two possible causes of this extinction have been suggested:

1. Collision with a near earth object (NEO) such as a comet or asteroid[5]

2. Volcanic eruption

There is fairly conclusive evidence that suggests a NEO caused the end of the Cretaceous period, as well as the dinosaurs. Therefore, it could also suggest that a NEO might have caused the P-T extinction as well. Unfortunately, most of any P-T extinction evidence has been swept into the earth by the plate tectonic subduction conveyer belt, and enough evidence does not exist to eliminate either proposal [54].

Regarding volcanic eruption, the obvious candidate is a vast outpouring of lava that occurred in Siberia about 250–251 MYA, exactly coincident with the P-T event [55]. This lava flow is termed the Siberian Traps. The word "traps" is derived from a Swedish word for stairs because the Siberian lava flows have a step-like appearance. The volume of lava was immense. Today the area covered is about 800 thousand square miles and estimates of the original coverage are as high as 2.8 million square miles. The original volume of lava is estimated to range from 250 thousand to 1 million cubic miles. The volcanism continued for about 1 million years [55].

Regarding the effect of this amount of lava flow, Dr. Norman Macleod of the Natural History Museum in London pointed out while discussing another large lava flow, the Deccan Traps: "We're talking about catastrophic effects in terms of changes in habitat, changes in rain fall patterns, changes in climate, all of the things you can think of that are going on in the modern world magnified many many times, many many orders of magnitude indeed" [62].

To date, questions surround each of these possible causes. It may have been a combination of the two theories. But one thing that is certain is the effects are still being felt today.

Triassic Period: The Triassic period opens on a rather different world and extends from the end of the P-T extinction to another extinction that occurred 199 MYA, the Triassic-Jurassic extinction [52].

In 1834, German geologist **Dr. Friedrich von Alberti (1795–1878)** published the results of his investigation of three apparently distinct sedimentary rock deposits that were found all over Germany and Northern Europe. The rocks were red sandstone overlain by chalk (calcium carbonate rock), which was in turn covered by black shale [53]. The fossils contained in the three layers were the same, therefore von Alberti realized that the three layers were actually part of the same formation, which he then termed "Trias" (from the Latin for triad). Von Alberti termed the time period occupied by these rocks the "Triassic" [56].

The Triassic landscape was dominated by the supercontinent Pangaea, which remained intact until the next geologic period, the Jurassic, at which point it began to break up [54].

Terrestrial animals of interest in the Triassic period are reptiles, some of which had survived the P-T extinction, and the archosaurs and cryodonts that had also survived the P-T extinction. Toward the end of the Triassic, the archosaurs evolved into dinosaurs (see next section). Cryodonts "laid low," avoiding the dominant dinosaurs and awaiting their turn as the first mammals during the Jurassic.

THE UNFORGETTABLE DINOSAURS.

Few extinct animals have captured the public's interest more than that most famous of reptiles, the star of stage, screen, and a gazillion books—the dinosaur.

As mentioned above, dinosaurs diverged from a group of reptiles known as Archosaurs [57] approximately 230 MYA during the Triassic period.

The first known dinosaur fossil was a tooth. It was found in 1822 in the United Kingdom by Mary Ann Martel. The tooth was huge and her husband, a physician, was determined to find out what kind of animal it came from. Dr. Martel eventually determined the tooth came from a giant lizard and named beast Iguanodon.

Eventually more and more bones were found, and the name was changed to dinosaur. The term "dinosauria" was first coined in 1842 by

British biologist and paleontologist **Sir Richard Owen (1894–1892)** [58]. It derives from Greek *deinos*, meaning "terrible, powerful" and *saura*, meaning "lizard." Dinosaurs dominated the earth for over 160 million years beginning in the Triassic period about 230 MYA and ending with the blinding flash of the asteroid that may have caused the Cretaceous-Tertiary extinction event 65 MYA.

Considerable confusion regarding what is and isn't a true dinosaur exists. For example, many older books include the flying pterosaurs and swimming ichthyosaurs as dinosaurs, but they are now considered to belong to their own classes.

Dinosaurs were an extremely varied group of animals. According to a 2006 study, over 500 dinosaur genera have been identified with certainty so far, and the total number of genera preserved in the fossil record has been estimated at around 1,850. It is therefore estimated that nearly 75 percent remain to be discovered [59].

Recent research beginning in the 1970s has demonstrated that dinosaurs are the most likely ancestors of birds, and most paleontologists regard birds as the only surviving dinosaurs [60].

K-T Extinction Sixty-five million years ago, another of the five major extinctions occurred. Because this ended the reign of the dinosaurs, more interest has been expressed in the K-T extinction[6] than any other.

The extinction effects varied. Almost 35 percent of the extant life forms perished. However, some of the fish survived, perhaps due to the protection of the oceans. Besides the demise of the dinosaurs, the K-T extinction marked the rise of mammals, which slowly came to dominate the earth.

As with other extinctions, the cause of the K-T is still contentious. Similar to the P-T extinction, there are two possible explanations: collision with a large NEO (perhaps an asteroid) or a large volcanic eruption.

The asteroid impact cause was originated by Nobel Laureate **Luis Alvarez (1911–1988)** who made a strong case for it [61]. The Alvarez explanation rests upon two observations. The first is a layer of the metal iridium at the K-T rock layer boundary. This layer is found all over the earth. Iridium is extremely rare on Earth, but it is commonly found in meteorites, hence the concept of an asteroid impact.

Alverez's second observation is an impact crater that was identified in 1990 based on the work of Glen Penfield in 1978. The crater was called Chicxulub Crater, and it was buried under the coast of Yucatan, Mexico. It is oval and has an average diameter of about 112 miles, approximately the size calculated by the Alvarez team. Material expelled when the crater was formed was dated to 65 MYA [55].

The alternate explanation to the K-T is a volcanic eruption known as the Deccan Traps, a large layer of flood basalt located in west central India. While not as extensive as the Siberian Traps (a contender for the P-T extinction) the Deccan Traps cover an area of 200,000 square miles and are more than 6,000 feet thick [62]. Also see Dr. Macleod's comments on the effect of the Deccan Traps, which were quoted earlier in this chapter.

As with the P-T extinction, experts can debate the causes, but for our purposes the events occurred and altered the course of Earth's creation, especially Earth's biological creation.

MAMMALS BECOME THE SUPREME SPECIES AND DOMINATE THE EARTH.

The main event of interest following the K-T extinction, which eliminated many of the top predators, was the rise of the mammals.

Mammals are bilaterally symmetric vertebrates with a head at the front of the body. The head contains eyes, ears, and a mouth, and an excretory anal pore is located at the back of the body, followed by a tail. Mammals differ from reptiles by sweat glands, including some sweat glands that are called mammary glands (hence the name mammal) that were modified for milk production [63].

The circulation system of mammals employs a four-chamber heart consisting of two atria and two ventricles [63]. Mammals are air-breathing, and they have lungs that transfer O_2 to the circulatory system. They also have a digestive system whereby food is ingested through the mouth, digested in a series of organs spread along the body, and excreted as waste at the back. The products of digestion are also passed to the circulatory system.

The mammalian body is supported by a skeletal system. Like reptiles, mammals have four legs attached to the skeletal system for locomotion. With the exception of the primate branch, all mammals walk on four legs. Mammals are warm-blooded and therefore have a more sophisticated thermal control than reptiles. Warm-bloodedness eliminates the lethargy that hampers reptiles when it gets cold.

As with reptiles, mammalian eggs are also fertilized within the protective female environment, and the resultant embryo is enclosed in the all-important protective amniotic membrane. However, mammals possess a significant improvement over reptiles in that they have an extended amniotic enclosure, termed the "placenta," in which an embryo develops until the mammal can give birth to live offspring. This obvious survival advantage is augmented by the mammary glands, which allow females to feed their young. This is an example of how evolution's natural selection filter continued to improve the odds of species survival.

The mammalian brain adds to the reptilian brain a new structure, the limbic system[7], which overlays the reptilian portion. The term "limbic" comes from Latin *limbus*, meaning "cap" [64]. The limbic system operates by influencing the endocrine system and is highly interconnected with the brain's pleasure center, which plays a role in sexual arousal and the "high" derived from certain recreational drugs. The limbic system is also tightly connected to the prefrontal cortex, the forward part of the brain [64]. We will encounter the limbic again when we discuss human behavior. The essential mammalian characteristics are included in appendix IV.

As mentioned above, mammals made their initial appearance in the Triassic, evolving from advanced cryodonts and developing alongside dinosaurs but, of course, avoiding them. However, more modern mammals appeared in a geological time interval termed the Eocene, or Dawn epoch, which was first identified by Lyell. The Eocene began 56 MYA and ended with another small extinction 34 MYA [65].

Eocene Extinction The Eocene extinction was termed the Grande Coupre in 1910 by Swiss paleontologist **Hans Stehlenn**, who observed a dramatic turnover of European mammals [66]. The Post Grande Coupre mammals included pigs, hippos, cattle, goats, sheep, and horses. This turnover gave European humans a significant advantage over humans in other parts of the world, as eloquently demonstrated by Jared Diamond in *Gun, Germs and Steel*, his interesting explanation of why Europeans came to dominate the planet [67].

PRIMATES—THE ULTIMATE MAMMAL.

We are the dominant members of the primate group. Primitive primates may have first appeared before the K-T extinction, but "real" primates, which include apes—the category that includes humans—appeared several million years after the K-T extinction [68].

Two divergences in the primate line have led to humans. The first was the split between the apes and human precursors, which included the chimpanzee. The ape-human split occurred approximately 10 MYA according to a study published in August 23, 2007 issue of *Nature* [69]. The next divergence, the human-chimpanzee split from our closest "relative" occurred between 5 and 7 MYA [70]. In the next paragraphs, we will briefly examine the developments that led to us

A. afarensis: The oldest fossil that is recognized as definitely human is the *Australopithecus* (southern ape) *afarensis*, an extinct hominid that lived between 3.9 and 2.9 MYA [71]. *A. afarensis* is represented by the celebrated fossil "Lucy," discovered in Ethiopia between 1973 and 1974 by paleontologist Donald Johansson. The specimen was apparently christened Lucy because Johansson was listening to the Beatles song "Lucy in the Sky with Diamonds" while he was examining the fossil elements [72]. *A. afarensis* was capable of walking upright, an improvement of significant value because it freed two "legs" to perform tasks. Upright walking also affords agility and improved ability to sense the environment.

Homo habilis: The next line is *Homo habilis* (Latin for "handyman" or "skillful person") is a species of the genus *Homo* that lived from approximately 2.2 to at least 1.6 MYA [73]. But this line is not significant as the next.

Homo erectus: *Homo erectus* (Latin for "upright man") is an extinct *Homo* species that appeared in Africa between 2 and 1.8 MYA. *H. erectus* is believed to have been the first of the genus *Homo* to leave Africa, possibly as a result of a dramatic climate shift. *H. erectus* fossils have been found in Europe (Spain), Indonesia, Vietnam, and China [74]. Recent findings in Northern Spain, reported in the Jan/Feb 2009 issue of *Archeology* [75] of an *H. erectus* jawbone dated to roughly 1.2 MYA. That is the earliest evidence of *H. erectus* outside of Africa. Only a small amount of the cave in which the jawbone was found has been excavated, so more findings are likely.

Homo erectus was probably the first early human species to fit squarely into the category of a hunter-gatherer society. The latest populations of *Homo erectus* were probably the first hominid societies to live in small-scale societies similar to modern hunter-gatherer band societies [76].

A site called Terra Amatam, located on an ancient French Riviera beach, seems to have been occupied by *Homo erectus* and contains the earliest least disputed evidence of controlled fire. It dates at around 300,000 BCE.

Evidence collected in 1984 indicated that despite *H. erectus's* human-like anatomy, they were not capable of producing sounds of a complexity comparable to modern speech. However, erectus may have communicated with pseudo language that lacked human language's fully developed structure, but it was more developed than the basic chimpanzee communication [74].

Homo Neaderthalensis: Since the first skeletal remains were found in a quarry near in the NeanderTal (valley of the Neander in Germany) in 1856, the Neandertal has proven to be an enigmatic member of our genus.

The Neandertal is an extinct *Homo* genus member whose remains have been found in Europe and Central Asia. In the August 8, 2008 issue of *Cell* magazine, an international research group reported a complete Neandertal mitochondrial DNA sequence [77]. Comparison with modern human DNA demonstrates that the two species diverged no later than about 350 thousand years ago (KYA). A more recent study reported in *Science* magazine updates the *Cell* data. This study also confirms that if Neandertal and modern humans had sex, they didn't have any babies [78].

But there is, of course, the question of Neandertals' disappearance. All evidence of Neandertal disappeared 22 KYA [79]. Since Neandertal and modern humans occupied essentially the same areas for approximately 15,000 years, did modern humans displace the Neandertal or did the Neandertal succumb to environmental changes? We'll address this question later.

Homo Sapiens Sapiens: *Homo sapiens* (Latin for "wise human" or "knowing human") walks upright and shares many of the characteristics with the previous members of species, but *H. sapiens sapiens* had an enlarged tissue layer overlying the limbic portion of the brain termed the cerebral cortex that provides improved survival benefits [80]. *H. sapiens sapiens* is associated with modern humans.

The limbic system was tightly integrated with the prefrontal cortex, the forward part of the cortex portion of the brain. The prefrontal cortex has been implicated in planning, complex cognitive behaviors, personality expression, decision making, and moderating correct social behavior. The basic activity of this brain region is considered to be the orchestration of thoughts and actions in accordance with internal goals. Some scientists contend that this connection between the limbic and prefrontal cortex is related to the pleasure obtained from solving problems [64].

The cerebral cortex, also called the cerebrum, plays a key role in memory, attention, perceptual awareness cognition (thought), language, and consciousness. In dead preserved brains, the outermost layer of the cerebrum is gray, hence the name "gray matter." The matter below it is white and is formed predominantly by myelinated (sheathed) nerves interconnecting different regions of the central nervous system (CNS) [81], [82].

DNA evidence indicates modern humans originated in Africa about 200,000 years ago. The superior mental capabilities of humans have allowed *H. sapiens sapiens* to occupy "every corner of globe". *H. sapiens sapiens* is the only surviving species of the genus *Homo*.

DEVELOPMENT OF MODERN HUMAN CULTURES.

Much of our knowledge regarding the development of modern European cultures[8] has come from explorations in France. Based upon tools and lifestyles and cultures, seven relatively distinct cultures have been identified. Most have taken their names from the first site, usually a cave, where artifacts were found. These are usually referred to as a "type site" because it can be used to ascertain which culture was present at a given location [83]. This section provides a brief overview of the development of human culture beginning with *Homo erectus* and ending with the beginning of civilization.

Here are some chronological subdivisions that are useful[9]:

- Paleolithic (Old Stone Age)
- Mesolithic (Middle Stone Age)

- Neolithic (New Stone Age)

These divisions recognize that the principal "tool" material employed by humans during these ages was stone. The later discovery and utilization of metals marks a major boundary in human development.
Another useful subdivision is:

- hunter-gatherer;

- farmer.

A final subdivision is:

- before writing;

- after writing.

Until relatively recently, most humans subsisted as hunter-gatherers. They hunted animals and gathered fruits and other edible plant material as sources of food. Due to the physical differences between males and females, males were usually the hunters and females the gatherers, which also allowed women to remain closer to camp and care for the children. The transition to farming marks another major boundary in human development and "before writing" as prehistoric (before written history) is also an important boundary.

The transition from stone to metal, from hunting and gathering to farming, and from no writing to writing have accompanied major changes in human culture, in particular the advent of writing because it led to civilization, as will be outlined below.

Lower Paleolithic—Acheulean Culture: The Acheulean culture represents the first evidence of deliberately manufactured "tools" and therefore is arguably the oldest human activity that can be considered a culture. The first artifacts identifiable as Acheulean were found at the St. Acheul type site in France[10] [84]. Acheulean artifacts are found in much of West Asian and European sites, usually in association with *Homo erectus* remains, which conclusively demonstrates that *Homo erectus* developed the Acheulean tools and was the first of the *Homo* species to leave Africa and colonize Eurasia [85].

Deposits containing Acheulean material have been dated with potassium-argon techniques and show Acheulean culture was the major culture spanning the time period of approximately 1.65 MYA to approximately 100 KYA [85]; [86]. That is an extremely long time period and coincident with the finding of *H. erectus* remains. But then, what happened to the Acheulean people? Why were they displaced by other cultures? This question appears in regards to all of the cultures that will be discussed later. I will save the discussion for the chapter 17 when a full explanation of the Territorial Imperative will be provided.

Middle Paleolithic, Mousterian Culture: Mousterian culture was identified by archaeologists as a style of predominantly flint tools and was associated primarily with *H. Neandertal*. It was named after the Le Moustier type site, a rock shelter in the Dordogn region of France [87]. Mousterian tools, associated with Neandertal, have been found in Europe that date from between 300 and 22 KYA.

Upper Paleolithic—Châtelperronian Culture: Châtelperronian is named for the La Grotte des Fées (fairies' grotto) in Châtelperron, France. It spanned the period 35–29 KYA [88]. Châtelperronian tools can be traced to earlier Neandertal, and Mousterian industry further indicates that modern humans and Neandertals co-existed in Europe. The Châtelperronian tools were an improvement (e.g., they were toothed-stone tools and a distinctive flint knife with a single cutting edge and a blunt curved back). This culture may also have produced jewelry. But, as will be discussed shortly, a coincident in time culture, the Aurignacian, produced vastly superior tools, most likely by the Crô Magnon people [80].

Upper Paleolithic—Aurignacian: Aurignacian is named for the Aurignac type site in the French areas of Haute Garonne and dates to between 32 and 26 KYA BCE, coincident in time with the Châtelperronian culture. Aurignacian toolmaking, however, was significantly advanced. They worked bone points with grooves cut in the bottom and produced some of the earliest cave art [90]. Their flint tools were more varied than those of earlier cultures, employing finer blades struck from prepared cores rather than from crude flakes. They definitely made jewelry (e.g., pendants, bracelets, and ivory beads). Perhaps most importantly, they made three-dimensional figurines depicting animal representations of the time period. These figurines are associated with extinct mammals, including mammoths and rhinos. Finally, the Aurignacians produced anthropomorphized animals such as the "Lion-Man" discovered in a German cave and dated at 32,000 years old. It is the oldest known anthropomorphic animal figurine in the world and could be inferred as one of the earliest evidence of organized religion[11] [89].

Crô Magnon Man: Since verifiable remains of the first truly modern human, the Crô Magnon[12], appear in Europe between 34 and 36 KYA, it would appear Aurignacian tools were made by the Crô Magnon, while Châtelperronian tools were a Neandertal improvement.

The first Crô Magnons remains were found in the cave of Crô-Magnon in southwest France and other remains are found until about 10 KYA. That means all humans after approximately 34 KYA were Crô Magnon [80]. To avoid confusion with naming conventions, in recent scientific literature the term "early modern humans" is used instead [80]. Some of the advances made by "early modern humans" are discussed below.

Upper Paleolithic—Gravettian Culture: The Gravettian is named for the La Gravette in the French Dordogne region and dates from between 28 and 22 KYA BCE [90]. The Gravettian people were definitely modern humans, not Neandertals, and they extended the Aurignacian tool repertory by adding a small pointed re-struck blade with a blunt but straight back, known as a Noailles burin (from the French word for "cold chisel") [91]. The Noailles burin was probably used for engraving or wood and bone carving .The Noailles burin is the identifying Gravettian artifact. In addition, they produced numerous "Venus" figurines, small statues of women with exaggerated breasts, hips, and other sexually oriented features [90].

Upper Paleolithic—Solutrean Culture: The Solutrean is named after the type site of Solutré in French Burgundy region and home of the famous Roche Solutré, a dominant sedimentary formation [92]. Solutrean culture features a relatively advanced flint toolmaking style and appeared about 19 KYA. The era's finds also include ornamental beads and bone pins as well as prehistoric art.

The Solutrean is probably a transitory stage between the Mousterian flint implements and the bone implements of the Magdalenian culture. Soultrean culture disappeared from the archaeological record around 15,000 BCE [92].

Upper Paleolithic—Magdalenian Culture: The Magdalenian is named after the rock shelter type site of La Madeleine located in the Vézère valley in the Dordogne region of France [93]. Magdalenian sites also contain extensive evidence for the hunting of red deer, horse, and other large mammals toward the end of the last ice age. Magdalenian sites have been found from Portugal to Poland.

The culture spans the period between circa 18,000 and 10,000 BCE toward the end of the last ice age. The later phases of the Magdalenian are also synonymous with the human re-settlement of northwestern Europe. Extensive research in Switzerland, southern Germany (Housley et al. 1997), and Belgium (Charles 1996) has provided detailed Atomic Mass Spectrometer (AMS) radiocarbon dating to support this.

The sea shells and fossils found in Magdalenian sites can be sourced to relatively precise areas of origin and have been used to support the hypothesis of Magdalenian hunter-gatherer seasonal ranges and perhaps trade routes. Cave sites such as the world famous Lascaux site or Altimira in Spain contain the best known examples of Magdalenian cave art [93].

Mesolithic: As its name (Middle Stone Age) implies, the Mesolithic was a transitional period that began at the end of the last ice age, circa 11.5 KYA and blended into Neolithic (New Stone Age) with the introduction of farming in the Neolithic [94]. The date of farming introduction varied considerably from region to region. In some areas, such as the Middle East, farming began shortly after the end of the last ice age because the area was not significantly affected by the ice age. On the other hand, farming came much later in Northern Europe.

Neolithic: The Neolithic "New" Stone Age was a period in the development of human technology that began about 10,000 BCE [95]. The name Neololithic was invented by British banker, politician, biologist, and archaeologist Sir John Lubbock (1834–1913) [96] in 1800. The term is more commonly used in Europe because it can be applied to cultures in the Americas that did not fully develop metalworking technology. But this does raise problems. The term "Neolithic" does not refer to a specific chronological period, but rather to a suite of behavioral and cultural characteristics including the use of crops (both wild and domestic) and the use of domesticated animals [95].

The Neolithic era that began with the rise of farming, ended when metal tools became widespread in the Copper Age (a.k.a. chalcolithic or Bronze Age) or developed directly into the Iron Age, depending on geographical region.

Neolithic culture appeared in the modern-day eastern Mediterranean area around 8500 BCE. By 8500–8000 BCE, farming communities arose in the eastern Mediterranean and spread to Asia Minor, North Africa, and North Mesopotamia. The Neolithic ended with the appearance of Chalcolithic culture [95].

Chalcolithic: The Chalcolithic is a word that combines the Greek word for copper, *khalkos*, and the Greek word for stone, *lithos*, and signifies the transition from stone tools to metal tools. Copper was the easiest and therefore the first metal employed and smelted from natural outcroppings [88]. The sole use of copper was short-lived because outcroppings of tin were readily available and the advantage of the copper-tin alloy bronze was soon discovered. That discovery launched the Bronze Age, which first emerged in the Fertile Crescent around 4000 BCE. The Fertile Crescent derives its name from the shape formed by the two main centers of Middle Eastern habitation, the Nile river and the twin rivers of Mesopotamia, the Tigris and Euphrates [97].

The Fertile Crescent is often termed "the Cradle of Civilization." We shall take up the topic of civilization in the next section.

SUMMARY OF CHAPTER 13.

We have spanned 4.5 billion years of history, beginning with the accumulation of matter from the rotating disc that surrounded the developing sun, totally devoid of life. The formation of planets from the rotating disc explains the observation that all planets lie essentially in the same plane and revolve around the sun in the same direction.

Formation of water: Although there was more than one source for water, most evidence points to water being deposited by asteroids and similar bodies. Pillow lavas provide firm evidence for the existence of water by 3.5 BYA because pillow lava forms only under water. During the formation of the oceans, iron sulphates were dissolved in the water.

Arrival of life: Life in the form of cyanobacteria found in stromatolites can be traced to 3.5 BYA; moreover, evidence of microbial life can be found in pillow lava tubes, also dated to 3.5 BYA.

The formation of O_2: O_2 has more than one possible source, but the most likely is the cyanobacteria/stromatolite combination that persisted for approximately 2 billion years. Initially the O_2 dissolved in the oceans and "rusted" the dissolved iron resulting in banded iron layers, the source of today's iron ore. Then the O_2 accumulated in the atmosphere.

Plate Tectonics: Evidence discovered in the Jack Hills of Western Australia demonstrated the existence of plate tectonic activity more than 4 BYA. Plate tectonic activity led to the formation and breakup of several supercontinents, which had extensive impacts on evolution.

First multi-celled life: Fossils discovered in the Ediacara Hills of South Africa demonstrated the existence of simple multi-celled life spanning a period of 650–540 MYA. The existence of multi-celled life during the Ediacaran period demonstrated that meiosis, the requirement for sexual reproduction, had evolved. This development set the stage for the "Cambrian explosion" between 540 and 500 MYA.

Cambrian Explosion: During the Cambrian, Most of the known phyla formed and began a sequence of steadily improving life forms:

- Primitive fish originated during the Cambrian, but modern fish did not appear until the late Silurian and Devonian periods, 443 to 395 MYA. Fish are constrained to water. Female laid large number of eggs which the male fertilized.

- The transition from oceans to land began about 382 MYA with the evolution of Tiktaalik, a lobe-finned fish that was capable of living on land. Tiktaalik is a transition to the next higher life-form, the amphibian. The amphibian's fossil record is poor, but it is found at the end of the Devonian, 360 MYA. Amphibians began life in the water, and the female laid large number of eggs that the male fertilized.

- The amniotic membrane evolved during this period, which added to an animal species' survival ability because the membrane protected the developing embryo. This capability was first employed by reptiles.

- Reptiles, the first animal capable of existing exclusively on land appeared around 320 MYA. Reptile eggs were fertilized in the female's body and the growing embryo was protected by the amniotic membrane. Upon maturity, the female laid eggs contained in a shell, which provided more protection than loose eggs in water.

- Five extinctions occurred over the period of 450–34 MYA. One of greatest extinctions occurred between the Permian and Triassic periods about 251 MYA in which about 95 percent of all species became extinct.

- After the earth recovered from the P-T extinction, some reptiles survived. This is when the precursor to mammals, the cryodont, appeared. Also, the famous dinosaur arrived.

- The next most severe extinction, the Cretatious-Ternary (K-T), probably caused by a large asteroid, occurred 65 MYA. It ended the age of dinosaur and ushered in the age of mammals. Mammals have the survival advantage of larger brains, warm-bloodedness, and live birth.

- About 34 MYA, a small extinction caused a dramatic change in the European animals with the appearance of pigs, cattle, goats, sheep, and horses. These animals provided improved diets and furs, which enabled Europeans to thrive and ultimately dominate the earth.

This chain of ever-improving life was the result of natural selection's filter and it is one of the more important demonstrations of evolution. Neither creationism or ID describes the development of life in this much detail.

The Arrival of Humans: Humans belong to the Hominidae family of the Primate order. Hominidae's include modern man and the extinct immediate ancestors of man [98]. Primates appeared several million years after the K-T extinction. True humans branched off from the closest primate ancestor, the chimpanzee, about 5 MYA. Human evolution steadily advanced beginning with *A. Afarensis* (3.9 to 2.9 MYA), then *H. Habilis* (2.2 to 1.6 MYA), and then *H. Erectus* (2.8 MYA to approximately 300 KYA).

Finally, *Homo Sapiens* appeared about 200 KYA. What is generally termed "culture" began with the Acheulean culture of the lower stone age approximately 1.6 MYA. Several different cultures developed, prospered, and disappeared from the archeological record. A possible reason for this will be discussed in a later chapter.

The apex of the Stone Age, the Neolithic (New Stone Age) began about 10 KYA with beginning of farming. It was gradually replaced by first the Bronze Age and then the Iron Age as humans developed metalworking technologies. These transitions occurred at different times. Civilization appears about 3,500 to 4,000 years BCE and is the subject of the next chapter.

Humans are thus the ultimate development of evolution's mandate to maximize the probability of species survival. Playwright Robert Ardry eloquently summarizes the arrival of humans in his seminal book *African Genesis* [99]:

"Not in innocence, and not in Asia, was mankind born. The home of our fathers was that African highland reaching north from the Cape to the Lakes of the Nile. Here we came about-slowly, ever so slowly-on a sky-swept savannah glowing with menace.

In neither bankruptcy nor bastardy did we face our long beginnings. Man's line is legitimate. Our ancestry is firmly rooted in the animal world, and to its subtle, antique ways our hearts are yet pledged. Children of all animal kind, we inherited many a social nicety as well as the predator's way. But most significant of all our gifts, as things turned out, was the legacy bequeathed us by those killer apes, our immediate forebears. Even in the first long days of our beginnings we held in our hand the weapon, an instrument somewhat older than ourselves.

Man is a fraction of the animal world. Our history is an afterthought, no more, tacked to an infinite calendar. We are not so unique as we should like to believe. And if man in a time of need seeks deeper knowledge concerning himself, then must explore those animal horizons from which we have made our quick little march. (p. 2)"

Ardry produced a second, companion volume to *African Genesis* five years later in 1966 titled, the *Territorial Imperative* [100], in which he developed the concept of the species most likely to survive, which can be summarized as: That species which instinctively forms groups which are internally cohesive and externally pugnacious.

I believe this is an accurate description humans as well as other mammals. However, the superior human brain has enabled humans to perfect this organization more successfully than other animals, hence human dominance of the planet. As mentioned above, we will investigate the *Territorial Imperative* in greater detail in chapter 17. Unfortunately, as will be explained, evolution has succeeded too well and we humans are unwittingly running the risk of destroying the blue paradise evolution has prepared for us.

Looking Ahead.

In this chapter, we have traced the creation of the earth and life upon the earth over a vast period of time through many trials and tribulations, ending with evolution's finest selection—at least to date—modern humans. In the next chapter, we develop the answer to the question, "Why are there three mutually exclusive origin explanations?

To do this, we will examine the evolution of humans in more detail, with emphasis on the development of the human mind. While it may not seem possible to follow the development of the mind, there are adequate pieces of evidence available to help us do this. One piece of evidence is the control of fire which required significant mental ability. We will use this and other pieces of evidence to arrive at a reasonable estimate of the mind's development. Of particular interest will be the reaction of the mind when it first becomes aware of itself and its surroundings. This reaction will provide the answer to the question, "Why are there three mutually exclusive origin explanations?"

This will, of course satisfy objective 1 by answering the question, "Why are there three mutually exclusive origin explanations?"

CHAPTER 14:
WHY THREE EXPLANATIONS OF CREATION?

We come now to what is probably one of the most interesting chapters—the examination and explanation of *why* there are three mutually exclusive origin explanations. As mentioned in the previous "Looking Ahead," this will satisfy objective 1.

In addition, this chapter will finalize the answer to which of the three origins explanations is true by conclusively eliminating both creationism and ID as contenders. This will, of course satisfy option 2.

The remaining option, option 3, establishing unequivocally that reason for the existence of evolution is the need for species survival, will be provided in chapter 17.

The previous chapter traced the basic evolution of humans, but there was not enough information presented to resolve the "why three explanations?" question. The answer to that question begins with the determination of the approximate time when humans first became aware of themselves and their surroundings (i.e., able to ask one or more of the six little friends questions related to gaining an explanation). Since it is the mind that creates *awareness,* our task is to determine when the mind had evolved sufficiently to become aware of itself and its surroundings, a task made a bit more difficult because the details of the emergence of the mind are obscured by "the mists of time." However, since the mind is merely the product of electrical interactions between neurons in the brain[1] [1], we can estimate the development of the mind by examining the development of the brain, assuming that mind development essentially paralleled brain development. The development of the brain can be traced through the availability of fossilized skull sizes [2]. This is, of course, somewhat of an over simplification, but it should suffice for our purposes.

Figure 14-1 crudely illustrates the roughly linear increase in brain volume over time for five hominids, from Australopithecus to modern humans. The squares represent average skull size.

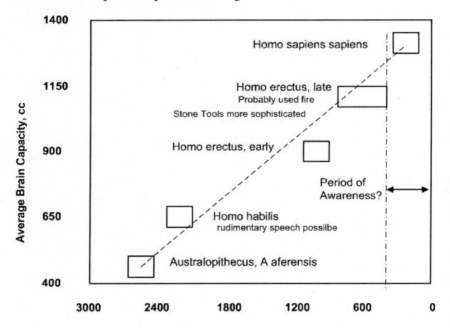

Figure 14-1. Brain capacity increase over time, time increment is KYA

Also shown are key capabilities each species is known or suspected to have achieved, as discussed in the previous chapter.

Referring to the graph, I believe it is reasonable to speculate that hominids, beginning with *Homo erectus*, who probably used fire, had some rudimentary speech capability and definitely possessed efficient stone tools, had also developed sufficient awareness of self and surroundings to ask questions about themselves and their surroundings[2]. This places the beginning of a limited amount awareness approximately 300 KYA.

Human awareness of themselves and their surroundings certainly existed by the arrival of *Homo sapiens sapiens* around 200 KYA, thus the attempt to understand ourselves and our surroundings has its origins sometime between 300 and 200 KYA.

It is difficult to imagine the reaction of early humans to this developing awareness, which presumably developed very slowly over perhaps thousands of years. This was a new development relative to other animals. Of course, survival requires that all animals have some awareness of their surroundings, but there is little evidence that other animals have the human capacity to envision what the effect of an event in their surroundings might have in the future, such as an approaching thunderstorm or the coming of winter.

FIRST EXPLANATIONS OF OURSELVES AND THE WORLD WE INHABIT.

It is common today for most people to desire explanations that are often associated with the "six little friends" questions: what, when, where, who, how and why (e.g., "What happened?" or "When did it happen?" etc.). Consequently, it is difficult to believe that this need for an explanation didn't develop in concert with developing awareness.

Accordingly, an obvious reaction to this growing awareness must have been an almost desperate need for explanations of their surroundings. "What is that strange light in the sky that appears on one side 'of the world' and disappears on the other?" or "What or who makes water fall from the sky?" and perhaps the perhaps the most fundamental question of all, "Where did humans and the world come from?"

Developing correct explanations was, of course, severely hampered by the illusions previously discussed. These illusions initially led to the incorrect belief that humans existed on a world located at the center of a small universe and to the equally incorrect belief that the world was apparently rather young.

It is impossible to know what explanations these early humans developed to answer these questions, but humans who lived more recently have left a strong clue.

This clue was discovered by British anthropologist **Sir Edward Burnett Tylor (1832–1917)**, who made in-depth examinations of early cultures in the mid-1800s. Tylor published the results of his examinations in several publications, but his most influential was *Primitive Culture*, published in 1871 [3].

Tylor's examinations revealed that all of the earliest cultures had strong beliefs that a soul or spirit existed in every object, even if the object was inanimate. Tylor coined the term animism as a name for these beliefs, where animism is derived from the Latin word *anima*, meaning "breath" or "soul" [4].

Tylor has been accorded the honor of the title "Founding Father of Modern Anthropology" by noted anthropologist Weston La Barre [5, p. 34]. Based on Darwin's explanation of evolution, Tylor believed there was a functional basis for the development of society and religion, which he determined was universal.

Considering Tylor's findings and the difficulties that must have been encountered by humans, approximately 200 thousand years in developng an explaination of themselves and their surroundings, their only rational explanation appears to have been the existence of undetectable supernatural entities we now call spirits or gods[3]. They apparently believed these spirits or gods created, inhabited, and controlled everything. What Tylor had discovered was the organization of these beliefs as it had developed over thousands of years into a belief system he termed animism. Tylor considered animism as the first phase of the development of religions [3]. This is a reasonably conjecture as animism and religion have much in common. Both are centered on the belief in a supernatural entity or entities that created the world and exerted considerable control over it.

From the viewpoint of this book, a key aspect of the development of animism is the fact that animism was essentially an explanation of the unexplainable created by proposing the existence of supernatural entities. It therefore follows that, since animism was the first religion, the need to explain the unexplainable is essentially the origin of religion.

While the first organized "religion" was animism, there were other early religious systems that should be mentioned for the sake of completeness. For example, there exists a loose association of individuals who follow the lead of a shaman. "Shamanism is a range of traditional beliefs and practices concerned with communication with the spirit world"[4] [6]. Finally, and more recently, we find individuals known as "master of animals." The master of animals is found in Mycenaen[5] mythology and could represent from the beginning of the Late Helladic period, approximately 1550 BCE [7] [8], a nature god who is related with hunting" [9].

In all cases, animism, shamanism, and master of animals involve a central belief in the existence of supernatural entities that created, inhabited, and controlled everything

I believe it is worth noting that, similar to the difficulty other investigators in the 1800s faced in developing explanations due to insufficient information, Tylor also labored with a paucity of information. While he had discovered animism, he would have difficulty understanding how it developed since figure 8-1 could have not been created much earlier than the mid-twentieth century, because *H. erectus* was not positively dated until the mid-1930s [10]. Hence Tylor could not have known that animism developed because early humans sometime between 300 and 200 KYA struggled to explain the observations about themselves and their surroundings that were being revealed by their growing awareness. As playwright Robert Ardry, who graduated in 1930 with a degree in anthropology points out in his seminal, *African Genesis*:

> "In the past thirty years [beginning in 1930] a revolution has been taking place in the natural sciences. It is a revolution in our understanding of animal behavior, and of our link to the animal world. In sum, therefore, the revolution concerns that most absorbing of human entertainments, man's understanding of man. Yet not even science, as a whole, is aware of the philosophical appraisal which must proceed from its specialists doings" [ch. 13, 100, p. 12].

In a more recent treatment of animism, Sara Wenner confirms that another term for "spirit" is "god." As she points out, "Animistic gods often are immortalized by mythology explaining the creation of fire, wind, water, man, animals, and other natural earthly things"[11]. This description could equally apply to the book of Genesis. So once again, it is clear that supernatural gods or god was the invention of the human mind that was desperate for explanations of the unexplainable. Voltaire is supposed to have remarked, "If God didn't exist, man would have invented him."

Regarding creationism, if we compare the dictionary definition of creationism—"Belief in the literal interpretation of the account of the creation of the universe and of all living things related in the Bible" [12]—with the descriptions of animism, it is clear that creationism is essentially a modern version of animism. As Tylor points out, this is the first phase of religion [3].

SUMMARY OF EARLY BELIEF SYSTEMS.

Before we progress farther into this chapter, I would like to take a moment to review and summarize the early belief systems

Awareness of ourselves and our surroundings can be traced at least to circa 200 KYA. As humans developed awareness of themselves and their surroundings, there was a great need for explanations. Having limited knowledge and coping with the illusions discussed in previous chapters, the only rational explanation was the existence of supernatural entities called spirits or gods that created, inhabited, and controlled everything.

Eventually anthropologists arrived on the scene and were able to identify three early belief systems in the historical or relatively recent prehistoric record: animism, shamanism and master of animals (though masters of animals was not strictly a belief system). While these systems differed in detail (e.g., animism was more organized than shamanism) all systems were similar in their beliefs in the existence of supernatural entities.

Thus, one can trace animism and probably shamanism back at least 2000 years, and since animism was organized, it was—as recognized by Sir Edward Burnett Tylor—"the first phase in the development of religion" [5]. Furthermore, since animism was recognized as the first religion, and since animism developed due to a need to explain the unexplainable, it follows that the need to explain the unexplainable is the origin of religion.

Regarding creationism, since creationism is a literal belief in the Bible—which contains beliefs essentially the same as animism—it follows that creationism is a direct descendent of animism.

EMERGENCE OF CIVILIZATION, ANIMISM EVOLVES INTO MORE ORGANIZED RELIGIONS.

As mentioned above, animistic beliefs were central to human life for thousands of years. Most likely, the transition from animism to more organized religions occurred during the development of civilization. While the placement of the emergence of civilization is almost as difficult as the placement of the emergence of awareness, Chester Starr presents one of the better explanations of what constitutes civilization:

> "The term "civilization" may have many meanings. At the present point we are seeking to detect its first appearance and early growth; and for this end certain fundamental characteristics of civilized society distinguish it from the "cultures" of earlier eras. Among these characteristics are the following:
> The presence of firmly organized states which had definite boundaries and systematic political institutions, under political and religious leaders who directed and also maintained society,
> The distinction of social classes,
> The economic specialization of men as farmer, trader, or artisan, each dependent upon his fellows,
> The conscious development of the arts and intellectual attitudes. In the last point are included the rise of monumental architecture and sculpture, the use of writing to keep accounts or to commemorate deeds, and *the elaboration of religious views about the nature of the gods, their relations to men, and the origin of the world*" [13, p. 27, emphasis added].

I emphasized the ending of the last characteristic because it is particularly pertinent to this book.

Development of Organized Religions. As civilization developed, animism developed into more organized religions in many places, with a pantheon of more human-like supernatural entities. Early religious systems developed in Greece, Egypt, the Mesopotamian region (e.g., Zoroastrianism, the religion of Persia [14]), India, (particularly Hinduism [15]), and the eastern Mediterranean (Judaism

273

and Christianity). Early religious systems also developed in the "New World" as was discovered when the first Europeans arrived. Moreover, these New World religions were essentially animism because they imbued spirits or gods with supernatural control over the earth and humans. They also often developed barbaric, at least in the eyes of the Europeans, rituals to propitiate these gods, including human sacrifice.

Animistic explanations were incorporated into all religions. The gods essentially created everything and made everything happen. Clearly, religion provided the first explanation of ourselves and our surroundings, an explanation based upon supernatural creation. Over time, many cultures developed elaborate stories explaining the creation of the universe and humans' relationship with the gods. We have discussed the book of Genesis in some detail earlier, but there were others, such as the Greek religions, that we now denigrate as paganism. However, they contain "reasonable" explanations based upon the knowledge available, for example, the Greek explanation of the seasons.

Greek Polytheistic Explanation of Seasons: The Greeks, as might be expected, developed one of the more elaborate systems of polytheistic based stories explaining all aspects of humans and the world around them. The story developed by the Greeks to explain the seasons is illustrative.

This story is embodied in the Eleusinian Mysteries, initiation ceremonies that were held each year for the cult of gods Demeter and Persephone, that apparently began in the Mycenaean age circa 1600 BCE[6] [16].

The story begins with a legendary "Golden Age," a time of the beginning of humanity [17]. The Golden Age was ruled by twelve[7] Titans, a race of powerful gods and goddesses who were divine children of Gaia (the earth) and Uranus (the sky) [18]. Two significant children of Gaia and Uranus were Cronos, the youngest Titan who ruled the Titans [19] and Rhea. Cronos married Rhea—apparently incest was not a problem for the Titans—and sired the twelve principal gods (twelve again). Therefore Rhea was known as "The Mother of the Gods" [20].

Chief among the twelve gods were Zeus, Hades, and Poseidon. They overthrew the Titans and "carved up" the universe among themselves. Zeus ruled the sky and became king of the gods[8], while Hades took control of the underworld, also called Hades [21]—an interesting confusion in terms—and Poseidon ruled over the sea. Because of his association with the underworld, Hades is often interpreted as a grim figure.

Cronos and Rhea also had a daughter, Demeter, whose name probably means "distribution mother." She was the goddess of grain and fertility, the pure nourisher of the green earth, and the health-giving cycle of life and death [22]. Demeter married her brother Zeus and had a daughter, Persephone. Persephone was also a fertility goddess [23]. One day, while gathering flowers with friends, Persephone was seized by her uncle Hades, who had gotten the consent of her father Zeus, and was taken to Hades's underworld kingdom [16].

Demeter, distraught by the loss of her daughter, searched for her in vain. While searching for Persephone, Demeter traveled many miles. Along the way, she encountered Tripolemus, a Greek warrior [25], and taught him agriculture, which explains origin of agriculture. When Demeter finally learned her husband, Zeus, had consented to Persephone's abduction, she created a terrible drought to force Zeus to permit Persephone to return. Finally, Demeter prevailed over Zeus, and she was reunited with Persephone. Joyful at Persephone's return, Demeter once more attended to the earth and life flourished.

While in Hades, Persephone had apparently been aware that the Fates, controllers of destiny, had decreed that anyone who ate or drank while there could not leave, so she had not eaten while under Hades's control. She was obviously hungry, and just before she left, Hades tricked her into eating four pomegranate seeds. Accordingly, Persephone was doomed to spend four months a year in Hades [24]. Since Persephone had to return to Hades for four months, when she was gone, Demeter ignored the earth, and the earth became barren. The alternating eight good months and four bad months accounted for the observed seasons.

There are numerous other examples of Greek mythological explanations for phenomena unexplainable at the time, such as the god Apollo, whose chariot carried the sun. However, there are no mythological explanations for the structure of matter other than the four principal elements. The reason for this is obvious. Until the development of modern instruments, as discussed above, humans were not aware solid matter was an illusion.

Decline and Conquest of Greece by Alexander the Great: For all its amazing accomplishments, the Greek civilization's position as the dominant civilization in the Mediterranean was relatively short-lived. The Athenians and Spartans quarreled among themselves. By circa 390 BCE, Sparta had defeated Athens. However, continued wars with other Greek city states left both Athens and Sparta decimated.

Seeing an opportunity, Phillip of Macedon began a campaign to conquer his Greek neighbor to the south. By 337 BCE, Phillip had succeeded and dictated terms at the council of Corinth. Phillip's celebration was short-lived, however. He was assassinated suddenly in 336 as he was mounting a campaign against Persia [26]. His nineteen-year-old son Alexander assumed command and continued his father's campaign. Alexander's campaign ranks as one of the greatest military campaigns in history. He completed the conquest of Persia, parts of India, and Egypt by 326 BCE. In ten years, he had conquered the better part of the known world [26].

Unfortunately, years of conflict and the weight of an enormous empire took its toll. Alexander the Great died at Babylon in 323 BCE. He was one of the most amazing humans to have ever lived. It is difficult to overstate his accomplishments. Alexander founded over seventy new cities. The Greek influence remained strong, and Alexander's successors continued the colonization process. The diffusion of Hellenic customs over Asia all the way to India was one of the most dominant effects of Alexander's conquests and undoubtedly influenced religious development.

He initiated the era of the Hellenistic monarchies and created a single market, if not politically at least economically and culturally, that extended from Gibraltar to India. It was a market open to trade, social, and cultural exchange. This vast territory had a common civilization, and Greek was in fact the common language of the time.

However, the divided Alexandrian kingdoms could not withstand the might of the next conqueror—Rome. The beginning of Rome can best be traced to the creation of the Republic circa 500 BCE. However, Rome's control of the Italian peninsula was only completed around 200 BCE [27]. In 146 BCE, Roman legions razed Corinth, and

Greece became a province of Rome [27]. With the defeat of Greece, the Romans absorbed Greek culture and religion. The Romans gods are essentially the Greek gods with different names (e.g., Zeus became Jupiter) [28].

Rome, like Greece and many other dominant civilizations, couldn't maintain the will and resources necessary to sustain their civilization. By the mid-400s CE, Gallic tribes from the north of Europe that had strengthened as Rome weakened sacked Rome, ending Rome's dominance of the Mediterranean. With the collapse of Rome, Christianity gained ascendancy in the western world. Weston la Barre presents an eloquent theory that Christianity is the Ghost Dance[9] of the Roman Empire [6], however, the evidence suggests Christianity had a more fundamental origin.

DEVELOPMENT OF MONOTHEISM—THERE IS ONLY ONE TRUE GOD,

The next significant religious development was the rise of monotheism, the belief in the existence of only one deity or the oneness of God [29]. As clearly and sternly stated in the Hebrew Bible, book of Exodus 20:2: "I am the LORD Thy God,… Thou shalt have no other gods before me." The concept of monotheism has largely been defined in contrast with polytheistic religions (e.g., animism or the Greek/Roman religions) and developed gradually in a number of locations.

Zoroastrianism is considered to be one of the earlier monotheistic beliefs [30]. It is the religion and philosophy based on the teachings ascribed to the prophet Zoroaster. Zoroastrianism explains that, "The supreme being is called Ahura Mazda (a.k.a. Ohrmazd), meaning 'Wise Lord.' Ahura Mazda is all good, and created the world and all good things, including people" [30]. It was once the dominant religion of much of Greater Iran (called Persia in the time of Zoroaster). Today, the number of adherents has dwindled, probably due to suppression by competing religions, to not more than 200,000 Zoroastrians worldwide, with concentrations in India and Iran [30].

There are other monotheistic religions such as Mithraism which has many similarities to Christianity [31]. For example, Mithras was believed by some to have been born on December 25, and Christianity merely subsumed the celebration of Mithras's birth. But Mithraism was completely displaced by Christianity.

For the purpose of this book, the important monotheistic religions are Judaism, Christianity, and Islam, all of which are largely based upon the Hebrew Bible [32]. Christianity developed out of Judaism, but as mentioned above, there are several similarities between Christianity and Mithraism, which some also trace to Zoroastrianism [30]. Judaism recognizes Yaveh as the one true deity, while Christianity recognizes God (which is an English word, *Deus* in French, *Got* in German, etc.) as the one true supreme deity.

While the narrative of the Hebrew Bible is the source of Judaism, there is no apparent consensus regarding the founding of Judaism. Christianity, the principal religion of interest for this book, diverged from Judaism about 30 CE.

It should be noted that Judaism may have received influences from various non-biblical religions present in Egypt and Syria. This can be seen by references in the Torah, Judaism's most sacred book, to Egyptian culture in Genesis in the story of Moses, as well as the mention of Hittite and Hurrian cultures of Syria in the Genesis story of Abraham [33]. Thus, there was considerable mixing of ideas and beliefs in Judaism and therefore also in Christianity. Moreover, there is considerable evidence concerning the connection between Christianity and the Greek mystery religions. For example, German religious scholar **Martin Hengel (1926–)** [34] argues that the religion of Dionysus existed in Palestine for centuries and influenced Judaism [36]. Dionysus was the Greek god of wine, the inspirer of ritual madness which freed people from daily cares, and the patron deity of agriculture and theater [35].

SUMMARIZING THE DEVELOPMENT OF THE RELIGIOUS EXPLANATION OF OURSELVES AND OUR SURROUNDINGS.

I would like to pause at this point and summarize the religious explanation of ourselves and our surroundings, which began with animism. From its earliest beginnings in perhaps 200 KYA, animism was a belief that a soul or spirit existed in every object, even if it was inanimate. As explained above Animistic spirits, or Gods, were often invoked to explain the creation of fire, wind, water, man, animals, and other natural earthly things.Considering the impediments to understanding early humans had to contend with, animistic beliefs in unseen supernatural beings that controlled and/or created the earth and everything on it relatively recently were a reasonable explanation.

As civilization developed in the fourth millennium BCE, animistic concepts—principally the belief in supernatural beings creating and controlling humans and everything around humans—developed into more organized religions, among which are Judaism and Christianity (which developed from Judaism). The concept of a single, all powerful God contributed by monotheistic religions such as Judaism and Christianity significantly modified animism.

The principal text of Judaism, the Hebrew Bible, was augmented by Christianity with the addition of the New Testament. However, the religious issues of interest to this book are the explanations of ourselves and our surroundings recorded in Genesis, which as explained above, can be traced directly to animistic explanations that are thousands of years old. Because Genesis is the basis of creationism, creationist explanations can also be traced to animism, thus we can conclude creationism is a direct descendent of animism.

REALIZATION THAT RELIGIOUS EXPLANATIONS ARE INCORRECT—THE DEVELOPMENT OF SCIENCE.

As chapters 1–12 discussed in some detail, the realization that early religiously oriented explanations were founded upon illusions and are therefore inherently incorrect began, in many cases, with the Greeks around 400–500 BCE. The Greeks were among the first to make observations that conflicted with the explanations like the geocentric Earth, and they began to develop correct explanations.

The resolution of the two illusions that have contributed most to the problem of developing correct explanations—the sun and stars around the earth illusion and the apparently unchanging earth illusion—marks the development of the explanation system listed third in the Gallup poll, evolution, which by easy extension includes all of science. In addition, the resolutions of these two illusions have been significant contributors to the science-religion conflict.

First Significant Science-Religion Conflict—the Heliocentric Solar System: Although short-lived, a few farsighted Greeks realized that the apparent sun and stars motion around the earth was an illusion and that the earth was not the center of the universe and just a planet revolving around the sun. The opposition of authorities, at first Aristotle and then later the Christian Church, delayed the announcement of proper explanation of these illusions until the publication of *On the Revolutions of the Celestial Spheres* by Copernicus in 1543. The Copernican explanation was of course met with vigorous and sometimes violent objections because it displaced man from his cherished position as the center of the universe. It began to undermine the idea of a supernatural creator (e.g., La Place's famous retort to Napoleon III, "I have no need of that hypothesis"). However, today the heliocentric solar system is well accepted.

Second, More Significant Science-Religion Conflict— Evolution: As will be discussed in the next chapter, the resolution of the unchanging biological and physical aspects of earth illusions, which revealed the extreme age of the earth and the process of evolution, was met with much more violent reactions from organized religions.

The discoveries beginning in the early 1800s by individuals soon to be known as geologists showed that the apparently unchanging physical features of earth was an illusion that had been produced by the extremely slow pace of geological phenomena. It led to a proper explanation of the creation of earth in which a supernatural creator played no role.

This, of course, was not well received by some. However, discoveries beginning essentially at the same time by individuals soon to be known as biologists proved the apparently unchanging biological aspects of earth was an illusion that had also been caused by the extremely slow pace of biological phenomena produced something much worse—the discovery of the process of evolution. This discovery was made public by the publication of Charles Darwin's magnum opus, *On the Origin of Species*, which provided the first reasonable explanation of the creation of life on Earth, also in which a supernatural creator played no role. *Origin of Species* collided with the fourth impediment, "Things are not what we would like," and ignited a firestorm of criticism.

While the displacement of man from the center of the universe by the heliocentric concept promulgated by Copernicus was tough to accept, the evolutionary concept was blasphemy. It held that man had descended from "lower forms of life" via natural selection and undermined humanity's most cherished belief in the existence of a supernatural God, a God who not only created the earth and ourselves but also watched over us, took care of us, planned our lives, and had a place for some of us—the "true believers"—in heaven when we died. Thus the battle between science and religion was joined and still very much alive today.

THE DEVELOPMENT OF THE "SECOND" EXPLANATION OF ORIGINS—INTELLIGENT DESIGN (ID).

While ID is listed second in the Gallup poll questionnaire, as implied by the discussion of the split off of science from religion above, ID was actually the third origins explanation developed.

In an article in the March 19, 2005 issue of the *Washington Post*, "The Origin of Intelligent Design," Frances Stead Sellers states, "It's an idea with a history as illustrious as its mere mention today can be contentious: that the order and complexity of the natural world are evidence of supernatural design" [37].

The concept that "the order and complexity of the natural world are evidence of supernatural design" has been promoted by William Dembski [38] and Micheal Behe [39], two leading proponents of ID who have played a significant role in formulating it.

Dr. William Albert "Bill" Dembski had an intelligent design epiphany at the conclusion of a 1988 Ohio State University conference on randomness. Dembski reacted to a concluding remark that "we don't know what randomness is" by concluding "that randomness is a derivative notion, which can only be understood in terms of design." Dembski cites his realization that randomness can only be understood in terms of design, as a catalyst for his subsequent work on design. [38]

Dembski formalized his ideas in *The Design Inference*. In the book, Dembski "attempts to establish a mechanism through which one could infer scientific evidence of (ID) in nature" [40]. To achieve this objective, he advances a concept he terms "specified complexity," which "is intended to formalize a property that singles out patterns that are both specified and complex." [41]. Dembski states that "specified complexity is a reliable marker of design by an intelligent agent" [38]. The specified complexity concept is generally viewed as mathematically unsound and has not been applied to any areas that might conceivably benefit from it such as complexity theory.

Dr. Micheal Behe developed a concept complimentary to specified complexity termed "irreducible complexity," which argues that "that certain biological systems are too complex to have evolved from simpler, or 'less complete' predecessors" [42]. Dr. Behe presented examples of supposedly irreducibly complex systems at the Dover, Pennsylvania trial regarding the teaching of evolution in public schools (see next chapter for a more detailed discussion of this trial). The trial judge, John Jones,

> "… in his final ruling relied heavily upon Behe's testimony for the defense in his judgment for the plaintiffs, noting that "Professor Behe has applied the concept of irreducible complexity to only a few select systems: (1) the bacterial flagellum; (2) the blood clotting cascaded; and (3) the immune system. Contrary to Professor Behe's assertions with respect to these few biochemical systems among the myriad existing in nature, however, Dr. Miller [testifying for the parents who had sued the school board] presented evidence, based upon peer-reviewed studies, that they are not in fact irreducibly complex"[39].

There is much additional information available on the Web regarding ID such as fact that that the phrase "Intelligent Design" first appeared in drafts of a 1989 high school textbook, *Of Pandas and People*, promoting intelligent design as a substitute for "creation science," which had been ruled unscientific by the U.S. Supreme Court in the 1987 *Edwards v. Aguillard* court case. The goal of the textbook was "to refer to the idea that there is scientific evidence that life was created through unspecified processes by an intelligent but unidentified designer" [38].

The Intelligent Design Network is one of the principal support organizations for ID [43]. Their Web site provides their basic philosophy, which reflects the contributions of Dembski and Behe:

> **"The theory of intelligent design (ID)** [bold text in original] holds that certain features of the universe and of living things are best explained by an intelligent cause rather than an undirected process such as natural selection. **ID** is thus a scientific disagreement with the core claim of evolutionary theory that the apparent design of living systems is an illusion.
>
> In a broader sense, Intelligent Design is simply the science of design detection -- how to recognize patterns arranged by an intelligent cause for a purpose. Design detection is used in a number of scientific fields, including anthropology, forensic sciences that seek to explain the cause of events such as a death or fire, cryptanalysis and the search for extraterrestrial intelligence (SETI). An inference that certain biological information may be the product of an intelligent cause can be tested or evaluated in the same manner as scientists daily test for design in other sciences."[43]

It is interesting that followers of ID view an "intelligent but unidentified designer" as superior to "an undirected process such as natural selection." Of course, as discussed in chapter 11, natural selection is not an undirected process; it is a process that is definitely directed toward the maintenance and ultimate improvement of species. It is also interesting the ID folks suggest that anthropology employs design detection. An internet dictionary, yourdictionary.com, defines anthropology as, "the study of humans, esp. of the variety, physical and cultural characteristics, distribution, customs, social relationships, etc. of humanity" [44]. This doesn't appear to have much to do with design detection.

Although a seemingly new idea, the concept of intelligent design has a long history. Of the many arguments for the existence of God, reference [45] lists ten. One that seems to be most often employed is the one championed by intelligent design: "the enormous complexity of life can only be explained by invoking an Intelligent Designer."

One of the most influential early proponents of the complexity of life implies design idea was British Christian apologist and philosopher **William Paley (1743–1805)** [46]. One of Paley's most famous arguments for an intelligent designer was his watchmaker argument, presented in 1802, in which he essentially argued that just as something as complex as a watch obviously required a designer, something as complex as a human also requires a designer, a supernatural designer [47].

A number of papers have been published demonstrating the flaws in Paley's argument, but most of them address aspects of the design process. As an alternative, I believe that a comparison of the watchmaking process with the "human-making" process[10] is a more effective means for displaying the flaws of the watchmaker argument. In both cases, there is much more to making something than design; you have to convert the design into something.

Watch Manufacturing Process: The manufacture of any item like a watch, which consists of a complex assembly of parts, involves four steps:

1. Design phase, including:

   a. Concept design—will the watch be small or large; elegant or plain, etc.?

   b. Detailed design, mostly mechanical, but with some electrical input if the watch is battery powered. Detailed design has two parts:

      i. Design of each individual part. This usually includes a parts drawing and perhaps some fabrication instructions and material requirements.

      ii. One or more assembly drawings which show how the parts are to be assembled, especially the order, and a companion set of assembly instructions.

2. Fabrication phase, where the individual parts are fabricated

3. Assembly phase, where the individual parts are assembled

4. Test phase in which the watch functions are verified and perhaps the timing of the watch is adjusted

Clearly, there is much more to watchmaking than mere design, and these general steps apply to almost any item that is manufactured. Finally, it is obvious that the application of human intelligence is involved in every step of the process. However, the human "manufacturing process" is entirely different.

Human "Manufacturing" Process: As detailed in chapter 11, the human "manufacturing" process involves the following steps:

1. The joining of two unique cells, an egg and a sperm, which creates a new, genetically unique cell.

2. An involved process in which a variety of signaling molecules control the mitotic cell divisions and differentiation of cells via a complex intracellular signaling process and into specialized organs, such as heart and lungs.

In humans, after approximately nine months all of this cellular signaling and cell division/differentiation yields a baby that is ready to enter the world and join the human race. This simplified discussion of the human "manufacturing" process, of course, doesn't even scratch the surface.[10]

In addition to cell division/creation within each cell, the complex process of protein manufacture is central to the entire "human-making" process. As discussed in chapter 11, protein manufacture requires two basic steps: (1) the selection of amino acids under the direction of genes, and (2) the assembly of amino acids into proteins. Contrary to the two-step fabrication and assembly watchmaking process in which separate drawings/instructions are required to make parts and then assemble them, in biological systems such as a human, DNA provides both the "fabrication drawing," (the gene) and the "assembly drawing" (the sequence of genes on the DNA).

Thus, it is quite obvious there is zero correlation between the making of a watch and the making of a baby. Therefore, the watch analogy and its other "biological complexity implies an intelligent designer," woodpile cousins are totally irrelevant.

But There is a Design Process at Work in Evolution.

In all of the smoke and heat but little light of the ID debates, one simple fact seems to get lost—there is a design process involved in evolution. It is one of the most common "design processes" and has been used for centuries: simple trial and error.

Almost everyone uses it, perhaps not really thinking that it is trial and error. Take for example the purchase of a pair of shoes. There are numerous cartoons showing a woman in a shoe store surrounded by shoes (no offense ladies). When we enter the store, we generally have some idea regarding the size and style we desire, but still the clerk brings out several types for us to try. After a few trials, we usually find a pair that fits and is the correct style. All the other pairs strewn about the floor are the "errors" in the trial and error process of shoe selection.

The process of evolution is no different. The variations produced by reproductive genetic mixing are the "trials" (forgive me for designating children as trials). Most children succeed to some degree. Occasionally a child is produced that has exemplary capabilities, and this child's survival probability is higher than contemporaries. So this child then has a greater chance of reproducing and passing along one-half of his or her genes, as well as the female or male with whom he or she has produced a child.

It is significant that no higher intelligence is required for this design process, which can indeed lead to very complex life-forms with no external assistance.

There are other arguments against intelligent design. But despite the defeat of ID in the court and the demonstration that the "manufacture" of human beings is completely different than the complexity requires, ID is advanced by such organizations as The Intelligent Design Network, Inc. There are a number of well-written books that do an excellent job of discrediting ID, among which are Kenneth R. Miller's, *Only a Theory*, and a book I found in France[11] (the isssued is not confined to the United States) [49].

Miller is quite outspoken regarding ID. He says, "those intellectual Vandals seeking to replace it [rational science] with their **delusional notion of pseudoscientific mendacious intellectual pornography known as Intelligent Design**" [48, emphasis in original].

SUMMARY OF CHAPTER 14.

I believe that, although the development of the three explanations did not take place in the order of the options posed by the Gallup poll, we have now accomplished the book's first object—answering the first question raised by the Gallup poll: "Why are there three mutually exclusive origins explanations?"

In addition, we have accumulated enough information to accomplish the book's second objective—answering the second question raised by the Gallup poll: "Since these explanations are mutually exclusive, which one is the true?"

The **first explanation of our origins, creationism**, developed slowly, beginning perhaps 100 to 200 KYA as humans gradually came to grips with their growing awareness of themselves and the world around them. While they had acquired raw intelligence, they lacked the means to cope with illusions such as the sun and stars around the earth and the apparently unchanging physical and biological natures of the earth. This limitation led first to animism, the belief that gods first created everything and that they then inhabited and controlled their creation—a reasonable explanation under the circumstances.

As humans developed civilization, animism slowly evolved into organized religions, initially with a pantheon of gods, usually with a chief god in charge. Then it evolved into the monotheistic religions, such as Christianity, which still exist today. Soon after writing developed, animistic ideas were written down, which resulted in literature like the book of Genesis, the first book in the Hebrew Bible. Today, these original religious explanations are generally known as creationism, However, as explained in chapter 9 with reference to the Web site, So just what is creationism trying to say?, creationism really doesn't explain anything.

Accordingly, creationism, which was essentially eliminated as a valid explanation of our origins in previous chapters, is further eliminated by the material presented in this chapter, which ends creationism as a viable explanation of our origins

The **second explanation of our origins**, which should be broadened to include all of science, began developing around 400–500 BCE as astute observers, armed with improved observation capability, started to realize that the religious explanations of ourselves and the world around us were incorrect. The sun and stars only appear to go around the earth; the earth actually revolves around the sun, and the stars around earth illusion is caused by the earth's rotation. Similarly, the apparently unchanging earth illusion was caused by the extreme slowness of physical and biological processes. Eventually, the true origin of humans via the process of evolution, which is the inevitable consequence of reproductive mixing, was discovered. Of course, the rise of scientific explanations represented a direct challenge to religious explanations, two of which are particularly significant.

The first significant challenge was Copernicus's heliocentric explanation of the solar system, which destroyed man's cherished view of being at the center of the universe. The second, more significant challenge came first from geologists who accumulated evidence that proved the earth was considerably older than the accepted religious-based explanation that the earth had been created 4000 years BCE. Next, biologists, especially Charles Darwin and associates, discovered the process of evolution, which proved life had evolved over thousands, perhaps millions of years. This destroyed man's most cherished view of being the special creation of a supernatural creator.

The heliocentric challenge initiated the science-religion conflict, but the process of evolution challenge inflamed it because the thought of being associated with the "lower animals" was just unacceptable to many. The conflict remains so today.

The **third origins explanation, intelligent design (ID)**, was developed in an attempt to retain a supernatural creator and a belief in God but avoid the restrictive dismissal of evidence required by creationism, such as a 10,000-year-old earth. ID accepts that evolution has occurred, but under the control of an intelligent, supernatural designer.

Many examples of complexity that implies a (supernatural) designer were put forward to support the ID hypothesis, one of the more famous being the famous Paley watch argument. Unfortunately, the ID concept fails because there is more to "manufacturing" a living being that design. Moreover, the manufacturing process employed in watchmaking is entirely different that the "manufacturing process" employed to make a human being.

Accordingly, ID, which was essentially eliminated as a valid explanation of our origins in chapter 12, is further eliminated by the material presented in this chapter and ends ID as a viable explanation of our origins.

Thus we have fully accomplished objectives 1 and 2 of this book by demonstrating that evolution is the answer to the question "Which explanation is true?" It is interesting to note that we have accomplished this by invoking Sherlock's theorem, which eliminates both creationism and ID as viable explanations of our origins.

As discussed below in "Looking Ahead," objective 3—establishing unequivocally that the reason for the existence of evolution is the need for species survival—will be accomplished in the next three chapters, especially chapter 17.

LOOKING AHEAD.

With the reason for the three explanations established, the final question remains: "Why does a belief in creationism and its offshoot intelligent design persist?" This question will occupy the next two chapters. First, we will discuss the religious reaction to evolution, which plays a large part in the persistence of a belief creationism and ID. This will then naturally lead into the answer to the question of why people still believe.

CHAPTER 15:
RELIGIOUS REACTION TO EVOLUTION IN THE UNITED STATES

CAUSE OF THE REACTION.

The previous chapter explained that the science-religion conflict originated with the development of the scientific explanation of ourselves and our surroundings. It was a significant departure from the long established religious explanation based on the Bible. Three critical events marked the beginning and exacerbation of the science-religion conflict.

The first event was the publication of the heliocentric solar system concept by the Polish monk Copernicus that explained why the sun appeared to go around the earth. The second event was the 1800s demonstration by geologists that the earth's age was millions of years rather than extant religious explanation, which based upon a literal interpretation of the bible (especially Genesis), which held that the earth was created by God only 6,000 years ago. The third and most significant event was discovery of evolution by Darwin and associates and the subsequent publication of Darwin's book, *On the Origin of Species.*

Evolution was by far the greatest challenge because it was totally at odds with the Genesis version of life's creation, which holds that man is a special creation of God. This provoked a strong religious reaction, and the church vehemently condemned evolution as "propaganda of infidelity."

Regarding the religious reaction in the United States, it is important to note that most of scientific discoveries related to biology and geology took place in Europe, especially the United Kingdom, France, and Germany. This resulted in two groups of individuals. The first was a small group of educated persons who either participated in the development of biology and geology or who understood and generally agreed with the findings and conclusions. However, there were doubters in this group, especially with regards to evolution (e.g., Charles Lyell). The second, much larger group included everyone else, especially people in the United States who were essentially unaware of the biological and geological revolutions[1].

The Victorian Web site [1] notes that in Great Britain, Rationalism and skepticism flourished in the latter half of the 1800s among the educated elite. The theory of evolution continued to win new converts, and by the end of the 1800s was accepted dogma at most institutions of higher learning."

As discussed above, there is some evidence that concern about religious reaction caused Darwin to be extra sure of his results before he published his obviously controversial findings. However, in the United Stated, as Douglas O. Linder points out in "Putting Evolution on the Defensive: John Nelson Darby, Dwight L. Moody, William B. Riley and the Rise of Fundamentalism in America" that "a reaction to Evolution and the challenge to religious dogma Evolution represented began to develop" [2].

The religious reaction to evolution provides an illustrative perspective for the persistence of the beliefs in creationism and ID, such as revealed by the Gallup poll. Another example is the recent anti-evolution action by the Texas School Board that I mentioned in the prologue of this book. As implied by Linder's title, the reaction to evolution was led by three men: Darby, Moody, and Riley.

John Nelson Darby (1800–1882) was an evangelist and an influential figure among the original Plymouth Brethren, which was a conservative Evangelical movement founded in Dublin, Ireland, in the late 1820s [3],[4].The Brethren derived their name from biblical references that designate believers as brethren.

Darby is considered to be the originator of dispensationalism [5];[6], a Christian theological biblical interpretation that became popular in the 1800s and is still believed today by many conservative Protestants. Dispensationalism derives from the idea that biblical history is best understood in light of a series of dispensations in the Bible, the dispensation of law, the dispensation of grace, and the dispensation of the kingdom. Dispensationalism advocates a form of premillennialism [7], a belief that Christ will literally reign on the earth for 1,000 years at his second coming. The doctrine of premillennialism views the current age as prior to the second coming and sees the past present and future as a series of dispensations. Dispensationalism places a heavy emphasis on prophesy and eschatology, the branch of theology dealing with last things such as death, immortality, resurrection, judgment, and the end of the world[2] [8]. Dispensationalism depends upon three core tenets:

1. The Bible is to be taken literally. In his book, *Prophecy in the New Millennium*, **John F. Walvoord (1910–2002)**, theologian, writer, teacher, seminary president, and defender of dispensationalism, provides this explanation:

   > History answers the most important question in prophetic interpretation, that is, whether prophecy is to be interpreted literally, by giving five hundred examples of precise literal fulfillments. The commonly held belief that prophecy is not literal and should be interpreted nonliterally has no basis in scriptural revelation. Undoubtedly, a nonliteral viewpoint is one of the major causes of confusion in prophetic interpretation. [9]

2. Dispensationalism teaches that the church consists of only those saved from the Day of Pentecost until the time of the rapture (which is the second coming)[2].

3. Dispensationalism teaches that Israel in the New Testament refers to saved and unsaved Israelites who will receive the

promises made to them in the Abrahamic Covenant, Davidic Covenant, and New Covenant.

Darby eventually separated from the Church of Ireland in 1832 and began to travel widely in Europe and Britain in the 1830s and 1840s, establishing many Brethren assemblies. In 1840, he gave a series of significant lectures in Geneva, which established his reputation as a leading interpreter of biblical prophecy.

The beliefs he disseminated then are still being propagated, in various forms, at such places as Dallas Theological Seminary and Bob Jones University. Darby's writings became the primary source of inspiration for the second theologian to figure prominently in the birth of the fundamentalist movement, Dwight L. Moody [2].

Dwight Lyman Moody (1837–1899) was an American evangelist possessed of great energy and foresight [10]. He founded the Moody Church, a huge structure that remains the largest non-pillared auditorium in the Chicago area. The curved balcony was one of the earliest examples of cantilevered construction, and its curvature—as well as the rest of the layout of the auditorium—was designed so that all lines focus on the pulpit. Designed in an era before modern sound systems, the building has almost perfect acoustics [11].

Moody also founded the Northfield Mount Herman School, which was originally intended to have a section for boys and a section for girls. Moody envisaged both these schools as parts of his dream to provide the best possible education for less privileged people. Indeed, Moody's schools matriculated students, even in their infancy, whose parents were slaves, Native Americans, and from outside the United States, something that was unimaginable in many elite private schools at that time.

In 1886, Moody founded the Moody Bible Institute (MBI) whose campus is located in near Chicago's North Side. The Institute has remained at the same location chosen by Moody 120 years ago. Then in 1894, Moody founded the Bible Institute Colportage (Book Missionary) Association (BICA) to facilitate the MBI's mission of affordable Bible education. In 1941 when BICA was renamed The Moody Press, almost 34 million copies of sermons, doctrinal books, and New Testaments had been published [11]. Moody's influence was enormous.

In the 1870s, Moody began an evangelical crusade on a scale never seen before in American history. He preached his ardent premillennialist message to large crowds in the British Isles for two years before returning to the United States. Moody's sermons drew multitudes; thousands were turned away at the gates and doors as Moody traversed the country. In New York in 1876, about 60,000 people a day filled halls at the Great Roman Hippodrome on Madison Avenue for the three to five rallies a day, held from February 7 to April 19.

Moody's last crusade started in Kansas City in November 1899. He fell seriously ill after delivering a sermon on "Excuses" and died a few weeks later [2].

As Moody's crusading career neared its end, the even more influential career of **William B. Riley (1861–1947)** was on its ascendancy. Riley was born in Indiana in 1861 and was known as "The Grand Old Man of Fundamentalism" [12]. Riley referred to Moody as his "hero," and adopted much of Moody's message. He conducted revival meetings from 1897 to the 1910s, attracting great crowds. Riley exhorted the crowds to "follow the Bible. God is the one and only author" [12].

Riley explained the writing of Bible by claiming, with no explanation of how it was done, that human writers "played the part of becoming mediums of divine communication"[12]. Stressing biblical inerrancy, the young Baptist preacher insisted that "every book, chapter, sentence, and even word" [12] came straight from God. His simple and forceful message resonated especially with persons on the bottom rungs of the middle class, who filled his rallies.

Riley invented the label "fundamentalist" and became the movement's prime mover. In May 1919, Riley founded the World Christian Fundamentals Association (WCFA) with crowd of 6,000 conservative Christians in Philadelphia. In the early 1920s, Riley identified the growing acceptance of evolution as the infidelity most threatening to Christian values. Riley made the teaching of evolution in the public schools his number one target. Evolution, he declared, was the "propaganda of infidelity, palmed off in the name of science" [12]. In 1923, Riley established an Anti-Evolution League in Minnesota, which soon became the Anti-Evolution League of America[3].

From 1923 to 1924, Riley pursued his crusade against evolution in Tennessee, which he viewed ripe for anti-evolution legislation.

Memphis was indeed full of ardent fundamentalists [13], and it was a Baptist "stronghold." Across the state, 50 percent of the population was Baptist. Riley's efforts easily made evolution one of the major issues of the 1924 state election, and he was instrumental in getting the anti-evolution bill, the Butler Act, introduced in the Tennessee legislature [14].

Although there was much anti-evolution sentiment for the anti-evolution bill, there was also strong opposition from the education and science establishment. But Riley was up to the task, and with his major allies—among whom was the renowned and popular William Jennings Bryan—he roused the faithful to write letters and send telegrams to undecided legislators, which resulted passage of the legislation.

The Butler Act was named for a Tennessee farmer named **John Washington Butler (1875–1952)**, who wrote the bill. The bill made it illegal for any state-funded educational establishment in Tennessee, "to teach any theory that denies the story of the Divine Creation of man as taught in the Bible, and to teach instead that man has descended from a lower order of animals" [14].

Butler specifically intended that the bill would prohibit the teaching of evolution. Ironically, he later admitted he had no knowledge of evolution when he introduced it. He had just read in the papers that boys and girls were coming home from school and informing their parents that the Bible was all nonsense, which greatly offended him [15]. However, as Linder points out, "Without the assistance of Riley, Bryan and persons of similar persuasion, the Fundamentalists would have failed" [2].

Evolution's proponents quickly mounted a challenge. The American Civil Liberties Union financed a test case, prevailing upon Dayton, Tennessee high school teacher John Scopes to intentionally violate the law. Scopes was charged on May 5, 1925, with teaching evolution from a chapter in a textbook that contained ideas consistent with Charles Darwin's book *On the Origin of Species*. The trial brought two of most superb legal minds in America to Dayton: William Jennings Bryan, a formidable orator for the prosecution, and prominent trial attorney Clarence Darrow for the defense. The trial spawned several books and a play and movie, both of which apparently took considerable "artistic license."

Scopes was ultimately convicted, as would be expected from an apparently biased jury, and on appeal, the Tennessee Supreme Court upheld the law under the Tennessee State Constitution because:

> "We are not able to see how the prohibition of teaching the theory that man has descended from a lower order of animals gives preference to any religious establishment or mode of worship. So far as we know, there is no religious establishment or organized body that has in its creed or confession of faith any article denying or affirming such a theory." [16]

SUMMARY OF SCIENCE-RELIGION RELATIONSHIP IN 1927.

It will be illustrative to pause and reflect upon the contrast between science and religion in the United States in 1927, which is starkly revealed by a quick review of scientific and religious activities in 1927:

- Edwin Hubble and associates were discovering and mapping the enormous size and age of the universe.

- Werner Heisenberg and Ernst Schrödinger were developing the correct formulation of quantum mechanics, the explanation of matter, which was known to be comprised of infinitesimally small atoms.

- Phoebus Levene, Fredrick Griffith, William Astbury, and Oswald Avery, among others, were unraveling the mystery of heredity which led to the solution of the mechanism of natural selection.

- John Scopes was found guilty of violating the religiously motivated Butler Law that forbade the teaching of evolution.

There was thus, a yawning gap between science and religion, which Maynard Shipley sums up succinctly in *The War on Modern Science*:

> "The forces of obscurantism in the United States are in open revolt!
> More than twenty-five millions of men and women, with ballot in hand, have declared war on modern

science. Ostensibly a "war on the teaching of evolution in our tax supported schools," the real issue is much broader and deeper, much more comprehensive in its scope.

The deplorable fact must be recognized that in the United States to-day there exist, side by side, two opposing cultures, one or the other of which must eventually dominate our public institutions, political, legal, educational, and social. On the one side we see arrayed the forces of progress and enlightenment, on the other the forces of reaction, the apostles of traditionalism. There can be no compromise between these diametrically opposed armies. If the self-styled Fundamentalists can gain control over our state and national governments, which is one of their avowed objectives-much of the best that has been gained in American culture will be suppressed or banned, and we shall be headed backwards toward the pall of a new Dark Age.

Centering their attacks for the moment on evolution, the keystone in the arch of our modern educational edifice, the armies of ignorance are being organized, literally by the millions, for a combined political assault upon modern science." [17]

Shipley's prose is perhaps a bit too strident. Many of his colleagues urged a more conciliatory approach to anti-evolutionists, but that was not Shipley's style.

As an historical footnote, The Tennessee Supreme Court ultimately overturned Scopes's conviction on a technicality; however, the law remained on the books until 1967 when a dismissed teacher challenged the law, stating that it violated his First Amendment free speech rights. Fearing another courtroom fiasco, the Tennessee legislature repealed the law [16].

PUBLIC SCHOOLS AND THE TEACHING OF EVOLUTION.

Until the 1840s, the U.S. education system was highly localized and available only to wealthy people. Reformers who wanted all children to gain the benefits of education opposed this inherent discrimination. Prominent among them were Horace Mann in Massachusetts and Henry Barnard in Connecticut. Mann started the publication of the *Common School Journal*, which took the educational issues to the public. The common-school reformers believed common schooling could create good citizens, unite society, and prevent crime and poverty. As a result of their efforts, free public education at the elementary level was available for all American children by the end of the nineteenth century [18].

With regard to the teaching of evolution, the expansion of public schools had an unintended effect. Until that time, children had only been exposed to the teaching of church and home, where the often unstated common belief was creationism and where evolution was never mentioned, probably because neither parents nor church knew much, if anything, about it. Suddenly these children were exposed to an idea that many felt was anathema to what they had been told and many rebelled, stoking to the fires of anti-evolutionism.

Further Thoughts on Teaching Teaching Creationism and ID in Public Schools. Along with creationists, proponents of intelligent design have been moderately successful in getting elected public school boards sympathetic to their ideas, which are then inserted into the curriculum. One of the more famous confrontations between parents who believed in intelligent design and parents who believe in science occurred in Dover, Pennsylvania.

> In 2004, the Dover Area School District added intelligent design to its science curriculum and mandated that teachers read a statement referring students to a creationist/ID textbook. Eleven parents then filed suit in federal district court against the school district on the grounds that the Dover policy violated the constitutional principle of separation of church and state. [19]

Many supporters of intelligent design, such as Michael Behe, testified at the trial. They viewed the trial as an opportunity to gain some validation of their position. However, the presiding judge, Judge John E. Jones III, saw through the flimsy intelligent design arguments and on December 20, 2005, wrote a strongly worded decision stating that, "ID is not science and cannot be adjudged a valid, accepted scientific theory as it has failed to publish in peer-reviewed journals, engage in research and testing, and gain acceptance in the scientific community" [19].

Regarding the school board, Jones wrote:

> We find that the secular purposes claimed by the Board amount to a pretext for the Board's real purpose, which was to promote religion in the public school classroom.... The citizens of the Dover area were poorly served by the members of the Board who voted for the ID Policy. It is ironic that several of these individuals, who so staunchly and proudly touted their religious convictions in public, would time and again lie to cover their tracks and disguise the real purpose behind the ID Policy. [19]

After such a stinging defeat and rebuke, one would think that the ID folks would back off a bit, but one of the lead witnesses for the intelligent design position, Steven Fuller, decided to take his case to the public with a book titled *Dissent over Descent Intelligent Design's Challenge to Darwinism*, which was reviewed in *Science* magazine by Steve Russ from the Florida State University Department of Philosophy in Tallahassee, Florida. In his review, Russ makes the following interesting observation:

> Fuller holds that Richard Dawkins[4] "arguably owes more to 18th-century secular theodicy than to Darwin's own 19th-century anti-theodicy." Thus it is not surprising to find him [Fuller] insisting a "literal reading of the Bible has done more to help than hurt science over the centuries. [20]

At the end of his review, Russ comments:

At Dover, the author supported the wrong side. Intelligent design theory is a form of Christianity made up to look like science. The judge correctly ruled that it has no place in science classrooms. Reading *Dissent over Descent* should not change anyone's verdict. As a historian and philosopher of science, I can only hope that the science community does not judge us all by Fuller's example.

But this will not deter the more ardent adherents like Michael Behe, whom I will discuss in more detail in the next chapter. At the end of the day, it is clear that intelligent design was invented by those who could not agree with the dismissal of evidence required by creationism but were unable and/or unwilling to relinquish a belief in God.

IMPORTANT DIFFERENCES BETWEEN SCIENCE AND RELIGION.

If we compare science and religion, we find a few stark differences. For example, religion has many branches caused principally by differing interpretations of the same sacred writings, while science also has many branches. However, the branches of science are not caused by differing interpretations of the same observation of phenomenon, but rather are due to the simple fact that in our modern times, no one individual or even group can know everything, hence the need for specialized fields such as astronomy, biology, geology, etc.

Besides the very different bases for the multiple branches in religion and science, I am unaware of any two branches of science, such as biology and geology for example, that have gone to war over their differences. However, religious wars between competing faiths have been one of humanities all too common failings. Some current examples are the conflicts between Christians and the followers of Islam and the often violent conflicts between Palestinians and Israelis, Sunnis and Shiites, and several in Africa. Relatively recently, a long-term battle between Christian Protestants and Christian Catholics raged in Northern Ireland. In each of these religious conflicts, there was considerable loss of life. Finally, an article in the *Turkish Daily News* on October 20, 2003, carried this headline: "Terror in the Name of God—what all religious terrorist organizations have in common, regardless of their religious affiliation, is the Manichean perspective of life, with the irreconcilable division of the World into good and evil" [22].

Religion generally relies upon one book for information: the Bible in Christianity (although there are variations due to differing interpretations as to what constitutes a proper sacred writing), the Koran in Islam, the writings of Zoroaster in Zoroasterism, etc. On the other hand, science relies upon libraries crammed with books, and unlike religious texts that rarely change except to adapt to changes in language, new science books appear continuously [23].

There are significant differences between scientific and religious explanations of ourselves and the world we inhabit. Religion provides a limited, unverifiable explanation of ourselves and our surroundings that was developed thousands of years ago and relies heavily upon a supernatural being. Moreover, this explanation has not significantly changed since it was originally written down. On the other hand, science provides a detailed, verifiable explanation of ourselves and our surroundings that was developed by some of the most brilliant persons who have ever lived. It is an explanation that is being constantly improved. The obvious reason for the persistence of the three origins explanations is the unwillingness and/or inability of many people to accept these simple facts.

SUMMARY OF CHAPTER 15.

The publication of Nicolas Copernicus's controversial heliocentric concept initiated the divergence between religious explanation and science explanation.

The publication of Charles Darwin's far more controversial *The Origin of Species* that set forth the process of evolution accelerated the divergence and caused a violent reaction against evolution, especially in the United States. Three theologians, Darby, Moody, and Riley, vigorously promoted this reaction, which resulted in a strong anti-evolution campaign, especially against the teaching of evolution in the public schools. The campaign is still very active today, as shown in the Dover, Pennsylvania court case.

Comparing science and religion, we find the following stark differences:

- Religion has many branches caused principally by differing interpretations of the same sacred writings.

- Science also has many branches. However, the branches of science are not caused by differing interpretations of the same observation of phenomenon.

- Different branches of science, such as biology and geology, have not gone to war over their differences, but religious wars between competing faiths have been one of humanity's all too common failings.

- Religion generally relies upon one book for information (e.g., the Bible in Christianity). Science, on the other hand, relies upon libraries crammed with books.

- Religion provides a limited, unverifiable explanation of ourselves and our surroundings that relies heavily upon a supernatural being. In contrast, science provides a detailed, verifiable explanation of ourselves and our surroundings. An obvious reason for the persistence of the three explanations provided by the Gallup poll is the unwillingness and/or inability of many people to accept these simple facts.

LOOKING AHEAD.

With this brief review of the religious reaction to evolution, it is will be easier to understand why, despite all the evidence supporting evolution, people still believe in either creationism or ID. We discuss this topic in the next chapter.

CHAPTER 16:
WHY DO PEOPLE STILL BELIEVE?

From the preceding chapter, it is clear that, within the United States, religious sentiment, inflamed against evolution by folks such as William B. Moody, has had a powerful influence, especially upon those who consider themselves fundamentalists. Thus, in spite of a mountain of evidence to the contrary and as the Gallup poll shows, many people still cling to either creationism or intelligent design. It is actually rather easy to see why, but there are also serious consequences, some of which will be discussed in this chapter, others of which will be revealed in chapter 18.

Although not explicitly mentioned yet, it should be clear from much of the previous discussion that there are two educational systems in our country: (1) the family and church; and (2) the secular public or private schools. These two educational systems often impart conflicting information to children and are one of the root causes for the persistence of the belief in creationism and intelligent design. While not stated as such, this situation is captured succinctly in **Paul Bloom and Deena Skolnick's** seminal study, *Childhood Origins of Adult Resistance to Science*. The abstract of the study states that:

> "Resistance to certain scientific ideas derives in large part from *assumptions and biases* [emphasis added] that can be demonstrated experimentally in young children and that may persist into adulthood. In particular, both adults and children *resist acquiring scientific information*

that *clashes with common-sense intuitions* about the physical and psychological domains. Additionally, when learning information from other people, both adults and *children are sensitive to the trustworthiness of the source of that information*. Resistance to science, then, is particularly exaggerated in societies where *nonscientific ideologies* have the advantages of being both grounded in common sense and *transmitted by trustworthy sources*" [1, emphasis added].

I have italicized the salient aspects of the abstract. A child's parents are his or her first teachers, and as alluded to above, they often impart (incorrect) assumptions and biases that are difficult to confront and linger long into adulthood . As stated earlier, one of the impediments to gaining a proper explanation occurs when there is conflict with common sense and intuition. Moreover, reliance on perceived trustworthy sources leads to further incorrect understanding.

In the main text of their study, Bloom and **Skolnick** point out, "The main source of resistance concerns what children know before their exposure to science." Of course, what a child knows initially is largely gained from parents or church. Accordingly, when a child enters school, he or she encounters a stranger standing at the front of the classroom who often tells them that what they have been taught either at home or perhaps in Sunday school may be wrong. This obviously places a young student in a difficult situation.

I encountered one of the more poignant examples of this problem in an article in the Palm Desert, California newspaper a couple of years ago. It advertised a sermon in a local church titled, "What do you say to your child when they come home from school and say 'our teacher said there was no God?'" It is doubtful that any teacher would be so foolish as to make a direct statement like that, but it is possible something a teacher might say could be interpreted that way. Perhaps the minister was only trying to catch people's attention, but for many children, much of what is taught in public school conflicts with what they were taught at home or in Sunday school. Sunday school teaches children biblical literalism, which has to conflict with the secularism that is taught in public schools.

I worked for NASA for a number of years in Huntsville, Alabama, back in the 1960s. I was honored to be invited to participate in the NASA speaker's bureau, which provided speakers for local schools to discuss the space program. Before any talk to either an elementary or high school, I was warned to avoid anything that might conflict with biblical teaching.

Another source of resistance, Bloom and **Skolnick** point out, is that a child's "intuitive psychology also contributes to their resistance to science. One important bias is that children naturally see the world in terms of design and purpose. For instance, 4-year-olds insist that everything has a purpose, including lions ('to go in the zoo') and clouds ('for raining')."

In their concluding paragraph Bloom and **Skolnick** state:

> These developmental data suggest that resistance to science will arise in children when scientific claims clash with early emerging, intuitive expectations. This resistance will persist through adulthood if the scientific claims are contested within a society, and it will be especially strong if there is a nonscientific alternative that is rooted in common sense and championed by people who are thought of as reliable and trustworthy. This is the current situation in the United States, with regard to the central tenets of neuroscience and evolutionary biology. These concepts clash with intuitive beliefs about the immaterial nature of the soul and the purposeful design of humans and other animals, and (in the United States) these beliefs are particularly likely to be endorsed and transmitted by trusted religious and political authorities. Hence, these fields are among the domains where American's resistance to science is the strongest. See also reference [2].

THE ODYSSEYS OF STEPHEN GODFREY AND ARLAN BLODGETT.

Godfrey and Blodgett are representative of those individuals who are inculcated with anti-evolutionism at home and in the church, but then begin to realize that anti-evolutionism may be wrong. Stephen Godfrey's personal story appeared in a recent *Science* magazine article [3].

Stephen Godfrey was raised by fundamentalist Christian parents in Quebec, Canada. His parents embraced an earth that was 6,000 years old as the good Bishop of Usher had proclaimed and accepted that Noah's flood had laid down every fossil. His father was a Sunday school teacher and every evening after dinner, he would lead the family in a reading of the Bible. Prayer followed afterward. The family attended church twice on Sundays, once in the morning and then again in the evening, and one parent or the other often dropped in on a Bible study class midweek.

As he matured, Godfrey developed a keen interest in biology, often collecting the skeletons of small animals to examine them. Godfrey's first encounter with evolution occurred in the first grade. A student teacher (perhaps not realizing the sensitivity of the issue) mentioned that apes were the ancestors of people. This statement bothered Godfrey greatly because he realized that, based upon what he had learned at home, it wasn't possible. At night he discussed the statement with his parents at the dinner table, it was concluded that the teacher must have been mistaken.

Godfrey entered Bishop's University in Sherbrooke, Quebec, where he majored in biology. He was still convinced scientists were engaged in a vast conspiracy to promote evolution. In general, his studies didn't influence his faith. He realized that "you can learn facts even if you don't believe them" [3].

But eventually his classes started to raise questions not easily answered by biblical literalism. A literal reading of Genesis implies that no animals died until Eve ate the apple, which then leads to the conclusion there were no carnivores preying on other animals. But this was just the opposite of what he was learning in his biology classes, where he examined predatory animals such as lions and tigers that were perfectly built to catch and kill other animals. What had God intended when he made them?

Godfrey's worldview collapsed during his first summer in graduate school. He joined a paleontological expedition in rural Kansas to help dig up pelycosaurs, 300-million-year-old animals. Quarrying through

308

thin layers of rock, Godfrey recounts that "we started to come across footprints of terrestrial animals. You can't imagine a global flood and animals finding ground to make footprints on." Godfrey had come upon the same problem Leonardo had encountered a few hundred years before. "That, more than anything, any other experience in my life, really shook me to the core" [3].

Godfrey often visited Dinosaur Provincial Park and observed layers of sedimentary rocks laid one on top of one another. One layer contained freshwater animals, the next seawater animals. He agonizingly realized that while "these animals all lived there at the same place, they all couldn't possibly have been there at the same time." This simple fact was completely irreconcilable with flood geology. It was then that "the rest of the young-Earth creationist ideas kind of exploded."

The *Science Magazine* article revealed that "Godfrey ran through bitterness, anger, and disappointment about having been deceived for so many years." It was through conversations during holiday meals that Godfrey's parents learned their son had changed. Deeply unhappy, they worried whether he could endorse an old earth and remain a Christian. Their message was, "It's all or nothing." The animosity raised by any mention of evolution caused Godfrey's father to tell Godfrey to stop mentioning the topic because it was too upsetting to the family, whose belief in their afterlife depended upon a belief in creationism.

Godfrey is not alone in having to cope with creationist believing families. Two other men are also mentioned in the *Science* magazine article: "'The day I had to tell my mother I wasn't a Young-Earth Creationist was the scariest day of my life,' says Denis Lamoureux, who teaches science and religion at St. Joseph's College in the University of Alberta in Edmonton, Canada [3].

Brian Alters is director of McGill University's Evolution Education Research Center in Montreal, and he testified for the plaintiffs in the 2005 trial in Dover, Pennsylvania, *Kitzmiller v. Dover Area School District*. Few of his friends or his enemies know that Alters abandoned his Fundamentalist Christian upbringing and rejected creationism in college. More than two decades later, he says, "I still have childhood friends and relatives who won't speak to me."

Godfrey believes that the scientific community needs to get more involved in the issue. While in college, he participated in an evolution-creationism debate and was incensed when no one from the biology department attended.

Arlan Blodgett's story is quite similar. I introduced Arlan in an earlier chapter about the biblical flood (ch 9, ref 18). Arlan Blodgett was raised in a conservative Seventh Day Adventist home that believed in creationism [5]. He became interested in archeology, in particular biblical archeology, and for many years, read the *Biblical Archaeological Review*. In 1997, Arlan decided he would like to spend his sixty-fifth birthday at a biblical archaeological dig, so he wrote to a number of the digs listed in the *Review*, listing what he believed would be useful abilities, such as fifty years of taking photographs.

A few weeks later, Dr. Yosef Garfinkel of Hebrew University offered him an opportunity to join a dig at Sha'ar ha-Golan in the upper Jordan Valley, investigating the Yarmukian culture that flourished between 6400–5800 BCE. Arlan was quite puzzled about the dates 6400–5800 BCE because 5800 BCE was 1800 years after the creation of the earth that Arlan had been taught.

During the three weeks of digging, Arlan kept wondering why the flood had not destroyed the town[1]. Anytime he had the opportunity when he met a new archeologist, Arlan would ask if the archeologist had seen any evidence of the flood. Most replied no, and some declined to discuss it all.

Not satisfied with the few archeologists he was able to talk with at the dig site, Arlan prepared a survey of members of the American School of Oriental Research. Of the hundred surveys he sent out asking if anyone had found any evidence of the flood, thirty-six replied; no one had seen any evidence of the biblical flood.

After much consideration of the results of his attempts to find someone who had evidence of the flood, Arlan was apparently forced by Sherlock' Theorem to realize that the flood probably hadn't happened and that perhaps the biblical account was a myth. As with Stephen Godfrey's experience, this called into question the entire concept of creationism, and Arlan abandoned it in favor of a more general theistic belief that I believe is similar to ID [4]; [5].

Reflecting upon Godfrey's and Blodgett's stories, I am convinced there are probably many who experience similar "awakenings." Based upon my own personal experience, I do not recall when I realized that what was taught in church and Sunday school were at odds with what we can observe. However, a person with a better memory than mine, the noted author and journalist Christopher Hitchens explains his reaction to the inconsistencies of religious belief in his controversial book, *god is not Great*:

> "I must not pretend to remember everything perfectly, or in order, after this epiphany, but in a fairly short time I had also begun to notice other oddities. Why, if god was the creator of all things, were we supposed to "praise" him so incessantly for doing what came to him naturally? This seemed servile, apart from anything else. If Jesus could heal a blind person he happened to meet, then why not heal blindness? What was so wonderful about his casting out devils, so that the devils would enter a herd of pigs instead? That seemed sinister: more like black magic. With all this continual prayer, why no result? Why did I have to keep saying, in public, that I was a miserable sinner? Why was the subject of sex considered so toxic? These faltering and childish objections are, I have since discovered, extremely commonplace, partly because no religion can meet them with any satisfactory answer"[6 p-3].

CURT SEWELL'S RETURN TO YOUNG EARTH CREATIONISM.

In fairness, it must be noted that some individuals revert to Creationism or Young Earth Creationism (YEC) as it is often termed, after they "see the light." One such person is Curt Sewell, who published his odyssey a few years ago [7]. Sewell was a Christian and a scientist who had worked on some complex problems, one being the Manhattan Project. Later, he became chief engineer of Isotopes Inc., a company that is involved in radioactive dating. Sewell had several conversations and discussions about the accuracy of radioactive dating with a PhD scientist associated with his company and eventually became convinced that there were serious flaws in carbon dating.

He eventually published a paper entitled, "Carbon-14 Dating Shows that the Earth is Young." In his paper, Sewell states that, "At least one of these [Carbon 14 dating] errors actually becomes a strong evidence that the Earth cannot be more than about 100,000 years old" [8].

Of course, Carbon 14 dating is only useful to about 30–45 KYA [9]. But Sewell commits the same error most creationists commit regarding the "problem of radio-dating." They find one apparent flaw in a radioactive dating system that contains many other ways to date objects and in a general dating system that has twelve different, independent methods of arriving at a correct date. Sewell's "strong beliefs" lead him to find data to support them. As he states in the end of his article:

> I was open to a YEC attitude, but still had never known there was such a thing as a "scientific creationist" who believed in YEC. When I finally met one, and he gave me one of the early copies of a Bible-Science Newsletter, I found there was actually such a thing as a scientist who really believed in early Genesis. I subscribed, and quickly became an active scientific creationist.[2]
>
> When someone asks me why I'm YEC now, one facet of my answer always includes the closer relationship to God the Creator, *and the comfort that gives.* No, you don't have to believe in YEC to be a Christian, *but it certainly does help one's relationship with God.* When I finally developed a complete Biblical and scientific world-view outlook, everything seemed to make sense, and it all fit together, perfectly. [8, emphasis added]

Poor Curt. He conveniently overlooks or disparages the work of others using the many other radiometric dating methods which demonstrate the earth is indeed 4.5 billion years old. Moreover, as discussed above, there are many other methods for determining the age of the earth. Curt appears to be a victim of matching the data to desired results: *the comfort that a relationship with God provides.*

Not surprisingly, former creationists believe that changing minds may not be worth the heartache it brings [3]. This belief is buttressed by people such as Michael Behe. As mentioned previously, Behe testified for the defense during the Dover, Pennsylvania. As a recent *Science* article explained [10], in an attempt to prove ID, Behe stated, "We can look high or we can look low in books or in journals, but the result is the same. The scientific literature has no answers to the question of the origin of the immune system" [10].

> The defense attorney was prepared and "began piling in front of the witness a large stack of recent journal articles, books, and book chapters, all relating research on the evolutionary origins of immunity." But Behe was intransigent.
>
> Eventually the defense attorney stated, "So these are not good enough?"
>
> "They're wonderful articles.... They simply just don't address the question that I pose," Behe responded.
>
> The presiding judge, Jones, found Behe's responses revealing. Behe "was presented with 58 peer-reviewed publications, nine books, and several immunology textbook chapters about the evolution of the immune system; however, he simply insisted that this was still not sufficient evidence of evolution," the judge wrote in his decision. Jones concluded that ID proponents set "a scientifically unreasonable burden of proof for the theory of evolution" [10].

Behe's response during the Dover, Pennsylvania trial is certainly revealing, and he has lots of company, particularly educated company. The consequences of this intransigent reluctance to accept "a mountain of evidence" will be discussed further in chapter 18 of this book.

SUMMARY FOR CHAPTER 16.

From this short review, it is clear that there are at least three the answers to the question, "Why do people still believe?"

1. Initial upbringing in an environment—family, friends, and church—that firmly believes in creationism and continually imparts this belief to the growing child

2. The rejection by the child of an explanation totally opposite to what they have been taught when it is presented by strangers of unknown trustworthiness. This rejection often includes refusal to examine the alternate explanation because they are afraid they might face eternal damnation or loss of an afterlife.

3. Wrenching decisions when conflicts between what the child, usually much older, observes in the "real" world and what the child has been taught

Some people like Stephen Godfrey or Arlan Blodgett, and I suspect there are more people like them (e.g., Christopher Hitchens) are able to overcome their doubts and the problems of rejection. For others such as Curt Sewell, the appeal of creationism and all it stands for is too great to reject.

Still others, I am certain, are caught in between what they have been taught from their earliest memory and what they see as they mature that does not agree with their earlier teachings. But they are afraid to "come out" for fear of ostracism. These persons I will designate as "closet questioners," those individuals who have been raised in a religiously fundamentalist family in which they were taught that the book of Genesis is literally true, but as they have matured, they find that a literal interpretation of the Bible is inconsistent with what they observe and learn, *but* they are afraid to voice any questions for fear of the kind retribution or ostracism from their family and friends that Godfrey experienced.

Finally, there are those like Michael Behe or Chris Sewell for whom no amount of evidence will change their beliefs.

Looking Ahead.

To this point, we have examined many of the answers provided by religiously oriented Web sites like Answers in Genesis or Christian Answers. In the next chapter, we examine answers to significant questions that are provided by our knowledge of evolution, which you will see are superior to the religiously oriented sites.

CHAPTER 17:
ANSWERS IN EVOLUTION.

The title for this chapter was motivated by the creationism-based Web site, Answers in Genesis, since I submit that "Answers in Evolution" are superior. Besides establishing the validity of answers in evolution, a fundamental objective of this chapter is the demonstration that evolution is the inevitable consequence of the need to ensure species survival. The accomplishment of this objective will involve the answers to or explanations of some of life's most perplexing questions, such as "Why must we die?" or "What is life's purpose?"

Because there are other intellectual disciplines that address these questions and there are apparently those who believe that there are dividing lines between the explanations that can be provided by various intellectual disciplines, it is reasonable to ask, "Are there limitations to evolution's or (more generally) science's explanatory purview regarding these and similar questions?"

Consequently, before we proceed, this issue must be addressed and resolved by examining the explanatory capability of what I submit are three of the more important relevant intellectual disciplines: philosophy, religion, and science (listed alphabetical to avoid any appearance of favoritism). These three disciplines are obviously quite complex, as has been demonstrated regarding science in previous chapters. They also may have different meanings for different people. Therefore for the sake of clarity, I have listed definitions provided by one of the better online dictionaries [1]. Each discipline has multiple sub-definitions, and I have selected those that seem most general and appropriate:

Philosophy is the

- "investigation of the nature, causes, or principles of reality, knowledge, or values based on logical reasoning rather than empirical methods."

- "critical analysis of fundamental assumptions or beliefs."

Religion is the

- "belief in and reverence for a supernatural power or powers regarded as creator and governor of the universe."

Science is the

- "observation, identification, description, experimental investigation, and theoretical explanation of phenomena."

As shown by their definitions, there are obviously some critical differences between these disciplines. First, the word "belief" appears in both definitions of philosophy and religion but not science. On the other hand, the word "explanation" appears only in the science definition. We have discussed explanation in considerable detail in previous chapters, but clarity will be aided by a review of the definition of "belief." Referring again to the dictionary, the word "belief" is defined as:

- "something believed or accepted as true, especially a particular tenet or a body of tenets accepted by a group of persons";

- "a principle, etc., accepted as true, often without proof" [1].

The "woodpile cousin" of belief is the word "faith"; they are often used interchangeably. Faith is defined as

- "firm belief in something for which there is no proof" [2];

- "belief that does not rest on logical proof or material evidence" [3].

In addition to an examination of the definitions of belief and faith, since philosophy eschews empirical methods and science embraces them, a review of empirical methods will also aid clarity: Referring again to the dictionary, "empirical method" is defined as

- "relying or based solely on experiment and observation rather than theory." [1]

The dictionary further clarifies empirical method via this example: "The law is *purely empirical*; it makes no attempt to explain the phenomenon." [1]

As discussed earlier, development of an empirical explanation is only the first, albeit important, step toward gaining a complete explanation of a phenomenon. While empirical methods provide a useful beginning, they usually do not explain *how* or *why* a phenomenon occurs. The *how* or *why* explanation is provided by theoretical methods (e.g., compare Kepler with Newton and Einstein in chapter 2, see also endnote 7 of chapter 1 regarding the use of the word theory).

Besides the presence or absence of belief or explanation, other critical differences are:

- Philosophy investigates causes but does not appear to offer any explanation of them.

- Philosophy employs logical reasoning rather than empirical methods.

- Religion relies almost exclusively upon belief in the supernatural power or powers.

- The science definition is consistent with this book's presentations.

Philosophy also shares a characteristic with psychology. There are many "schools" of psychology (e.g., structuralism, behaviorism, psychoanalysis and humanism [4]) all of which explain human behavior from different viewpoints. Likewise, there are many philosophical systems, each investigating *principles of reality* based on logical reasoning without offering objective proof of their versions. Accordingly, at the risk of offending philosophers, philosophical answers don't satisfactorily explain the questions to be examined in this chapter.

Regarding religious "answers," I believe this book has satisfactorily demonstrated that religious "answers" based upon supernatural causation are really not answers; they are merely beliefs which, as explained above, are "accepted as true, often without proof" and thus are not valid explanations[1].

In view of the above, I submit that the only reliable answers to the questions that face the world today (which this chapter will examine) are found via the use of the scientific method. Moreover, I submit that nothing of value worth knowing is outside the purview of science, in particular the branch of science known as evolution.

CANDIDATES FOR ANSWERS/EXPLANATIONS IN EVOLUTION.

Regarding candidates, there are many, but the following questions are reasonable candidates:

1. Why must we die?

2. What is our life's purpose?

3. Why are we here?

4. Why do we engage in sex?

5. Why is there so much variation in humans, especially behavior?

6. What is the origin of our propensity for violence, especially war?

7. Evolution's fatal flaw, the issue of population growth—can we control it?

We will begin with the first four questions because they are related to the demonstration that evolution is a consequence of the need to ensure species survival. Beginning with death may seem a bit macabre, but as will be seen, the ordering of these first four is intentional. The question regarding death, "Why must we die?" instead of "Why do we die?" was also phrased intentionally.

The Bible contains one of the early explanations regarding the question of death. In his letter to the Romans, Paul writes in Romans 5:12, "Wherefore, as by one man [presumably Adam] sin entered into the world, and death by sin; and so death passed upon all men, for that all have sinned" [5]. Unfortunately, we have here another biblical non-answer. Paul only informs us of a convoluted reason for our death, because of Adam's original sin. Nothing is said about the mechanism of death—the *how* and *why* we die. Of course, we cannot blame Paul. No one in his day knew much about what caused death except wounds, disease (which was not well defined at the time), and old age.

From a modern viewpoint, the previous chapters indicate that the short answer to the "why must we die?" question is that life begins with a single fertilized cell that slowly grows into an adult via continued mitotic cell divisions, and then due to further cell divisions, each of which produces a slightly inferior cell due to mutations and other flaws, the adult slowly degrades until one of more of his or her "vital" organs fail, making life impossible, or his or her body succumbs to an onslaught of bacteria or virus. However, this really just explains what evolution has currently provided. But there is an interesting aspect to the "must" portion of the question: "Will death always be inevitable?" We will address this intriguing question in a later section.

Since, our lives are finite, we must be replaced in order for our species to survive, hence the short answer to our life's purpose is to respond to evolution's central mandate—the need to ensure species survival—by having sex[2] at a time when a child can be conceived. On the other hand, as with the previous question, there is much more to the life's purpose question, which we will also address in a later section.

Finally, the short, and really only answer to the "why are we here?" question is that our parents, responding to evolution's mandate, had sex when our mother was able to conceive. This leads to what may appear to be an odd question, "Why must we have sex?" which we answer in the next section[3].

ESTABLISHMENT OF SEXUAL ACTIVITY TO MEET EVOLUTION'S SPECIES SURVIVAL MAXIMIZING GOAL.

To place the satisfaction of maximizing species survival in perspective, it will help to exhibit the comparison of individual and species survival needs modelled after Table 12-1

Need	Individual Survival	Species Survival
Oxygen	Y	N
Food	Y	N
Access to food	Y	N
Digestion of food	Y	N
Suitable Environment	Y	N
Reproduction	N	Y
Defense of Territory	Y	Y

Table 17-1. Comparison of individual and species survival needs

This table demonstrates that, with the exception of defense of territory, none of the activities required for individual survival are required for species survival; moreover, the greatest species survival need is not necessarily of interest to individuals.

Regarding life in general, I believe it is relatively safe to say that your parents probably lived relatively comfortable lives compared to the lives of Stone Age peoples such as the Aurignacians and Gravettians discussed earlier, for whom life was probably "Nasty, Brutish and Short" to quote from *The Leviathan*, published in 1651 [6] by British philosopher **Thomas Hobbes (1588–1679)** [7].

Consequently, under Stone Age circumstances, engaging in sexual intercourse was an extremely inconvenient interruption in the daily struggle to meet the needs of individual survival such as food, shelter, and possible protection from enemies, especially since individuals had no need for sexual activity for their survival. Accordingly, unless the need for sexual activity was raised to a level above all other needs, at least once and perhaps more often in an individual's life, species survival would have been impossible due to insufficient reproduction. This was of course true of all multi-celled life.

Evolution's main tool, the natural selection's filter solved the species survival need problem very effectively by selecting only those animals for whom, at some time in their life, sexual activity becomes more important than anything else[4].

Natural selection's filter has also selected only animals whose sexual activity is constrained to that time during the year that is optimum for offspring survival. Thus, animals with short gestation periods such as fish and birds and have offspring in the spring, which provides the best chance of survival during the relatively warm summer months. Mammals, on the other hand, whose gestation period is relatively long, mate in the fall so that the offspring will be born in the spring, also providing the best chance of survival during the relatively warm summer months.

Finally, natural selection has selected many animals who return to the place where they were born to mate, which presumably further enhances survival of offspring. We observe this all the time, (e.g., salmon expending all their energy in an attempt to return to the place where they were born to lay eggs and fertilize them and birds flying thousands of mile to their favorite mating ground).

Humans are a bit different and engage in sexual activity many times during the year. But they have shelter and other aids that allow them to safely have children at any time. Clearly, reproduction has many competitors; therefore, satisfying evolution's need was not simple. One of reproduction's significant characteristics is its cyclical nature, which provides a clue to how evolution accomplished its objective; hence, let us briefly examine the basic bodily cycle.

Basic Bodily Cycle: In addition to the cyclical nature of sex, many bodily activities that satisfy the needs listed in the table are cyclical (e.g., breathing O_2) and can be depicted in the (obviously simplified) diagram shown in figure 17-1.

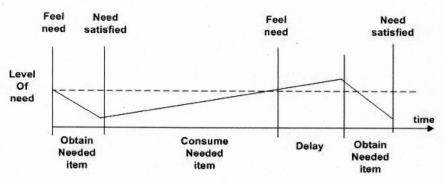

Figure 17-1. Basic bodily cycle

Beginning at the left, a need is felt, which initiates action to obtain that which will satisfy the need. Once the needed item has been obtained, the need is satisfied. Consumption of the item then begins until the item is consumed and the need is felt again. As shown, there is sometimes a delay between the time a need is felt and when it is satisfied (e.g., we feel hungry, but we need to acquire food which requires time). A brief examination of the breathing and eating cycles will provide some examples.

The breathing cycle occurs in response to physical activity. As we perform physical activity, we breathe O_2 into the lungs, which transfer the O_2 to the circulatory system, which then carries the O_2 to the cells where it is employed for a variety of tasks such as metabolism. One of the principal waste products of metabolism is CO_2 [8], which must be eliminated via gaseous exchange in the lungs. The rate of breathing is controlled by the amount of CO_2 in the circulatory system. When the amount of CO_2 reaches the "need more O_2 level," we need to breath. As our physical activity increases, rate of breathing increases.

This simple cycle operates continuously throughout our lives, usually automatically. It is possible to interrupt it by holding our breaths, but we can only do this for a short time.

The eating cycle is essentially the same. It begins with a time when we feel hungry, and we eat. About thirty minutes later, we feel full[5]. As our bodies consume the food, we eventually feel hungry again, and we eat. Under normal conditions, there may be a delay between feeling hungry and eating since time is required to acquire and/or prepare food.

HUMAN FEMALE AND MALE SEX CYCLES.

As mentioned above, most animals conform to an annual sexual cycle, with sexual activity occurring either in the spring or fall. The human sex cycle, at least the female human cycle that controls the probability of reproduction, is approximately twenty-eight days.

The human sex cycle is similar to, but more complex than, the breathing or metabolic cycles because it is actually two intertwined cycles, one for the female and one for the male, and it is governed by complex hormonal interactions[6]. We begin with a brief review of the female cycle, termed the menstrual cycle [9].

The Female Menstrual Cycle: Averaging twenty-five to thirty-four days, the menstrual cycle begins with first day of menstruation and ends at the start of the next period, which averages between three to five days. The menstrual cycle has two distinct phases:

1. **Egg Preparation:** Each month, responding selectively to the female sex hormones, egg cells begin to ripen within the ovary until one egg is ready to exit and enter the adjacent fallopian tube to be either fertilized or expelled from the body. The exit of the egg from the ovary is termed "ovulation" and is the time of the menstrual cycle when a female is usually most receptive to sexual activity with the male.

2. **Fertilization/No Fertilization:** Fertilization usually occurs during the egg's passage through the fallopian tubes. If fertilization occurs, the fertilized egg is placed in the lining of the uterus, which grows and eventually forms the placenta in which the embryo develops. If fertilization does not occur, all the tissues that had been created to support embryonic development are expelled.

Regarding the female cycle, it is important to note that the cycle functions automatically, and the female takes no overt action for it to occur.

The Male Sex Cycle: The male sex cycle does not operate on a relatively fixed cycle like the female. Rather the male cycle varies with age and other variables such as general emotional state, which can shorten or lengthen it. However, it is usually relatively short at the onset of puberty and then lengthens as the male ages. The male cycle does have, in some sense, two "parts." The basic "cycle" conforms to the general cycle depicted in figure 17-1.

Beginning with puberty, the male experiences a growing need for sexual activity. This normally involves copulation with a female. Over time, the need grows until a threshold is crossed in which a male "feels the need." The "need" is fulfilled by achieving an orgasm, which accompanies the release of sperm. If the need is not fulfilled, the male typically becomes increasingly agitated. Once a male satisfies the need with sexual activity, the need "goes away." However, the need accumulates until the male feels it again.

The second phase of the male sexual cycle is achievement of an orgasm. This involves a number of steps: finding a receptive female, appropriately stimulating the female to increase the female's receptiveness, and performing the act of coitus [1]. Once he fulfills these steps, the tension associated with need subsides.

From the viewpoint of species survival, the sperm-releasing male orgasm that occurs during coitus is the ultimate goal of male sexual activity, since this may lead to fertilization and another human to carry on the species. Lasting but a few seconds, the male orgasm is one of the most euphoric sensations a male can experience, which is an obvious reason why males will do almost anything to achieve it and why evolution's natural selection filter has provided for it. While not as strong, females can also achieve orgasm, which assists in creating a bond between male and female, which is necessary to assure survival of the children.

However, in its desire to assure reproduction occurs, evolution has caused the female orgasm to be initiated by a different motion than the male orgasm. The thrusting motion employed by a male is relatively ineffective for the female, as the female requires continuous stimulation of the clitoris, which is usually achieved by rubbing against the male. The result is the male usually achieves orgasm first, which achieves evolution's objective but can be frustrating for the female.

Complications Associated with Human Sexual Activity: As with all other animals, engaging in sexual activity is complicated by the fact that individual humans do not require sexual activity for survival, hence something must be added to the human sex cycle to assure coitus occurs. The obvious addition, as suggested above, is the arrangement for the human sexual need to be greater than any individual survival need a sufficient number of times, preferably early in a human's life to assure that enough humans are born to maintain species survival, and that is what we observe.

The diagram in figure 17-2 is a simplified depiction of the relationship between sex and eating, for example, with sex occasionally becoming more important than eating.

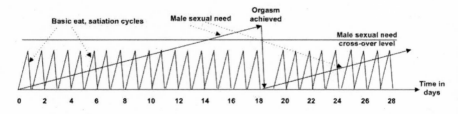

Figure 17-2. Periodically placing sexual need above all other needs to assure reproduction

We see growing male sexual need superimposed on a simplified daily eating-satiation cycle. For obvious reasons, the male sexual need crossover is set relatively high so as not to interfere too much with the need to eat, but eventually the growing male need reaches the crossover, and an eating cycle may be interrupted or delayed so that coitus takes place. This is, of course, a highly oversimplified depiction, but this behavioral pattern will achieve evolution's need; otherwise, species would not survive.

Although not always true, in general the male need for sex is greater than the female need, which from an evolutionary view point is reasonable. The male must seek a receptive female and then perform coitus with the female, all the while being mindful of the need for hunting animals or protecting his group from an another group.

While males generally initiate sexual activity, females also play a role in assuring optimum reproduction. As mentioned above, natural selection's filter has constrained sexual activity in animals to certain times during the year. Animals with short gestation periods such as fish and birds mate in the spring, while mammals—whose gestation period is relatively long—mate in the fall. This constraint of mating to optimal times during the year is largely controlled by female receptiveness.

SUMMARIZING THE PROOF THAT EVOLUTION IS THE INEVITABLE CONSEQUENCE OF THE NEED TO ENSURE SPECIES SURVIVAL.

The preceding sections have shown how evolution's natural selection filter functions. In particular this filter has selected those multi-celled animals that occasionally experience a need for sex that exceeds all other needs. This need occurs often during the individual's lifetime; thereby, guarrenteeing that enough offspring are produced. to assure species survival

If we review the development of life as described in chapter 13, we observe that initially only single-celled plants and animals, those without a nucleus, populated the earth. These animals reproduced by simple binary fission (they just split in half). Gradually, single-celled animals with nuclei developed and they reproduced by mitosis. Then, approximately 650 MYA, during the Edicarran Period, sexual reproduction appeared in the fossil record.

The appearance of sexual reproduction initiated the "Cambrian Explosion" in which all major life types appeared. At the conclusion of the Cambrian, the dominant animal was the fish, which appeared about 500 MYA and dominated the Devonian Period. The fish was followed by the amphibian which was in turn followed by the reptile. Each new animal type had improved survival capability, thereby enhancing the animals survival potential.

This gradual improvement of life, facilitated by evolution's natural selection filter, continued until the arrival approximately 200 KYA of modern man.

This entire process of life's continued improvement, beginning with a simple single-celled animal and culminating in modern man, we have named evolution. It is a process born out of the need to assure species survival; hence, without the process of evolution there would be no life on earth today except possibly simple single-celled animals. Evolution is truly the inevitable consequence of the need to ensure species survival.

But Regarding Our Life's Purpose, Is Sex all there Is?: Before addressing this question, I would like to make an observation and raise a controversial issue. We may observe that purpose and meaning are similar; hence, in addition to life's purpose, we might ask, "What is life's meaning?" However, the existence of both questions implies positive answers to a more basic question "Does life have purpose and/or meaning?" While most would answer, "Why, of course life has purpose and meaning!" this exclamation is usually followed by, "If not, then why are we here?"

We have answered the "why are we here?" question and it has nothing to do with fulfilling some lofty purpose other than to reproduce ourselves once we are here. So there is at least one answer to the "what is life's purpose?" question; however, it's not clear that the "what is life's meaning?" question is so easy to answer. But I will leave that for the philosophers to debate.

Regarding life's purpose, there definitely is more to life than sex. Referring to the table above, in addition to reproducing ourselves, we need to keep ourselves alive by meeting all of the needs listed in the table.

Of course, our current lives are quite different than those of our Stone Age ancestors, therefore it is reasonable to assume that our ancestors fulfilled their "life's purpose" by staying alive long enough to reproduce themselves. This was made particularly difficult by a relatively low life expectancy (e.g., in the Stone Age, life expectancy was approximately twenty years, rising all the way up to about twenty-five in classical Greece and Rome [10]). In the early twentieth century, life expectancy had improved to approximately thirty-five years, but it was not until the middle of the twentieth century when modern sanitation and medical science really became sufficiently effective that life expectancy began to climb to its present approximate seventy to eighty years. Accordingly, concern about life's purpose, other than staying alive, is a relatively recent phenomenon.

One current answer to life's purpose and the "why we are here?" question is provided by the Reverend Rick Warren's extremely popular book, *The Purpose Driven Life* [11], which has sold over 30 million copies. The book lists five purposes for a *purpose-driven life*, which basically involve "serving God" such as:

> "1. We were planned for God's pleasure, so your first purpose is to offer real worship" or "5. We were made for a mission, so your fifth purpose is to live out real evangelism."

While 30 million people apparently found this of value, I find it repugnant that my basic purpose in life is to give an invisible entity pleasure and to go about being an evangelist for this invisible entity, since I have no idea what this invisible entity finds pleasurable except for "real worship," whatever that is.

I hope those who read Warren's book get something out of it, but I would like address the purpose of life from what I believe is a more useful perspective. This perspective will be presented along the lines I sketched above, bearing in mind that evolution deals each of us a different hand, some better than others, some worse. In addition, realizing that unlike the invisible entity of religion, which supposedly has everyone's interest in mind, evolution unfortunately has no specific interest in any particular individual, only that enough survive to continue the species.

Considering the dictionary definition of purpose[7], perhaps when we talk of having a purpose, we may really mean achieving a goal or an end result of value. Then achieving life's purpose will obviously depend upon many things: a person's ability, ambition, interests, etc., all of which vary considerably among individuals. Accordingly, one of the answers to the question, "What is life's purpose?" depends significantly upon human variability.

Before discussing human variation specifically, a few words about the measurement of variation in general are in order. Examination of coin flipping provides a simple way to view variation. Suppose we flip a coin ten times and record the number of heads (H) and the number of tails (T). Then we repeat this experiment several times, recording H and T each time. If we plot the ratios of H to T at the end of the flipping exercise, we will get a curve that looks like this (see figure 17-3):

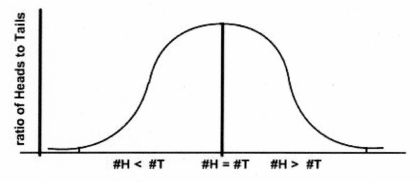

Figure 17-3. Results of flipping a coin ten times multiple times

This curve is called a normal distribution curve and shows that the H-T combination that occurs most often (i.e., the occurrence with the highest probability) is an equal number of heads and tails. Moreover, the ratio of H to T or T to H will occur less often, the wider the difference between the number of heads and tails. A bit of reflection will show that this is obvious. Flipping a coin ten times and getting no heads would intuitively seem to be a rare occurrence. This chart also illustrates what most people know when flipping a coin, a head will normally appear 50 percent of the time.

As another example, if you measure the heights of a large group of men and then plot the number having a given height, you will find that the heights will follow the above distribution curve as shown in this chart (see figure 17-4), which uses data taken from a survey of British men [12].

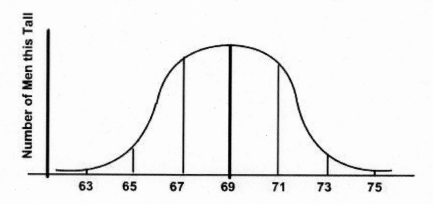

Figure 17-4. Distribution of men's heights in centimeters

It should be clear that the greatest number of men are 69 centimeters tall, or we might say that the average man is 69 centimeters tall. Also, the curve illustrates that the number of men 65 centimeters tall is much less than the number of men 69 centimeters tall. In general, we can surmise that a man picked at random will probably be 69 centimeters tall. The vertical lines associated with the specific heights will be explained below.

Of particular interest in coin toss experiments is each coin toss is independent of the other; hence then a distribution curve is produced by a random phenomena. In the same manner, measurements of men's heights are a random phenomenon and based upon the previous chapters, we might suspect that the variation in men's heights is due to random genetic mixing.

Regarding the vertical lines in the diagram, the normal distribution curve has another interesting property. It is possible to calculate a characteristic of the curve known as the standard deviation, which we will designate as SD as is depicted in figure 17-5.

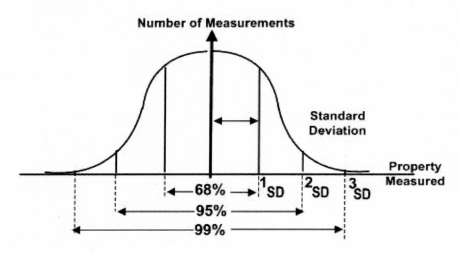

Number of Measurements

Standard Deviation

Property Measured

←—68%—→ ¹SD ²SD ³SD

←———95%———→

←————99%————→

Figure 17-5. Normal distribution showing standard deviation increments

As shown, if measurements result in a normal distribution, then 68 percent of the items measured are included in one standard deviation or 1SD, while 99 percent are included within 3 SD. Thus, if you have a normal distribution and calculate the value of SD, which is easily done on almost any hand calculator, you can arrive at a reasonable estimate of how much variation can be expected in 99 percent of items measured. In the case of men's heights, the numbers 63, 65, 67, 71, 73, and 75 represent the one, two, and three SD values; hence we can, with fair confidence, say that 99 percent of men's heights range from 63 inches to 75 inches and that almost 70 percent vary between 67 inches and 71 inches.

Regarding humans, if you measure almost any physical human parameter in a large enough population, you will find that the measurements, if plotted, will produce a normal distribution curve regardless of whether the parameter measured is height, weight, bra size, strength, blood pressure, etc.

Mental properties also fall into a bell-shaped distribution [13], but that is more controversial and invokes the impediment, "Things are not what we desire."

While this curve is officially termed a normal distribution curve, it is more often referred to as a bell curve due to its shape. The bell curve is one of the most widely used distributions for the analysis of the

330

variation of a random parameter, such as most human properties [14]. For the purposes of this book, we will not be interested in detailed values, just the general picture. Hence, we could show men's heights as the curve in figure 17-6 depicts.

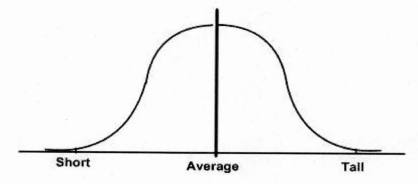

Figure 17-6. General distribution of heights

The measurements would, of course, range from short to tall, with the average height being in the middle.

One aspect of the human variation bell curve is those individuals who lie beyond the three SD value. Obviously, there are not many who do, but there are individuals who might be five or six SD from the mean. Such individuals include Mozart, Shakespeare, or Einstein. They are individuals that "do not fit the mold," so to speak.

Considering individuals whose abilities lie beyond the three SD value, there is the intriguing debate regarding the authorship of Shakespeare's plays. The paucity of data regarding William Shakespeare of Avon, especially any indication of great intellectual prowess, has led to considerable speculation that someone else wrote the plays. Reference [15] contains one of the better summaries of the controversy, but most candidates for "Secret Shakespeare" don't withstand close scrutiny, and as one who has researched the issue, I believe Shakespeare was fortunate to have received, via the reproductive genetic mixer, a prodigious intellect. He was at least six SD from the mean, which enabled him to overcome a modest upbringing and write his famous plays.

WHY DO HUMAN PROPERTIES CONFORM TO A BELL CURVE?

Due to its importance, we will investigate why so many measurements of human properties or phenomena seem to fit the bell curve.

The answer lies in the reproductive process, which as discussed in chapter 12 is a completely random process. The particular genes provided by the male and female are picked at random. The obvious result of this random selection is the simple fact that human genes are distributed at random. Therefore, human characteristics will, in general, fit a normal curve. This also explains the occurrence of a physically or mentally gifted individuals or physically or mentally challenged individuals. The variation which results from reproductive genetic mixing can produce a wide variety of outcomes.

But of course, relative to human behavior, which is obviously an important part of achieving life's purpose—whatever it may be—at present we are just beginning to understand human behavior genetics. Constance Holden points out that very fact in a November 7, 2008, *Science* article:

> ...."As scientists are discovering, nailing down the genes that underlie our unique personalities has proven exceedingly difficult. That genes strongly influence how we act is beyond question. Several decades of twin, family, and adoption studies have demonstrated that roughly half of the variation in most behavioral traits can be chalked up to genetics. But identifying the causal chain in single-gene disorders such as Huntington's disease is child's play compared with the challenges of tracking genes contributing to, say, verbal fluency, outgoingness, or spiritual leanings. In fact, says Wendy Johnson, a psychologist at the University of Edinburgh, U.K., understanding genetic mechanisms for personality traits 'is one of the biggest mysteries facing the behavioral sciences'.
>
> All we really know so far is that behavioral genes are not solo players; it takes many to orchestrate each trait. Complicating matters further, any single gene may

play a role in several seemingly disparate functions. For example, the same gene may influence propensities toward depression, overeating, and impulsive behavior, making it difficult to tease out underlying mechanisms" [16].

One of the reasons for the "several seemingly disparate functions" that genes can produce lies in the simple fact that gene "expression" (the term applied to gene function) can be significantly affected by a phenomenon known as epigenetics, which is also discussed in chapter 12 and [17]. As mentioned in chapter 12, epigenetic means "in addition to genes" (i.e., factors external to a gene that modify gene expression without modifying the gene). Hence, the fact that the same gene may have different effects is to be expected.

Effect of the Brain on Behavior: While we may at present lack a detailed understanding of how genes affect behavior, we know, as Constance Holding points out, that genes must have significant influence. Recognizing that genes play a significant role in brain structure, we can gain considerable insight into human behavior variation by examining the structure of the human brain, which is the main determinant of human behavior and capabilities.

As mentioned in the chapter 12, the brain is actually composed of three sections, each with very different characteristics that are presumably governed by genes:

1. The reticular formation, sometimes termed the "reptilian brain" because we inherited this section from the reptiles

2. Limbic, which is part of our mammalian heritage

3. Neocortex (new layer), which is strictly human

I have assembled a brief review of these three sections to facilitate the remaining portions of the chapter.

The Reptilian Brain, a.k.a. the R-complex: As described in chapter 12, the reptilian portion of the brain is composed of various clumps of cells that determine the brain's general level of alertness and controls basic life functions such as autonomic nervous system, breathing, heart rate, and the fight or flight mechanism. Lacking language, its impulses are instinctual and ritualistic. It's concerned with fundamental needs such as survival, physical maintenance, hoarding, dominance, preening, mating, and perhaps most importantly, acquisition and control of territory. It is located at the base of your skull and emerged from your spinal column [18]. The basic ruling emotions of love, hate, fear, lust, and contentment emanate from this first stage of the brain.

Over millions of years of evolution, layers of more sophisticated reasoning have been added upon this foundation [19]. However, when we are out of control with rage, it is our reptilian brain overriding our rational brain components. If someone says that they reacted with their heart instead of their head, what they really mean is that they conceded to their primitive emotions (the reptilian brain base) as opposed to the calculations of the rational part of the brain [19].

The Limbic: The term "limbic" comes from Latin *limbus*, meaning "cap" [20]. The limbic, the emotional center of the brain, is a set of brain structures including the hippocampus [21], which facilitates short-term memory, the amygdala (which controls arousal), the autonomic responses associated with fear, emotional responses, and hormonal secretions.

The limbic system also operates by influencing the endocrine system and is highly interconnected with the brain's pleasure center, which plays a role in sexual arousal and the "high" derived from certain recreational drugs. The limbic system is also tightly connected to the prefrontal cortex, the forward part of the brain. This brain region has been implicated in planning complex cognitive behaviors, personality expression, decision making, and moderating correct social behavior. The basic activity of this brain region is considered to be orchestration of thoughts and actions in accordance with internal goals. Some scientists contend that the connection between the cortex and limbic is related to the pleasure obtained from solving problems.

The Cerebral Cortex: The cerebral cortex, also termed the cerebrum, plays a key role in memory, attention, perceptual awareness cognition (thought), language, and consciousness. In dead preserved brains, the outermost layer of the cerebrum is gray, hence the name "gray matter." The matter below is white matter and is formed predominantly by myelinated (sheathed) nerves that interconnect different regions of the central nervous system (CNS) [22].

THE THREE BRAINS IN ACTION.

An interesting hypothetical situation suggested by Richard Restak places the action of the three brains into perspective:

"You've accidentally bumped into an old lover at a weekend ski resort. While the Wife and kids are up in the room readying for dinner, you step into the bar for a drink and-there she is. You smile and begin the usual pleasantries: 'How long it's been,' 'Small world,' 'Hope things are well with you,' etc. Your cerebral cortex can ramble on almost indefinitely with pleasant chitchat, all the while taking in countless observations about the person before you. At the same time, you begin to feel the rumbling of the limbic system, which doesn't deal in chitchat but remembers the old drives and the old feelings. It too speaks, but through its connections to the hypothalamus and down to the brainstem. Soon you may be embarrassingly and painfully aware of a racing heart, sweaty palms, a feeling of constriction around the neck, stomach churnings, and perhaps even the beginnings of an erection. Or perhaps the emotion felt is annoyance, the desire to get out of an uncomfortable, no longer 'relevant' situation. In this case, too, there are limbic accompaniments as your face flushes, or your eyes dilate enough to make you uncomfortably aware of the glare of the light flashing in your face from over the bar.

While all this is going on, the R-complex, our old reptilian brain, is active with various forms of body language, spelling out for astute observers the contradictions between your pleasant verbal messages and the inner turmoil you are experiencing. Perhaps you are shaking your head a little too often, shifting position from one foot to another, engaging in expansive gestures that would be more appropriate to a large theater than a small cocktail lounge. But finally, after what seems an eternity, the former lover remembers she must be dressing for a dinner date, wishes you the best and is gone.

The relief you feel is almost immediate as your 'three brains' come back into harmony. Your stomach is settled, the heart is languidly flip-flopping, and your hands are steady even before picking up 'the double' you just ordered" [19, pp. 65, 67].

As Restak points out:

"Such a hypothesis of brain function is, of course, difficult to prove. But the feelings are all there in each one of us, requiring only the appropriate trigger to release wide disharmony between our behavior and how we're feeling at the time. Paul MacLean [noted psychologist who coined the term limbic [23]] and others have demonstrated that sensory input-what we see and hear-can be transmitted not only to our cortex but to the limbic system as well. Therefore, such a cocktail lounge scenario is explicable, since all three brains would be experiencing the same scene, each in a different way"[19, p. 67].

From this brief review, it can be seen how evolution provides a clear explanation of potential variations in humans by dint of three different sections of the brain that have evolved over millions of years. Variation in these different sections of the brain will clearly produce considerable human variation. For example, the variation in cognitive ability as depicted in figure 17-7.

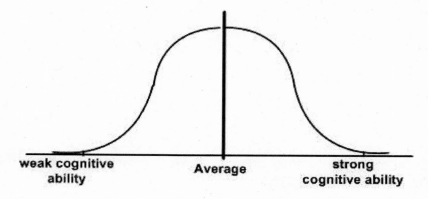

Figure 17-7. General distribution of cognitive ability

WHAT DOES/CAN EVOLUTION SAY ABOUT LIFE'S PURPOSE?

Returning to life's purpose, let's recall the dictionary definition of purpose was "achieving a goal or an end result which has value."

Since reproductive genetic mixing results in considerable human variation, evolution exerts considerable influence over life's purpose. Besides the most important purpose, reproduction, evolution has imbued each of us with a set of capabilities and interests, no two of which are the same. Therefore after satisfying the most important purpose plus the other basic needs related to staying alive, our lives are what we make of them within the variability attendant with our evolutionary heritage. We are not puppets controlled by some unseen supernatural deity; we are free to do as much or as little with our lives as we desire or can.

I would therefore define a "purpose driven life" as one in which a person strives to achieve all that evolution has granted them, maximizing their abilities directed toward achieving a desirable goal and perhaps endeavoring to leave the world a little better for having been there rather than merely essentially being a slave to an invisible supernatural deity who needs our worship.

WHY MUST WE DIE? REVISITED.

As discussed above, the fact that we do die is relatively easily explained by evolution. However, having discussed life's purpose, we can revisit the question, "Why must we die?"

For many, death is one of the more dreaded aspects of life. In view of this dread and the apparent finality of death, early civilizations seemed to have held out some hope for an afterlife, particularly for the more important members of the group (e.g., the elaborate burial practices of the ancient Egyptians, especially the pharaohs).

The allure of Christianity, with the offer of everlasting life in Heaven, has proven irresistible to many even though not a shred of evidence exists that this will actually occur. However, the allure is so strong that families are torn apart over the "afterlife" as revealed in Stephen Godfrey's *Odyssey*, as discussed in chapter 13: The animosity raised by any mention of evolution caused Godfrey's father to tell Godfrey to stop mentioning the topic as it was too upsetting to the family whose belief in their afterlife depends upon a belief in creationism.

One of most perceptive comments on our ignorance of the afterlife was provided by the Persian astronomer and philosopher Omar Khayyám, who was born in Naishápúr, India circa 1170 CE. He summed up the situation with this quatrain from "The Rubáiyát of Omar Khayyám," translated by Edward Fitzgerald:

> "Strange, is it not? That the myriads who
> Before us pass'd the door of Darkness through
> Not one returns to tell us of the Road,
> Which to discover we must travel too" [24].

Based upon our current understanding of our biology, upon our death our brain's cells disintegrate and with them all of our knowledge and our personalities. How this is then recreated in another location such as heaven is as difficult to comprehend as is the location of heaven itself.

As with many aspects of our existence, evolution explains the mechanism of death. Beginning with the simple fact that reproduction is required for multi-celled species survival, we recall that all multi-celled life begins with a single cell, a fertilized egg, which divides relatively rapidly by mitosis until the fertilized egg has reached mature size. But since cells are not immortal, cell division must continue, albeit at a slower pace. On average, approximately 50 billion cells in our body die every day from a process termed "apoptosis"—a word derived from the Greek: *apo* (from) and *ptosis* (falling) —from traumatic cell death [25].

It is important to note that cell death is a completely normal process in living organisms that was first discovered by scientists over 100 years ago. However, it was not until 1965 that John Foxton Ross Kerr at University of Queensland was able to distinguish apoptosis while studying tissues using electron microscopy [25].

There are approximately 50 trillion cells in the average human body [26]; therefore if 50 billion die every day, the life of an average cell is 1000 days or approximately three years. However, some cells have shorter lives. For example, the average life of red blood cells is about 100–120 days, while nerve cells live considerably longer.

When a cell dies, it must be replaced; so over the course of a lifetime, the cells in your body are replaced on average of every three to four years, or approximately twenty-two times for a person who lives ninety years, an age more and more people are reaching. Unfortunately, the cell replication process, which involves mitosis, is imperfect. Mutational defects occur, and the body slowly degrades. Or as we more politely say, it ages. Eventually the defects accumulate until a vital organ ceases to function. The top three leading causes of death in America are heart disease, cancer (especially of the lung), and stroke (death of the brain). Clearly we cannot exist without a heart, lungs, or brain.

From an evolutionary viewpoint, cancer is a particularly pernicious villain because it is essentially a disease of the DNA. As new cells are created, mutations that occur are related to DNA copying errors, as discussed chapter 12. If sufficient errors accumulate, especially in genes known as tumor suppressors, or if cancer-causing genes known as oncogenes become active, the cell ceases to function properly. Instead of dying, the cell begins uncontrolled replication, a process we call cancer. Hence, the very function that makes multi-celled life possible, cell replication via cell division, can ultimately bring death via cancer. Of course, all of the other problems leading to death are also related to cell degradation associated with cell division. Thus, evolution first grants multi-celled life and then it takes it away.

This leads to another question. Is death an inevitable attribute to life, or is it just another fatal disease that is as amenable to intervention as any other disease? Until relatively recently, the answer was the first choice, but now more and more biomedical researchers and other relatively enlightened individuals are beginning to lean toward the latter.

In 2004, computer genius Ray Kurzweill and physician Terry Grossman wrote an intriguing book titled, *Fantastic Voyage, Live Long Enough to Live Forever* [27]. Kurzweill and Grossman propose that biomedical advances are proceeding at such a rapid pace that it is reasonable to believe some of the major causes of death such as heart disease will be significantly reduced within the lifetimes of many persons. This will lead to significant life extension. Of course, there are significant implications in this thesis, such as the effect on population size, but we'll address that later.

The bottom line is that death is not necessarily preordained because of the act of a mythical "First Man" approximately 6,000 years ago. While death is still the fate of all humans today, there may come a day when significant life extension, if not immortality, will become a reality.

Why are We so Prone to Violence? One of the more perplexing aspects of the human condition is our propensity toward violence, including robbery, murder, and most importantly, war. Of course the Bible has a rather simple explanation. It reads, "Whoever sheds the blood of man, by man shall his blood be shed; for God made man in his own image" (Genesis 14:1–10).

On the other hand a more reasonable origin of violence is rather easy to establish. As multi-celled animals evolved, they divided into two groups: predators and prey. The predators killed and ate prey—a rather violent act. But one should not be too harsh on predators. The interaction between predators and prey keep populations in control and forestalls the inevitable overpopulation that would occur.

Predators and prey appear in all animal groups. For example, the barracuda and shark are predators in the group of animals that swim in the ocean. In the amphibian family, frogs are generally carnivorous, dining on spiders and other insects. Reptiles, of course, are voracious predators, and the reptilian brain is quite violence prone. Mammalian predators such as lions and tigers are well known and feast on a variety of prey.

In addition to the acquisition of food, violence is also associated with the acquisition or defense of territory in many species. There are numerous pictures of two male animals fighting with a female standing nearby. The pictures are often titled, "Males fighting for female." However, as Robert Ardry pointed out in his seminal *Territorial Imperative* [ch. 7, ref. 100], if you look closely the female is not really paying any attention to the fighting. Moreover, observations in the wild demonstrate that the female will stay with whichever male wins the fight. Therefore, the males are actually fighting over territory. Now admittedly, not too many males are killed in these combats because once the outcome is clear, the loser will withdraw before being killed. Humans on the other hand, have a much better developed sense of pride and need for control; hence humans will actually kill to defend or acquire territory if necessary.

Thus, it is easy to see how humans carry violence in their genetic heritage. It is primarily associated with the reptilian portion of our brain. Of course with the advent of the emotional limbic and logical cerebral cortex, human violence becomes much more complex. Humans are known to kill for pleasure, although fortunately it is not particularly common. There is also anecdotal evidence that other animals kill for pleasure.

So violence in humans is a complex phenomenon, to a large extent associated with defense, although murder for profit (e.g., killing someone to obtain what they have) and murder for revenge are staples of television crime shows, the daily newspaper, and the nightly news.

Since all facets of human behavior conform to a bell curve, it is only natural that violence would also. Violence, however, does not appear to be a continuous aspect of human behavior. If it was, there wouldn't be many of us left. Accordingly, there must be a control mechanism that modulates violence. The two curves pictured in figure 17-8 depict control variation and violence variation.

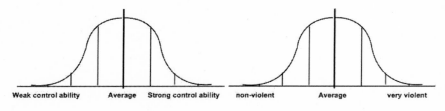

Figure 17-8. General distribution of control ability and violence

The propensity toward violence is controlled by at least two intersecting variance distributions: (1) the general propensity toward violence primarily initiated by the reptilian brain, and (2) the ability to control this propensity, which is centered in the cortex. Those possessed with a strong control ability are much less likely than those with weak ability to "fly off the handle," even if they have a propensity to violence. We see this all the time. "Such and such person has a short fuse," is a phrase we often hear to in reference to someone with a weak control ability. In view of the above, there is a genetic component to our violent ways that has roots millions of years old.

War—the Ultimate Expression of Violence: War is perhaps man's most bewildering activity. Although there appears to be almost universal desire for peace, we seem to be eternally engaged in war somewhere on Earth. The obvious question is why?

Numerous authors have tackled this question. In his well-researched book, *On the Origins of War*, Yale historian Donald Kagan presents one of the more thorough treatments of war's origins. He examined four conflicts spanning over 2000 years from the war between Athens and Sparta in the fifth century BCE to the Second World War. Kagan begins his book with the following observation:

> The collapse of the Soviet Union put an end to the dangerous rivalry between the great powers that threatened the peace and safety of the world for almost half a century .For many, the victory of the West over the East, of the free market over command economies, of democracy over Communist dictatorship, promises a new era of security, prosperity, and peace. [28, p. 1]

However, Kagan is skeptical that this peace will last. He comments:

Past theories of war's obsolescence were much the same as today's... It has been characteristic of our time to seek the causes and origins of war in impersonal forces: monarchy and aristocracy and the military ethos of an earlier age that surrounds them; atavistic reversions in the modern age to these outmoded ways; the class struggle; imperialism; arms races; alliance systems, etc.

Kagan also comments that these are merely symptoms and that "the wisest modern students of war have concluded that something more fundamental produces wars: the competition for power" (p. 15). Moreover, he claims, "The 5th century BCE historian Thucydides wrote 'In the struggle for power, whether for a rational sufficiency or in the insatiable drive for all the power there is,' Thucydides found that people go to war out of "honor, fear, and interest" (p. 17).

However, Kagan observes, "To many in the modern world the word *power* has an unpleasant ring. It seems to imply the ability to impose one's will upon another, usually by the use of force. Power is felt inherently to be bad" (p. 16). Presumably, we would term these people "pacifists" and many would note that they harbor unrealistic dreams given the realities of history. Since, as Kagan points out

[p]ower and honor have a reciprocal relationship. It is obvious that when a state's power grows, the deference and respect in which it is held are likely to grow as well. But the opposite is also true: even when its material power appears to remain the same, it really declines if in some manner these attitudes toward it change. This happens most frequently when a state is seen to lack the will to use its material power. (p. 18)

Finally, Kagan notes:

The Chinese sage Sun Tzu said: "The art of war is of vital importance to the state. It is a matter of life and death, a road either to safety or ruin. Hence it is a subject of inquiry which can on no account be neglected." No less vital is the art of avoiding war, and no more may the attempt to understand its origins and causes safely be neglected. (p. 19)

While there is some evidence that violence seems to be diminishing noted cognitive scientist, Steven Pinker states in a perceptive speech that "Contrary to conventional belief…. Global violence has fallen steadily since the middle of the twentieth century" [29]. However, Pinker doesn't seem to know why, as he laments, "It would be nice to know what, exactly, it is."

WAR—THE UNFORTUNATE RESULT OF THE NEED TO ENSURE SPECIES SURVIVAL.

While, as Kagan and others have pointed out, there are numerous causes of war, I believe one of the principal causes of war is the need for increasing amounts of territory. The need for increasing amount of territory grows as we heed evolution's mandate for species survival, which is similarly manifested by an occasionally overwhelming need for sexual activity that provides the offspring needed for species survival. Unfortunately, we tend to provide too many people, but that again is evolution's rather blind objective: more people increases the probability of species survival. However, more people also increases the probability of conflicts between groups as increasing amounts of territory are required, which ultimately invokes the "Territorial Imperative."

The Territorial Imperative and the fate of cultures: In the *Territorial Imperative*, playwright Robert Ardry develops the concept that territory, especially possession of it, as human's strongest emotion, stronger even than sex[8].

However, Ardry is only partially correct. While the need for territory may be the strongest overall emotion, at some point sex must overcome even the need for territory to assure continuation of the species. On the other hand, sexual activity requires relatively little time when compared with the defense of or acquisition of territory. Accordingly, there is no obvious conflict between the two.

While Ardry didn't specifically discuss evolution's role, as explained in chapter 13, he defined one of the more important human behavioral characteristics as the instinctive need to form groups that are internally cohesive and externally pugnacious (e.g., violent toward anything that incurs on their territory or that prevents expansion of territory as population grows). These characteristics Ardry termed the "Territorial Imperative."

Regarding reproduction, there are three possibilities for the relationship between birth rate (B) and death rate (D):

$$B < D$$
$$B = D$$
$$B > D$$

Obviously, B < D is a dead end, and while B = D would be optimum, the variability of D required that B > D be the only practical solution. Unfortunately, B > D generally leads to population growth, as proposed by Malthus. From the viewpoint of the Territorial Imperative, as a groups population grows there is need for increasing amounts of territory. In relatively uninhabited areas, population can grow without consequence. But since all groups of humans will grow, eventually one or more groups will reach the point where their territorial needs overlap.

In the case of early humans, who were apparently not very sophisticated in the ways of negotiation, survival of the fittest was applied to the contesting groups. Take for example the Aurignacian and Gravettian cultures. As discussed in chapter 13, the Aurignacian culture dates from about 32–26 KYA when the culture disappeared from the archeological record. The Gravettian culture dates from about 28–22 KYA. Clearly there was a short overlap (2,000 years) between the two. As noted in chapter 13, the Gravettian culture produced superior stone tools (a great euphemism for weapons), thus it is reasonable to assume that the Gravettian people won the survival of the fittest struggle with the Aurignacians.

Regarding groups, as Gary Marcus points out in chapter 11 of this book survival of the fittest does not mean survival of the optimum; it merely means survival of the most fit at the time. Clearly the natural selection filter had favored the Gravettians, and their application of the Territorial Imperative was superior to the Aurignacians. Accordingly, the Aurignacians were displaced from their territory and ultimately from the archeological record.

There is much speculation regarding the fate of the Neandertal, especially modern humans' involvement in the Neandertal extinction. After all, extinction is an ugly concept. Recall Yale historian Kagen's comment: "Power is felt [by many] inherently to be bad"[28].

On the other hand, if we examine recorded history, which does not require the reading of archeological "tea leaves," there are many examples of one group of humans effectively eliminating or absorbing another group. Take for example Alexander's treatment of the Persians and Rome's treatment of the Carthaginians, and the Romans suffered the same fate at the hands of the Germanic tribes. In more modern times, we have the Holocaust of WWII.

Thus, it is reasonable to suggest that while the Gravettians may not have exclusively ended the Neandertal's existence, they certainly contributed by taking Neandertal territory and driving them to their last known location in Spain[9].

In chapter 11, we discussed the fact Thomas Malthus's most dire predictions regarding the human race have been averted so far by improved food production technology. We also commented that Malthus really didn't know *why* excessive population growth occurred. We now know that the evolutionary driven tendency of humans to expand their populations causes excessive population growth such that groups of humans eventually come into territory conflicts. Those groups that are most prone to violence (i.e., their ability to control their violence is less than the other group) will tend to clash violently with the other group and often either annihilate or subjugate them. As mentioned above, history is filled with examples of this.

Thus, evolution has provided humans with a triune brain and a variation in violence. When that tendency to violence is combined with population growth problems that are also caused by the evolutionary need to ensure species survival, it leads inexorably to the need to gain more territory and ultimately leads to war or conflict. Regarding this issue, an informative article regarding the relations between nations, groups of individuals with internal cohesiveness and external pugnacity, appeared in a recent *Wall Street Journal* article regarding then President Obama's approach to unfriendly nations:

> 'We will listen carefully,' Mr. Obama said with a view to Tehran, 'we will bridge misunderstandings, and we will

seek common ground.' Some 500 years ago, Francis I of France was asked what misunderstandings had fueled his constant wars with the Habsburg Empire's Charles V. He replied: 'None, we are in complete agreement. We both want control over Italy.'

Conflict between states is made from sterner stuff than bad manners or bad vibes, past grievances or imaginary fears. International politics is neither psychiatry nor a set of "see me, feel me" encounter sessions. It is about power and position, about preventing injury and protecting interests. Love and friendship move people, not nations. [30]

POPULATION GROWTH—IS IT GOOD OR BAD?

We come now to the last item in the answers in evolution list, an item I believe is perhaps the most important because the control of our population is key to our long-term survival. If we were to control our populations, the pressure for more territory would presumably be reduced. Unfortunately as will be seen, this is one question evolution can't answer because evolution is the culprit.

While there is great controversy regarding whether population growth is good or bad, I am unaware of anyone who claims that population growth can continue indefinitely. However, as discussed above, indefinite population growth is our evolutionary heritage.

While the survival of multi-celled species requires the birth rate to exceed the death rate, for millions of years, the birth rate was just barely greater than the death rate. In some cases, the death rate greatly exceeded the birth rate during the occurrences of extinctions. Keep in mind that 95 percent of all the species that have evolved on this planet are now extinct.

Somewhere between 200–100 KYA, *Homo sapiens sapiens* appeared on the planet with a nice big brain. Since then, the diseases, plagues, and food supply problems that kept populations in check have been slowly solved. The result is an expansion of the human population that many term an explosion, as depicted figure 17-9. Figure 17-9 illustrates the rapid growth in population that came with the Industrial Age (all dates are approximate) [31].

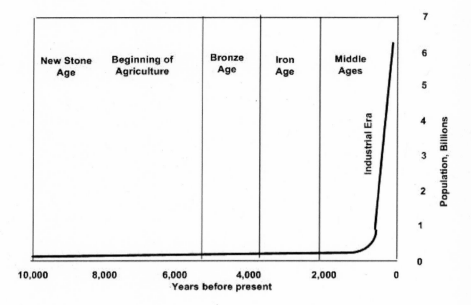

Figure 17-9. Human population growth over time

As pointed out in chapter 11, this is an exponential growth curve that is a correction to the geometric growth Thomas Malthus predicted.

In 1972, an organization called The Club of Rome that was founded in 1968 by an Italian industrialist and a Scottish scientist [32], published a book titled, *The Limits of Growth*. The book was prepared by Donnella H. Meadows and Dennis L. Meadows, students in the MIT Systems Engineering Labs [33]. The MIT group had created a software "model" of the earth's economic system that allowed them to make crude predictions of the consequences of unlimited growth.

The program generated a number of curves the one shown in figure 17-10, which was adapted from [33] Figure 35, p. 124.

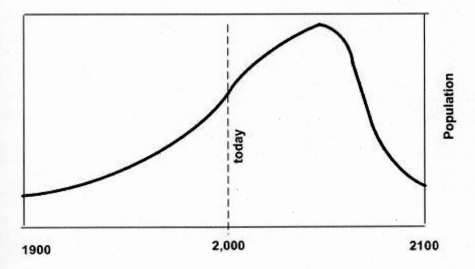

1900   2,000   2100

Figure 17-10. Estimate of population trends, 1900 through 2100

Not surprisingly, reaction to the study was swift and not very generous. There are many sources for information regarding this study, both pro and con, so I won't dwell on it other than to point out what I believe is obvious: growth cannot continue indefinitely, and there are those that believe we have already passed the point of no return. For example, this excerpt from the article,

> "New studies predict record land grab as demand soars for new sources of food, energy and wood fiber.
> Escalating global demand for fuel, food and wood fibre [*due to rapidly expanding world population*] will destroy the world's forests, if efforts to address climate change and poverty fail to empower the billion-plus forest-dependent poor, according to two reports released today by the U.S.-based Rights and Resources Initiative (RRI), an international coalition comprising the world's foremost organisations on forest governance and conservation" [34].

The phrase in brackets is my addition because population growth is not mentioned in the article. In fact, population growth is the "elephant in the room"[10] in many articles regarding Earth's ecological problems.

Any reasonably observant person can see that population growth is *the problem*, but it is rarely mentioned. Perhaps that is because population control requires objectionable measures such as birth control, which will be discussed in the next chapter.

It is interesting that Gary Marcus apparently realizes that "we must recognize evolution's limits," but he doesn't elaborate about the limits as he remarked in his article regarding the survival of the fittest: "If we are to move past perpetual cycles of fantasy-driven booms followed by devastating busts, we must recognize evolution's limits, and confront them head-on" [20, ch 11].

EVOLUTION'S FATAL FLAW—A PROBLEM EVOLUTION CANNOT SOLVE.

At this section's beginning, I noted that the population control issue is one question evolution can't answer. The reason is quite simple. The tendency to excess population growth is largely the result of evolution's selection of those multi-celled animals, including us, that occasionally experience a need for sex that exceeds all other needs, thereby causing said animals to produce enough offspring to ensure species survival.

Evolution is blind to the ultimate consequences of this selection, which resembles the famous "Sorcerer's Apprentice" problem of an operation that apprentice started with good intentions but then couldn't shut off. Evolution initiated the occasional need for sex to exceed all other needs to ensure species' survival, but while this solved the problem, it also sowed the seeds of evolution's fatal flaw. Since evolution cannot deselect this need or else place species survival in peril, the solution to the population control problem lies with evolution's greatest achievement—the human mind.

Consequently, unless we humans accept the existence of the process of evolution, recognize that our occasionally excessive need for sex is the result of evolution's prime need to assure species survival, abandon our outdated religiously motivated anti-population control attitudes, and take effective population control measures, evolution's greatest achievement may not survive.

SUMMARY FOR CHAPTER 17.

One of this chapter's fundamental objectives is the demonstration that evolution is the consequence of the need to ensure species survival. Accomplishment of this objective requires an examination of some of life's basic questions:

1. Why must we die?
2. What is our life's purpose?
3 Why are we here?
4 Why do we engage in sex?

In addition to these, two other important questions were examined in this chapter:

5 Why is there so much variation in humans, especially behavior?

6 What is the origin of our propensity for violence, especially war?

Finally, one of the more import facets of evolution, "evolution's fatal flaw," which adversely affects the issue of population growth, ended the chapter.

Since these questions might be considered more properly addressed by other intellectual disciplines such as philosophy and religion, the chapter began with a brief review of the definitions of philosophy, religion, and science to ascertain whether evolution does indeed have purview. We made the following observations:

- Philosophy investigates causes but does not appear to offer any explanation causes for them.

- Philosophy also employs logical reasoning rather than empirical methods.

- Religion relies almost exclusively upon belief in supernatural power or powers.

- The science definition contains the word "explanation" and is general consistent with this book's presentations.

351

Accordingly, evolution does indeed have purview. From there, the discussion began with the first three questions because of their relation to each other.

THE FIRST FOUR QUESTIONS.

One answer to the first question is provided by the Bible. We must die because the first man (presumably Adam) brought sin into the world and with it, death. That is not much of an explanation. Evolution has a better "short answer" to the first question : from the moment we are conceived, continuous cell division ultimately degrades our body until either a vital organ fails or we succumb to a viral or bacterial infection.

Since we die, species survival requires us to replace ourselves; hence the answer to the second question is that our life's primary purpose, which is evolution's mandate, is to engage in coitus at a time when the female partner can conceive, which leads to the answer to the third question, "Why are we here?" We are here because our parents heeded evolution's mandate to reproduce, and had sex when our mother was able to conceive.

This finally leads to what may seem to be a strange question: why do we engage in sex? The answer is provided in the next section which discusses the difference between individual survival and species survival needs.

ENSURING REPRODUCTION.

We now live in relative comfort whereas for our more ancient Stone Age ancestors, life was most likely "Nasty, Brutish and Short" [7]. Consequently for them, engaging in sexual intercourse was an extremely inconvenient interruption in the daily struggle to meet the needs of individual survival, especially since *reproduction provides no survival benefit to individuals.*

Accordingly, to ensure that sexual activity occurred often enough, evolution selected—via the natural selection filter operating over millions of years—those males who placed the need for sexually activity above all other needs at the time when the female was receptive and fertile. Concentrating on male need was not an act of evolutionary male chauvinism; it merely recognized that the male typically initiates sexual activity.

But females also play an important role. They must be sexually receptive at a time when there is an optimum chance of offspring survival. For example, those animals with short gestation periods tend to mate in the spring, while those with longer gestation periods tend to mate in the fall. Because humans had some shelter, they were not constrained to mate at only certain times of the year. However, there is evidence that women are more sexually receptive just before ovulation [36].

The preceding paragraphs have shown how evolution's natural selection filter has chosen only those multi-celled animals that experience a need for sex that exceeds all other needs sufficiently often during the individual's lifetime to assure sufficient offspring to assure species survival.

A review of the development of life, as described in chapter 13, will illustrate how this selection evolved. We observe that initially, only single-celled plants and animals—those without a nucleus—populated the earth. These animals reproduced by simple binary fission; they just split in half. Gradually, single-celled animals with nuclei developed and they reproduced by mitosis. Then, approximately 650 MYA, during the Edicarran Period, sexual reproduction appeared in the fossil record.

The appearance of sexual reproduction initiated the "Cambrian Explosion" in which all major life forms appeared. At the conclusion of the Cambrian, the dominant animal was the fish which appeared about 500 MYA and dominated the Devonian Period. The fish was followed by the amphibian which was in turn followed by the reptile. Each new animal type had improved survival capability, thereby enhancing the animals survival potential.

This gradual improvement of life, facilitated by evolution's natural selection filter, continued until the arrival of modern man approximately 200 KYA.

This entire process of life's continued improvement, beginning with a simple single-celled animal and culminating in modern man, we have named evolution. It is a process born out of the need to assure species survival; hence, without the process of evolution, there would be no life on Earth today, except possibly simple single-celled animals.

Therefore, Evolution is truly the inevitable consequence of the need to ensure species survival.

Why is there so Much Variation in Humans? Before revisiting life's purpose, the question of human variation was addressed because it has an impact on fulfilling life's purpose. Human variation is the direct result of reproductive genetic mixing in which the assignment of genes to sperms and eggs is random. Therefore, when they meet to create a genetically unique being, the resultant characteristics of a large number of offspring can be expected to follow a bell curve.

Returning to Life's Purpose: There is obviously more to life's purpose than sexual activity because this occupies a relatively small amount of a human's time. The Reverend Rick Warren asserts in his book, *The Purpose Driven Life*, that the basic human purpose is to serve God, but this seems extremely limiting since God, being undetectable, is difficult to know.

Therefore, since variation in human characteristics is an obvious outcome of reproduction, besides replacing ourselves our more general life's purpose is the maximization of the abilities that evolution has granted us.

But Must We Die? Is death an inevitable attribute to life or is it just another fatal disease that is as amenable to intervention as in any other disease? More and more biomedical researchers and other relatively enlightened individuals are beginning believe that death is just another fatal disease and that significant life extension will be possible in the not too distant future.

For example, in 2004, computer genius Ray Kurzweill and physician Terry Grossman wrote an intriguing book titled, *Fantastic Voyage, Live Long Enough to Live Forever* [27]. Kurzweill and Grossman propose that biomedical advances are proceeding at a rapid pace that it is reasonable to believe that some of the major causes of death, such as heart disease, will be significantly reduced within the lifetimes of many persons, which will lead to significant life extension.

TERRITORY, POPULATION GROWTH, VIOLENCE, AND EVOLUTION'S FATAL FLAW

The first three concepts are inextricably intertwined and result from Evolution's Fatal Flaw. As mentioned earlier, evolution has selected those humans whose need for sex occasionally exceeds all other needs and reproduction occurs; thereby ensuring species survival. Unfortunately, the birth rate generally exceeds the death rate and population expands with a concomitant requirement for more territory.

Evolution has also selected human characteristics in which humans instinctively form groups that are internally cohesive and externally pugnacious (i.e., violent toward anything that incurs on their territory or prevents expansion of territory as population grows). These three characteristics are generally called the "Territorial Imperative."

The result of the interaction of these characteristics is easy to predict. As populations expand and with them the need for more territory, eventually two or more groups come in contact, the Territorial Imperative takes over, and war ensues. In general, the group that is most fit and the most violent will "win the war." Much of human history is a chronology of these activities.

With regard to population growth, we have seen how evolution has produced the human race over millions of years of continued improvement via the tendency of the most fit to survive and pass one their superior genes. This is of course desirable, but ensuring species' survival by a very strong reproductive need leads inevitably to unsustainable population growth, which I now identify as evolution's fatal flaw.

On the other hand, evolution has also provided us with a cerebral cortex with which to reason and hopefully control evolution's blind dictum to have as many children as possible. However, if humans do not accept that evolution has occurred and continues to function, and learn what evolution is and how it works so that evolution's propensity for unsustainable population growth can be counteracted, the human race may be doomed.

LOOKING AHEAD.

There is one more area that should be explored briefly because it is extremely important: the science-religion conflict and its consequences such as unsustainable population growth. This is the subject of our last chapter.

CHAPTER 18:
CONSEQUENCES OF THE SCIENCE-RELIGION CONFLICT

While this book is directed primarily toward demonstrating that evolution is a consequence of the need to ensure species survival, it is obvious that the divergent views regarding evolution contribute significantly to the science-religion conflict, especially—as pointed out in the prologue of this book—when one searches for the reason for three mutually exclusive origins explanations.

As summarized in chapter 13, the origin of the conflict can probably be traced to the ancient Greeks, who were the first to realize that religious explanations conflicted with an increasing number of observations that did not seem to fit into religious explanations. However, the Greeks were a bit ahead of their time and it was actually the 1543 CE publication by Nicolas Copernicus of his ground breaking opus, *On the Revolutions of the Celestial Spheres*, that most appropriately marks the beginning of the science-religion conflict.

While Copernicus's publication caused vehement religious reaction, the 1859 publication of Charles Darwin's research and explanation of evolution in another earth shaking opus, *On the Origin of Species by Means of Natural Selection*, really fanned the fires of the science-religion conflict which, as this book has amply demonstrated, continues today.

We have already seen examples of the consequences of fundamentalist beliefs in the stories of persons such as Stephen Gregory and Arlan Blodgett. On another level, the dichotomy reflected in the Gallup poll findings was succinctly captured in an interview with a member of the town of Dover, Pennsylvania after the famous trial discussed above. He asked, "How can we go to church on Sunday, profess a belief in God as the creator of ourselves and our world, and then go to school on Monday and teach the opposite?[1]"

During a recent winter while escaping the cold Washington weather, I found this question framed slightly differently in a religious ad in a Palm Desert, California newspaper. It read in effect, "What do you tell your child when they come home from school and says 'my teacher says that God did not create the world.'"

These questions go to the heart of the question, "Why are there three explanations of origins?" which I believe we have satisfactorily answered. However, if this were merely a debate between the correctness of the religious or scientific worldview, one could just decide whether or not to participate in the debate. After all, the conflict "does no harm." But a closer examination of the various points of conflict demonstrates that the science-religion conflict has serious consequences.

Some examples of the more important harmful science-religion conflicts include:

- religiously motivated opposition to vaccination;
- belief in the efficacy of prayer;
- belief in divine intervention in medical problems;
- opposition to abortion and contraception;
- religiously motivated opposition to the teaching of evolution in public schools.

Each of these will be discussed in term.

RELIGIOUSLY MOTIVATED OPPOSITION TO VACCINATION.

Vaccination has been one of the most effective medical intervention programs in history, virtually eliminating such scourges as poliomyelitis, first introduced in 1951, plus small pox and measles to name just a few. However, seemingly beyond all reason, there is a well-organized religiously based opposition to vaccination without any verifiable evidence of harm caused by vaccination and considerable evidence of harm if people are not vaccinated. Of the many anti-vaccination program problems, I will address three:

- The belief that vaccination in general causes autism
- Measles vaccination opposition
- Human *papillomavirus* (HPV) vaccination opposition

Vaccination in General Causes Autism: Regarding autism, the Friday, February 12, 2009 *Wall Street Journal* contained an article titled, "U. S. Court Rejects Vaccine Connection to Autism" [1]. Although the case involved only three plaintiffs, 5,000 families are seeking damages claiming that vaccination caused their children's' autism. In a strongly worded decision, the court found that "The numerous medical studies concerning these issues, [vaccines cause autism] performed worldwide, have come down strongly against the petitioners contentions [that vaccines cause autism]."

At the conclusion of the case, one parent, apparently active in the vaccination caused autism campaign, complained, "It's tough when you're taking parent support calls and you hear same story [presumably a claim that the callers child has autism that was caused by vaccination] day after day. When does anecdotal evidence become enough" [1]? Of course, this parent apparently doesn't understand that anecdotal evidence isn't evidence, it's merely coincidence and there can never "become enough."

The debate regarding whether vaccines cause autism can be traced to a 1998 study published in the prestigious English medical journal *Lancet* by researcher Andrew Wakefield and some of his colleagues [2]. While the study was small, it generated considerable interest and fanned the anti-vaccination flames. A few years after the study's publication:

> "Wakefield's associates withdrew their endorsement of the paper and cited as part of the reason an undisclosed potential conflict of interest for Wakefield, namely that at the time of its publication he was conducting research for a group of parents of autistic children seeking to sue for damages from MMR vaccine producers"[2].

Unfortunately "true believers" accused the British Medical Establishment of covering up "the truth."

Opposition to vaccination based upon religious convictions is very well organized with Web sites such as the Frequently Asked Questions About Religious Exemption (to vaccination) site. The site states, "A religious exemption for vaccination is a written form certifying that the parent's objection to immunization for religious reasons exempts the parent and child from state vaccination requirements" [3]. The Web site further states that "a conflict arises if you believe that man is made in God's image and the injection of toxic chemicals and foreign proteins into the bloodstream is a violation of God's directive to keep the body/temple holy and free from impurities."

Anti-Measles Vaccination Efforts: Another example of the deleterious effects of these religiously motivated anti-vaccination efforts is the effect on the elimination of measles, as attested by a headline that appeared in a January 2008 San Diego, California newspaper: "Outbreak of Measles --- San Diego, California" [4].

The news article stated that "Measles, once a common childhood disease in the United States [now controllable by measles vaccine], can result in severe complications, including encephalitis, pneumonia, and death." However, the article went on to say:

> "In January 2008, measles was identified in an unvaccinated boy from San Diego, California, who had recently traveled to Europe with his family. After his case was confirmed, an outbreak investigation and response were initiated by local and state health departments in coordination with CDC, using standard measles surveillance case definitions and classifications. This report summarizes the preliminary results of that investigation, which has identified 11 additional cases of measles in unvaccinated children in San Diego that are linked epidemiologically to the index case and include two generations of secondary transmission".

And this is not an isolated case of measles anti-vaccine problems, as indicated in a new article in the Olympia, Washington paper, the *Olympian*, on March 17, 2009. The article carried the headline, "Diseases may come back due to vaccine worries" [5]. It contained no new information but did feature another parent whose child appeared to become sick as a result of vaccination. The parent could not be convinced it was a coincidence and refused to vaccinate other children. Thus the problem is spreading. But measles is not the only viral infection of concern.

Anti-HPV Vaccination Efforts: Human papillomavirus (HPV) is a potentially deadly infection that can, in some cases, lead to cancer of the cervix. It is transmitted by sexual contact [6]. Fortunately, a vaccine to protect women against HPV has recently been approved. With over 70 percent of the population infected, protecting young girls would appear to have high priority. However, religious objection to HPV vaccination has surfaced for a "moral reason" on a Web site called Religious Objections to HPV Vaccine, which points out:

> 'To prevent infection, girls will have to be vaccinated before they become sexually active, which could be a problem in many countries. In the US, for instance, religious groups are gearing up to oppose vaccination, despite a survey showing 80 per cent of parents favor vaccinating their daughters. 'Abstinence is the best way to prevent HPV', says Bridget Maher of the Family Research Council, a leading Christian lobby group" [7].

The article closes with:

> "Religious people who oppose this because there is a *risk* that people *may* have more sex (unlikely, because the threat of things like HIV is far more immediate than the chance of cancer decades later) are putting their religious ideology over the others' lives. They would rather see women *die* [emphasis added] than possibly have extra-marital sex".

Of course, abstinence-only programs "fly in the face" of the powerful sexual urge evolution has imbued in us to ensure species survival.

Finally, due to the liability associated with anti-vaccination lawsuits like the one described above, more and more pharmaceutical companies are abandoning vaccine manufacture and development. Vaccine research, development, testing, and manufacture are expensive and difficult. And markets are small since vaccines are used only a few times during an individual's lifetime, unlike medicine such as cholesterol-lowering medicines. Thus a vital weapon in the fight against disease is being eliminated by well-meaning but superstitious people.

BELIEF IN THE EFFICACY OF PRAYER.

One of the more powerful religious beliefs is the belief in the power of prayer, despite the lack of any evidence to support the belief.

Due to the lack of interest by trained, objective observers, few tests of the power of prayer have been conducted. However, Dr. Herbert Benson of Harvard Medical School has led one double blind study. The study involves about 1800 persons undergoing heart bypass surgery. The results of the study were published in an article titled, "Study of the Therapeutic Effects of Intercessory Prayer," in the *American Heart Journal*, April 2006 issue [8].

Three Christian groups were recruited to pray for particular patients, starting the night before their surgeries and continuing for two weeks. Volunteers were given a patient's first name and last initial and were asked to pray for "a successful surgery with a quick, healthy recovery, and no complications."

The patients were divided into three groups.

> The first set of patients was being prayed for and knew it.

> The second group was also the subject of prayers but only knew it was a possibility. Patients in the third group weren't prayed for, although they were told they might be.

All patients were then monitored for any complications for thirty days following surgery, and no effects of prayer on the patients' recovery were detected. However, the study did find that, surprisingly, 59 percent of the patients who knew they were being prayed for developed medical complications while only 52 percent of those who thought they might be prayed for had complications.

The researchers said they had no explanation for the higher complication rate among patients who knew they were being prayed for. In addition, researchers emphasized, presumably for political reasons, that their study could not address questions such as whether God exists or answers prayers. However, prayers were definitely not answered in this case.

A question that is relevant to this study but which seemed to go unasked is, "What did the people who prayed expect to happen?" There are many complications that can occur in heart bypass surgery, as listed in this site [9]. Most involve surgical problems such as leakage due to a faulty suture, additional blockages, etc. If a prayer for a "healthy recovery" were to be "answered," then somehow supernatural intervention would have been required (e.g., to repair the faulty suture or remove the blockage), all presumably without the patients or anyone else's knowledge. It is rather difficult to image this happening.

While the study reported above was an apparently unsuccessful attempt to establish the ability of prayer to affect medical outcomes, numerous demonstrations of the futility of prayer to alter medical outcomes have been reported in the daily press with much more serious consequences. An example occurred recently in Wausau Wisconsin. On March 23, 2008, Dale Neuman's 11 daughter Kara died due to lack of proper medical attention. Mr. Neuman claimed he was unaware that his daughter was diabetic and, being a "Full Gospel Christian," was convinced his prayers would cure his daughter. Despite his daughter's deteriorating condition, Mr. Neuman refused medical assistance and persisted in prayer until his daughter finally stopped breathing [10]. Mr. Neuman has been convicted of second-degree reckless homicide in has daughter's easily preventable death. In a statement similar to, incorrect "there is no evidence" statements reported in other chapters of this book, Neuman's lawyer stated that "There's also not 'a shred of evidence' Neumann knew his prayers would fail to help his daughter or cause her death." [10]

Of course, there are those who are certain their prayers were answered. "We prayed for (desired outcome) and (desired outcome) occurred. There are two possible explanations for this "proof that God answers prayers"

1. Occurrence of (desired outcome) was simple coincidence and not the result of prayer

2. God is capricious and only answers selected prayers based upon an unknown selection process

Even if true, explanation 2 provides little comfort since one cannot rely upon a prayer being answered and thus must act on the assumption the prayer will not be answered. If Dale Neumann had followed this advice, his daughter would still be alive.

On a larger scale, in view of the amount of prayers being said, if pray really worked, the world would probably be a very different and probably much better place. All wars would have ended because praying to end wars is probably high on everyone's prayer list. Other items that would be affected would be disease in general, poverty, and many other major afflictions of the human race. It's too bad prayer doesn't work.

BELIEF IN DIVINE INTERVENTION IN MEDICAL PROBLEMS AND END-OF-LIFE DECISIONS.

Approximately 57 percent of 1000 adults in a randomly selected survey stated they believed that God intervenes in serious illness [10]; [11]. While this would seem to be no less benign than the belief in prayer, strong believers interfere with medical care, especially in situation when the case is definitely terminal with no known effective medical intervention. Loved ones will insist that physician "do whatever they can" to prolong the person's life with hope that "God will intervene." Despite the ultimate letdown, valuable medical facilities are usurped in these futile attempts.

Similar to the belief in divine intervention, persons who have strong religious faith and are terminally ill typically seek more aggressive end-of-life treatment than those of lesser faith [12]. This has the same effect on valuable medical facilities as those who believe in divine intervention and perhaps is motivated by similar beliefs.

RELIGIOUS OBJECTION TO ABORTION AND CONTRACEPTION.

Objection to abortion and contraception exacerbates one of the most serious problems facing the earth today: the expanding world population, as discussed in detail in the previous chapter. While abortion and contraception are interrelated issued, religious objection to abortion is more prevalent because religiously inclined individuals acquaint abortion with murder, even if the "murder" is only of a few cells. These are well-known aspects of religious intolerance, but religious objection to contraception is relatively new and growing [13].

In a related incident, on Pope Benedict XVI's April 2009 visit to Africa, which has one of the most serious AIDS problems on Earth, it was reported that: "Pope Benedict XVI said. condoms are not the answer to the AIDS epidemic in Africa and can make the problem worse, setting off criticism Tuesday as he began a weeklong trip to the continent where some 22 million people are living with HIV" [14].

Contrary to the Pope's ill-timed remarks, condoms have proven to be extremely effective in preventing the spread of HIV/AIDS. A USAID study reports:

> "The body of research demonstrating the effectiveness of latex condoms in reducing sexual transmission of HIV is both comprehensive and conclusive. Scientific studies of sexually active couples, where one partner is infected with HIV and the other partner is not, have demonstrated that the consistent use of latex condoms reduces the likelihood of HIV infection by 80 to 90 percent" [15].

And yet one of the most powerful persons on Earth disparages the use of condoms, presumably because condoms are anathema to the Catholic religion's anti-artificial contraception position. So rather than risk abdicating his responsibility to conform to Catholic dogma, the Pope places the lives of Africans in peril.

Apparently the prospect of global catastrophe is not as important as clinging to ancient religiously based beliefs. This is similar to those refuse to protect their daughters from cervical cancer due to the mistaken belief that premarital sex is somehow more harmful than cervical cancer when there is ample evidence that it is not.

RELIGIOUS OPPOSITION TO THE TEACHING OF EVOLUTION IN PUBLIC SCHOOLS.

One of the most influential educational institutions in the country, the Texas School Board, is attempting to limit the teaching of evolution in public schools and substitute creationism or ID [16]. These attacks are part of a continuous assault on the public educational system not unlike the Scopes Trial.

Stephanie Simon reported in a March 28, 2009 *Wall Street Journal* article, titled "Texas Opens Classroom Door for Evolution Doubts," that:

> "The Texas Board of Education approved a science curriculum that opens the door for teachers and textbooks to raise doubts about evolution.
>
> Critics of evolution said they were thrilled with Friday's move. 'Texas has sent a clear message that evolution should be taught as a scientific theory open to critical scrutiny, not as a sacred dogma that can't be questioned,' said Dr. John West, a senior fellow at the Discovery Institute, a Seattle think tank that argues an intelligent designer created life" [16].

On the other hand, the article said, "Kathy Miller, president of the pro-evolution Texas Freedom Network, said, 'The board crafted a road map that creationists will use to pressure publishers into putting phony arguments attacking established science into textbooks.'"

Ms. Simon added that, of particular concern,

> "It isn't just evolution at issue: The board also approved an earth-science curriculum that challenges the widely accepted Big Bang Theory. Students are expected to learn that there are 'differing theories' on the 'origin and history of the universe.'
>
> Board members also deleted a reference to the scientific consensus that the universe is nearly 14 billion years old. The board's chairman [Dr. Don McLeroy] has said he believes God created the universe fewer than 10,000 years ago"[16].

Dr. McLeroy and his minions, of course, merely trot out all the anti-evolution ideas that have been discredited earlier in this book with the usual lack of any evidence to support his views. However, the Texas school system has great influence due to the volume of books they order.

In a related article La Shawn of La Shawn Barber's Corner offers this comment in defense of the Texas position:

> Darwinian evolution, the absurd theory that magnificent life is based on *random chance* and developed through genetic mutations (which are *harmful* to the organism, by the way —X-Men is only entertainment), is taught in government schools as if it were the gospel. In my opinion, it takes much more faith to believe in macroevolution than to believe a Creator designed the universe with purpose. [17]

As with similar creationist discussion, this is just a rant. Barber offers no proof of his assertions.

A couple of years ago, recognizing the growing anti-evolution movement in the public schools, I addressed this issue in a paper given at an AAAS meeting [18]. I was probably just preaching to the choir because it doesn't seem to have had much effect, but a key point in the paper emphasized that the anti-science forces have well-organized programs to elect persons like Dr. McLeroy who are sympathetic to their beliefs to school boards where they have great influence on the educational system. This was, of course, what happened in Dover, Pennsylvania.

SUMMARY FOR CHAPTER 18.

This short chapter has pointed out five (of many) consequences of the science-religious conflict, consequences that can affect life and death:

- opposition to vaccination;
- belief in the efficacy of prayer;
- belief in divine intervention in medical problems;
- opposition to abortion and contraception;
- religiously motivated opposition to the teaching of evolution in public schools.

There is zero verifiable evidence of a relationship between maladies like autism and vaccination however, desperate but ill-educated or uninformed persons are suing vaccine manufacturers for damages in the mistaken belief that vaccine caused their child's autism or some other deadly problem. In addition to being ineffective, these actions threaten one of the best medical intervention programs in history. Moreover, opposition to vaccination is not limited to incorrect beliefs about autism; measles vaccination is also under attack.

Belief in the efficacy of prayer is perhaps harmless, but it leads people down the wrong path, and cases where "it worked once" but doesn't work a second time can cause unnecessary pain and the occasional serious diversion of medical facilities. The same is true with the belief in divine intervention, except it is the insistence that divine intervention is possible that unnecessarily burdens medical facilities at a time of rising costs and facility scarceness.

Opposition to abortion is not only misguided and an interference with a woman's life but can place the woman's existing children in jeopardy by the addition of another person to an already over-burdened family. Furthermore, the opposition tends to exacerbate the growing population problem. The same is true of opposition to contraception although this is even more misguided as no "life" is at stake, only a probably unwanted future life.

Finally, religiously motivated opposition to the teaching of evolution in schools threatens the foundation of democracy—an educated populace.

At the end of the day, there appears to be little good to be said about religion. It's an outdated and essentially discredited explanation of who we are and of our surroundings. It inflicts great hardship due to the refusal to abandon religious dogma. For example, there are those who oppose the HPV vaccine which prevents a potentially fatal cancer or condoms which can greatly reduce one of mankind's greatest scourges, because these preventative measures conflict with religious dogma regarding sexual activity. Religion's one redeeming virtue appears to be the comfort it brings to those who cannot find any other source of comfort, especially those who are convinced that they will somehow be reunited with loved ones that have "gone before them."

LOOKING AHEAD.

I believe we have accomplished the book's objectives as posed in the prologue:

1. Answering the question "Why are their three mutually exclusive origin explanations?"

2. Answering the question "Since these explanations are mutually exclusive, which one is true?"

3. Establishing unequivocally that the reason for the existence of evolution is the need for species survival.

4. Validating the book's title: *Evolution: The Inevitable Consequence of the Need for Species Survival.*

Accordingly, we should summarize what has been covered.

EPILOGUE.

As stated in the Prologue, the primary objective of this book has been to validate at least 2 controversial claims: the first claim being that evolution is the correct explanation for our origins, while the second claim is that evolution was the inevitable consequence of the need to ensure species survival. Unfortunately, ensuring species survival led to Evolution's Fatal Flaw, uncontrolled population growth. As suggested in Chapter 17, only the human brain can counteract Evolution's Fatal Flaw by consciously recognizing the Flaw's existence and countering it by exercising control over population growth – a supernatural entity is not going save us.

This book recounts the long journey of inquiry which produced the evidence necessary to accomplish this book's objective.

Also stated in the Prologue, this book owes its existence to two chance encounters at sea. The first encounter occurred on a trans-Atlantic voyage aboard the Queen Mary, while the other occurred one lovely summer day while I was on a whale watching excursion on Puget Sound.

This first encounter involved a chance dinner companion who happened to be an author who encouraged me to try self publishing. The second encounter led me to a series of eight Gallup polls regarding the evolution controversy. The survey encapsulated the essence of this controversy in a simple, three-part question:

> "Which of the following statements comes closest to your *explanations* of the origin and development of human beings? 1) God created human beings pretty

much in their present form at one time within the last 10,000 years or so, 2) Human beings have evolved over millions of years from other forms of life and God guided this process, 3) Human beings have evolved over millions of years from other forms of life, but God had no part in this process" [Prologue, ref. 2].

Table E-1 (summarized from the table P-1) averages the eight years of Gallup poll results and demonstrates a considerable divergence in beliefs regarding our origins.

Eight Years of Poll	God created humans in present form: Creationism	Humans developed, with God guiding: ID	Humans developed, but God had no part in process: Evolution	No Opinion
Average	46	37	11	6

Table E-1. Summary of Gallup poll results

The existence of these three origins explanation leads to two rather fundamental questions:

1. Why are there three mutually exclusive origin *explanations* extant?

2. Since these *explanations* are mutually exclusive, which one is the true one?

Answering these two questions and conclusively demonstrating that evolution is the inevitable consequence of the need for species survival, established these objectives for the book:

1 Answering the question "Why are there three mutually exclusive origin explanations?"

2. Answering the question "Since these explanations are mutually exclusive, which one is true?"

3. Establishing unequivocally that reason for the existence of evolution is the need for species survival.

Once these objectives had been accomplished, then the primary objective of this book, the validation of the book's title *Evolution's Fatal Flaw: The Inevitable Consequence of the Need to Ensure Species Survival* would have been established.

The first fourteen chapters of the book effectively accomplished objective one by demonstrating that the three explanations of our origins are a result of the historical development of our explanation of ourselves and our surroundings. These chapters also eliminated creationism and ID as viable origins explanations; thereby accomplishing objective two.

The third objective was accomplished in chapter 17, "Answers in Evolution," which also satisfied the primary objective.

It should be noted that the examination of the historical development of our understanding of ourselves and our surroundings involved the realization that "understanding" means the ability to answer one or more of six questions—*what, when, where, who, how,* and *why*—with *how* and *why* being the most important.

As explained in the prologue and demonstrated throughout the book, the historic gaining of a correct answers to these six questions was made particularly difficult by five impediments:

1. Illusions—things are not what they seem.

2. The unaided human eye has limitations.

3. Explanations often conflict with "common sense" and "intuition."

4. Explanations often conflict with authority, especially with family and religious authority.

5. Unrealistic expectations—things are not always what we desire.

Because of these impediments, the development of a correct explanation of ourselves and our surroundings proceeded in two phases.

In the first phase, the earliest explanations were developed. These were usually incorrect because they were based upon limited and/or incorrect observations and incorrect interpretations of those observations.

In the second phase, improving observational capabilities slowly convinced the more astute observers that these improved observations were at odds with the early explanations and ultimately led to correct explanations. This second phase often involved two kinds of explanations:

1. Empirically based explanations, which typically required laborious examination of large numbers of observations from which explanatory information was "distilled".

2. "Theoretical" explanations based upon a proper understanding of the underlying fundamental principals regarding the phenomenon being observed.

As discussed in detail in the book, illusions are the most serious impediment to gaining a correct understanding of ourselves and our surroundings, but the others contributed their share of difficulty. The most serious illusions are:

1. The apparently solid earth;

2. The apparent motion of the sun and planets around the earth;

3. The apparent same size of the sun and moon and the apparent closeness of both the sun and moon;

4. The apparent motion of the stars around the earth, and the apparent closeness of the stars;

5. The apparently unchanging physical and biological features of the earth.

Due to the complexity of the five illusions, resolving them has involved many disciplines and many investigators, therefore chapters 1 through 5 were devoted to the solid earth illusion, chapters 6 through 8 were devoted to the three astronomical illusions, and two chapters each were devoted to the apparently unchanging physical and life illusions. Chapters 9 and 10 focused on the apparently unchanging physical features illusion, and chapters 11 and 12 on the apparently unchanging animal and plant structures illusion. The significant information revealed by the resolution of these illusions is summarized in the next section.

SUMMARY OF THE SIGNIFICANT INFORMATION REVEALED BY ILLUSION RESOLUTION.

We began with the solid earth illusion because its resolution provided important material for the resolution of the other illusions.

The earth is, of course, not "solid," as demonstrated by Thompson's 1897 discovery of the electron, which disclosed that the atom had structure, thereby overturning centuries of belief that atoms were indivisible. Further atomic investigations by Rutherford revealed the existence of a tiny positive nucleus at an atom's center, around which electrons moved in fixed orbits.

The discovery of the atom enabled Niels Bohr to prove three significant experimental findings regarding "solid" matter: (1) atoms form unique, identifiable compounds such as H_2O; (2) atoms can be grouped in a table according to their properties; and (3) when placed in an evaluated glass bottle and subjected to an electric current, gaseous atoms such as hydrogen give off light that can be shown to have a unique spectrum, which provided a technique for detecting atoms in places like the sun. One spectrum in particular, the hydrogen spectrum, proved to be spectacularly useful in charting the universe.

Next, Becquerel's 1896 studies of heavy metals such as radium revealed that heavy atoms spontaneously emit positively charged alpha particles, negatively charged beta particles, and high-energy gamma rays. The three types of nuclear decay, or radioactivity, are very predictable and aid in determining the earth's age. Finally, Röntgen's 1895 studies revealed the existence of X-rays, which was then extended by Von Laue, who discovered X-ray diffraction, which is used to determine molecular structure (e.g., DNA).

It is rather remarkable that three of the most important discoveries relative to the structure of matter occurred with two years.

Further investigations of the atomic nucleus led to the discovery that light atoms combine, or fuse, to create heavier atoms, thereby releasing enormous amounts of energy which is source of energy for the sun and all stars. This discovery also solved a mystery that had perplexed early geologists: how old is the sun? Until the discovery of nuclear fusions showed the sun had probably existed for about 5 billion years and would "burn" for another 3 billion years, geologists couldn't devise an explanation for the sun's age greater than a few million years.

Atomic investigations culminated in quantum mechanics, which explained *why* atoms behave as they do and *how* various phenomena such as the emission of light or radioactivity were accomplished.

Of particular interest to this book, studies of "solid matter" led to the development of twelve independent methods for determining the age of the earth and the universe.

The resolution of the three astronomical illusions proceeded in three steps. In 1543 CE, Nicholas Copernicus rediscovered the 2000-year-old findings of fifth century BCE Greek philosopher Aristarchus, that the apparent motion of the sun and planets around the earth illusion is caused by the earth's axial rotation and that the planets actually revolve around the sun—a concept known as the heliocentric planetary organization.

The Copernican explanation was met with great hostility from the church. An outspoken proponent, a monk named Bruno was burned at the stake, and the venerable Galileo, who supported the Copernican explanation, was forced to recant. Copernicus's heliocentric concept displaced man from his favored position at the center of universe and was the first serious conflict between science and religion and can be considered as the initiation of the science-religion conflict.

The heliocentric explanation was satisfactorily confirmed via the efforts of Danish astronomer Tycho Brahe, followed by Brahe's assistant, Johannes Kepler, who demonstrated that planets orbit the sun in elliptical orbits. In 1636, Sir Isaac Newton theoretically derived Kepler's orbital equation from fundamentals. Newton was able to answer *why* planets orbited as they do; however, it remained for Albert Einstein to explain *how*, with his theory of special relativity. Einstein's theory showed that a large mass like the sun warps space-time, forcing objects orbiting around it into elliptical orbits.

The resolution of the apparently same size of the sun and moon illusion and the apparent closeness of both the sun and moon illusion led to an explanation of the correct size of the solar system via the development of triangulation methods. These methods were first successfully employed by Cassini, who measured the Earth-Mars distance and then the Earth-sun distance in 1672. Determination of

Earth's orbit size led to the discovery of the finite light speed by Danish astronomer Olaf Roëmer, also in 1672.

Using the Earth-sun distance as a baseline, astronomers were able to measure the distance to the closer stars, thereby resolving the apparent closeness of the stars illusion. Additional astronomical measurements by Hubble using Cepheid variable stars, discovered by Harriet Leavitt, revealed that the Andromeda galaxy is over one million light-years distant. Continued measurement of star distances and the discovery of the "red shift" in the hydrogen spectrum led to the conclusion that the universe was expanding. If it was expanding, it must have begun as a much smaller object. This led to the Big Bang hypothesis, which was conclusively confirmed by the Wilkinson Microwave Anisotropy Mapping (WMAP) satellite, which measured the background radiation left over from the Big Bang to unprecedented accuracy and revealed *why* the universe is 13 billion years old. Further astrophysical studies revealed *how* stars (including the sun) and planets (including the earth) form. An understanding of *how* stars are formed led to an understanding of *how* elements are created—in exploding stars..

The resolution of the apparently unchanging physical features of the earth was initially slow because no one realized it was an illusion. Early explanations such as Genesis, the basis for creationism, proposed supernatural creation, which explained *who* created the earth but failed to answer *how* or *why* the earth was created. Also, the genealogy from Adam to Jesus allowed the Bishop of Armagh to estimate Earth's age as approximately 6,000 years, and finally, the biblical flood was employed to explain all of Earth's physical features.

Gradually, geologists such as James Hutton, Charles Lyell, Louis Agassiz, and others—after examining such identifiable but dissimilar processes as sedimentary rock formation, erosion, and glaciations—began to realize that the Genesis explanation did not fit their observations. Moreover, it was clear that these processes must have operated over thousands, if not millions of years.

Examination of fossils by the legendary Leonardo da Vinci demonstrated that the biblical flood could not account for his fossil investigations. The discovery of a true mega flood in western Washington State demonstrated what traces a true mega flood left behind, but nothing like it has been found where the biblical flood is supposed to have occurred. Therefore the biblical flood is another myth.

Studies of the apparent similarity between the east coast of the Western Hemisphere with the west coast of the Eastern Hemisphere led geologists to discover one of the strangest phenomena of all—plate tectonics. The granitic continents "float" on a layer of basalt and are in constant, although excruciatingly slow motion. This motion, propelled by the heat generated by radioactive decay, caused the two major hemispheres to part millions of years ago.

Finally, radiometric dating based upon the discoveries of Becquerel using of some of the oldest rocks on Earth and meteorites remnants of the formation of the solar system demonstrated Earth is approximately 4.5 billion years old.

The resolution of the apparently unchanging animal and plant structures of the earth illusion provides an important part of modern biology, especially the process of evolution. The process of resolving this illusion proceeded in two parts: before and after Charles Darwin. Prior to Darwin, several investigators suspected the existence of evolution; however, Darwin was the first person to assemble enough evidence to demonstrate it had occurred. He also developed part of the explanation regarding *how* it had occurred, via natural selection's "filter," which selected the most fit to survive. However, Darwin didn't know *how* the variable input to natural selection provided by reproduction worked. Darwin published the results of his investigations in 1859 under the title, *On the Origin of Species by Means of Natural Selection, or the Preservation of Favoured Races in the Struggle for Life.*

Publication of *The Origin of Species* triggered a much stronger religious reaction than Copernicus's heliocentric explanation. Whereas Copernicus displaced man from the center of the universe, Darwin displaced man from the center of creation. This was unacceptable to the religious community, and the conflict between science and religion that had begun with Copernicus intensified and is still going strong today.

Approximately 150 years after Darwin's death, meiotic cell division, which is required for sexual reproduction of multi-celled animals; the discovery of genes, the blueprint for assembling amino acids into proteins; the discovery of DNA; and the location of genes on DNA all completed the explanation of *how* natural selection functions.

One set of genes from the male and one set genes from the female join to create a genetically unique offspring. Due to the random selection of genes provided by the male and female, the offspring exhibit significant variability of characteristics. Some have superior survival ability while others do not. Accordingly, there is a reasonable probability that evolution's natural selections filter will be able to select those offspring that have improved survival fitness and other characteristics and thus have a higher probability of passing along their genes. This will result in a gradual improvement in life, which is what we observe. As stated in chapter 17, "The entire process of life's continued improvement, beginning with a simple single-celled animal and culminating in modern man, we have named evolution."

Resolution of the fifth illusion is of particular interest because it revealed the details of the physical and biological processes that have shaped the earth. Of particular importance, at the conclusion of chapter 12 creationism and ID had been largely eliminated as an explanation of our origins. Creationism was essentially eliminated due to the existence of twelve independent dating methods, all of which conclusively prove that the earth is approximately 4.5 billion years old, and the proof that the biblical flood, one of the lynch pins of creationism, never occurred. ID was essentially eliminated due to the lack of any evidence of the existence of an intelligent designer in the explanation of the process of evolution.

Building on the material presented in the first twelve chapters, chapter 13 was devoted to the correct explanation of Earth's formation, including both the formation of Earth's physical features and the evolution of life on Earth, which demonstrated the how evolution's natural selection slowly developed improved forms of life.

The first thirteen chapters then set the stage for chapter 14, which provides the final answers the two questions: why are there three origin views and which one is true? It answered those questions by tracing in greater detail the development humans from the earliest known member of our line—*A. Afarensis* to modern humans. A significant aspect of this development was the gradual growth of awareness of ourselves and our surroundings, which appeared approximately 200 KYA. The first result of this awareness was the development of animism, the belief that spirits or gods inhabited, created and controlled everything. It was shown that animism morphed into modern religions and is the basis of creationism—the first option in the question, "Which is the true explanation of our origins?". The further discussion of creationism in chapter 14 completed the elimination of creationism as a viable explanation of our origins.

Chapter 14 next showed how science, the second explanation of ourselves and our surroundings, split off from religious explanations commencing with the Greeks in 500 BCE and developing more extensively, around 1500 CE. Chapter 14 demonstrated that the science-religion split was caused by the growing realization that religiously based explanations, influenced mainly by the five illusions, were in direct conflict with the growing observations and explanations being made by persons who became known as scientists.

The third explanation of our origins, ID, was also discussed in detail in chapter 14 and was completely eliminated as a viable explanation of our origins. A principal reason was the failure of the many examples of complexity "which implies a (supernatural) designer" that were put forward in support of the ID hypothesis, to actually support the hypothesisFurthermore, the intelligent design concept fails as a method for "manufacturing" improved life forms because there is more to manufacturing a living being than design. Moreover, the manufacturing process employed in manufacturing items such as a watch (a "complex item" put forward as an exampled of complexy implies Intelligent Design) is entirely different that the "manufacturing process" employed to make a human being.

Thus, at the end of chapter 14 the first two objectives delineated in the prologue—(1) answering the question of why there are three mutually exclusive explanations of our origins and (2) determining which one is true—had been achieved.

The book then proceeded to related issues such as the short overview presented in chapter 15 of the religious reaction to evolution, which has seriously impeded acceptance of evolution in America and is a significant reason for the Gallup results.

Chapter 16 extended the discussion in chapter 15 with an examination of why, in spite of a vast quantity of evidence against creationism and ID, many people still believe in them. The principle problems are:

- Initial upbringing in an environment—family, friends, and church—that firmly believes in creationism and continually imparts this belief to the growing child.

- The rejection by the child when presented with an explanation totally opposite to what they have been taught by strangers of unknown trustworthiness. This rejection often includes refusal to examine the alternate explanation because the child is afraid he or she might face eternal damnation or loss of an afterlife.

- Wrenching decisions when conflicts between what the child observes in the "real" world when he or she is older and what the child has been taught. Most believers seem to resolve these conflicts by rejecting them because, as discussed above in chapter 16, it is too upsetting to those whose beliefs in their afterlife depends upon a belief in creationism.

However, I suspect there are many others who have also been raised in a religiously fundamentalist family that find as they mature that a literal interpretation of the Bible is inconsistent with what they observe and learn, *but* they are afraid to voice any questions for fear of retribution or ostracism from their family and friends, as discussed in the case of Stephen Godfrey. These persons I designated in chapter 16 as "closet questioners," and are one of the principle audiences for this book.

Chapter 17, "Answers in Evolution," was motivated by the creationism-based Web site "Answers in Genesis." It discussed such topics as why we must die, what life's purpose is, and why we must have sex, which then lead to the demonstration of why evolution is the inevitable consequence of the need to ensure species survival, which unfortunately resulted in Evolution's Fatal Flaw.

Chapter 18 addressed the consequences of the science-religion conflict, which is largely due to the rise of science, especially the discovery and subsequent understanding of the evolutionary process. If there was merely a debate between the correctness of the religious or scientific worldview, one could just decide whether or not to participate in the debate. After all, the conflict "does no harm." However, a closer examination of the various points of conflict demonstrates that the science-religion conflict can have significant impact. Some examples which are religiously based are:

- opposition to vaccination;
- belief in the efficacy of prayer;
- belief in divine intervention in medical problems;
- opposition to abortion and contraception;
- religious opposition to the teaching of evolution in the public schools.

In summary, the facts relevant to the Gallup poll question of which of the three origins options are true—creationism, intelligent design, or evolution?—have been irrefutably established:

- The earth is 4.5 billion years old.
- The biblical flood never occurred.
- No evidence of a creator has been found.
- Reproductive genetic mixing—the input to evolutions natural selection filter—explains the gradual improvement in life on Earth, thereby eliminating the need for an intelligent designer.

Regarding the religious reaction to evolution, first the Copernican heliocentric concept and then Darwin's discovery and promulgation of evolution led to a strong religious reaction, especially to evolution. The most virulent form of this reaction occurred in the United States with the rise of fundamentalism in the mid-1880s. Fundamentalism is still alive and well today.

As I researched and prepared this book, I began to realize that, besides answering the Gallup poll and related questions, evolution also provides a better explanation of and possible solutions to the most perplexing problems confronting humanity than religion, and that it is evolution, not God, that controls our lives and our destinies. It should be noted that there is an interesting difference between God and evolution. Whereas God supposedly somehow cares for everyone, evolution has no particular interest in any specific individual. Evolution is only interested in having the maximum number of individuals survive, thereby providing the highest probability of species survival.

FROM ALPHA TO OMEGA.

Given the basic premise of this book, that evolution is the inevitable consequence of the need to ensure species survival, I believe it is illustrative to briefly trace the need to ensure species survival to the surprising conclusion—evolution's fatal flaw.

We begin with the fact that all multi-celled animals die; hence, species survival requires that enough of the species be replaced to ensure species survival, which requires reproduction. Reproduction requires sexual activity, but since individuals don't require reproduction to survive, sexual need must at times become more important than any other need in order to ensure the production of enough offspring to satisfy the need to ensure species survival. However, the strong sexual need unfortunately leads to excessive offspring, which then results in overpopulation, which ultimately thwarts evolution's need to ensure species survival due to lack of resources. This is evolution's fatal flaw.

Unfortunately, evolution cannot deselect the strong sexual need because that would place species survival in peril. Consequently, the solution to the population control problem lies with evolution's greatest achievement—the human mind.

Accordingly, unless we humans accept the existence of the process of evolution, recognize that our occasionally excessive need for sex is the result of evolution's prime need to assure species survival, abandon our outdated religiously motivated anti-population control attitudes, and take effective population control measures, evolution's greatest achievement may not survive.

ARE SCIENCE AND RELIGION COMPATIBLE?

At the end of the day, the question of whether science and religion are compatible is to some extent one of the more important questions regarding the issues addressed in this book.

There is considerable desire to achieve a compromise such that science and religion could coexist in relative **equanimity.** The compatibility of science and religion has been the subject of many books, e.g., Paul Kurtz, Ed., *Science and Religion, Are They Compatible?*. [23, Chapter 16] and much discussion. However, as Cosmologist Lawrence M. Krause pointed out in a recent Wall Street Journal Article, "A scientist can be a believer. But professionally, at least, he can't act like one." [1].

Dr. Krause's article contains an illustrative quote attributed to British biologist, scientist, professor J. B. S Haldane (1892- 1984) fromHaldane's 1934 book, *Fact and Faith*: "My practice as a scientist is atheistic. That is to say, when I set up an experiment I assume that no god, angel or devil is going to interfere with its course; and this assumption has been justified by such success as I have achieved in my professional ca*reer...*" [1]

Dr Krause also quotes Sam Harris, author of *The End of Faith*, that "a reconciliation between science and Christianity would mean squaring physics, chemistry, biology, and a basic understanding of probabilistic reasoning with a raft of patently ridiculous, Iron Age convictions." [1]

Dr. Krause concludes with "Finally, it is worth pointing out that these issues[compatibilty brtween Science and Religion] are not purely academic. The current crisis in Iran has laid bare the striking inconsistency between a world built on reason [i.e Science] and a world built on religious dogma" [1].

As this book has amply demonstrated Science and Religion are essentially two totally incompatible explanations of ourselves and our surroundings arrived at using two diametrically opposite investigative methods, one based upon rational observation and reason, the other on a belief in the supernatural. Accordingly, they are unfortunately not compatible.

PUTTING IT ALL TOGETHER.

Considerable material has been covered, but there are some particularly salient points of interest. We have established that creationism is not a viable explanation because the earth and the universe can be irrefutably demonstrated to be 4.5 billion years old and 13.7 billion years old, respectively, which is far too old for Young Earth creationism. Intelligent design is an intriguing possibility, but there is no verifiable evidence that an intelligent designer has played a role in the evolution of the earth, and strong arguments have been advanced which show that ID is not a viable concept. Hence, by virtue of Sherlock's theorem, we have an alternate means of establishing evolution is the only viable origins concept; however, I believe the material presented in the first thirteen chapters establishes the viability of evolution in its own right, which firmly validates claim one. Finally the second claim, the basic premise of this book, that evolution is the inevitable consequence of the need to ensure species survival, which is implied by its title, has also been validated. Finally the basic premise of this book, that evolution is the inevitable consequence of the need for species survival, which is implied by its title, has been demonstrated.

This book has indeed presented a long journey of inquiry in which some of the more complex findings the human mind is capable of and some of the most perplexing problems facing this wonderful human mind have been discussed. We exist today because of the great power of the evolutionary process to maximize the probability of species survival and to provide gradual improvement through the input of the results of reproductive genetic mixing into evolution's natural selection filter, which gives the most fit at the time the highest probability of surviving.

Unfortunately, the principal driver of reproductive genetic mixing is an occasionally overriding desire for sexual activity, also provided by evolution's natural selection filter. While the overwhelming desire for sexual activity cannot be realistically controlled, there are readily available means for preventing the results of sexual activity from producing too many offspring. If we are unable and/or unwilling to exercise this option, the devastating population growth plaguing the earth today will continue and, as mentioned above, ultimately doom evolution's greatest achievement.

APPENDIX I:
EXTRACTS FROM GENESIS, CHAPTER 1

(Taken fromthe **BELIEVE Religious Information Source web-site**, J. J Roberts author.)

CREATION OF THE EARTH AND BY IMPLICATION, THE UNIVERSE.

1 (P) In the beginning God created the heaven and the earth.

2 And the earth was without form, and void; and darkness was upon the face of the deep. And the Spirit of God moved upon the face of the waters.

3 And God said, Let there be light: and there was light.

4 And God saw the light, that it was good: and God divided the light from the darkness.

5 And God called the light Day, and the darkness he called Night. And the evening and the morning were the first day.

6 And God said, Let there be a firmament in the midst of the waters, and let it divide the waters from the waters.

7 And God made the firmament, and divided the waters which were under the firmament from the waters which were above the firmament: and it was so.

8 And God called the firmament Heaven. And the evening and the morning were the second day.

9 And God said, Let the waters under the heaven be gathered together unto one place, and let the dry land appear: and it was so.

10 And God called the dry land Earth; and the gathering together of the waters he called Seas: and God saw that it was good.

CREATION OF LIFE.

11 And God said, Let the earth bring forth grass, the herb yielding seed, and the fruit tree yielding fruit after his kind, whose seed is in itself, upon the earth: and it was so.

12 And the earth brought forth grass, and herb yielding seed after his kind, and the tree yielding fruit, whose seed was in itself, after his kind: and God saw that it was good.

13 And the evening and the morning were the third day.

14 And God said, Let there be lights in the firmament of the heaven to divide the day from the night; and let them be for signs, and for seasons, and for days, and years:

15 And let them be for lights in the firmament of the heaven to give light upon the earth: and it was so.

16 And God made two great lights; the greater light to rule the day, and the lesser light to rule the night: he made the stars also.

17 And God set them in the firmament of the heaven to give light upon the earth,

18 And to rule over the day and over the night, and to divide the light from the darkness: and God saw that it was good.

19 And the evening and the morning were the fourth day.

20 And God said, Let the waters bring forth abundantly the moving creature that hath life, and fowl that may fly above the earth in the open firmament of heaven.

21 And God created great whales, and every living creature that moveth, which the waters brought forth abundantly, after their kind, and every winged fowl after his kind: and God saw that it was good.

22 And God blessed them, saying, Be fruitful, and multiply, and fill the waters in the seas, and let fowl multiply in the earth.

23 And the evening and the morning were the fifth day.

24 And God said, Let the earth bring forth the living creature after his kind, cattle, and creeping thing, and beast of the earth after his kind: and it was so.

25 And God made the beast of the earth after his kind, and cattle after their kind, and every thing that creepeth upon the earth after his kind: and God saw that it was good.

26 And God said, Let us make man in our image, after our likeness: and let them have dominion over the fish

of the sea, and over the fowl of the air, and over the cattle, and over all the earth, and over every creeping thing that creepeth upon the earth.

27 So God created man in his own image, in the image of God created he him; male and female created he them.

28 And God blessed them, and God said unto them, Be fruitful, and multiply, and replenish the earth, and subdue it: and have dominion over the fish of the sea, and over the fowl of the air, and over every living thing that moveth upon the earth.

29 And God said, Behold, I have given you every herb bearing seed, which is upon the face of all the earth, and every tree, in the which is the fruit of a tree yielding seed; to you it shall be for meat.

30 And to every beast of the earth, and to every fowl of the air, and to every thing that creepeth upon the earth, wherein there is life, I have given every green herb for meat: and it was so.

31 And God saw every thing that he had made, and, behold, it was very good. And the evening and the morning were the sixth day.

On the seventh day God rested.

APPENDIX II:
THE GEOLOGIC TIME SCALE

EON	ERA	PERIOD	MYA
Phanerozoic	Cenozoic	Quaternary	1.6
		Tertiary	66
	Mezozoic	Cretacious	138
		Jurassic	205
		Triassic	240
	Paleozoic	Permian	290
		Pensylvanian	330
		Mississippian	360
		Devonian	410
		Ordovician	435
		Silurian	500
		Cambrian	570
Proterozoic		Ediacarran	630
Archean			2500
			3800?

Appendix III:
Time Line for the Appearance
of Fossils in Rocks

Species	First appeared	Became extinct
Humans	4–5 MYA	
Primate	30 MYA	
Mammals	100–65 MYA	
Dinosaurs	230 MYA	65 MYA
Reptiles	320 MYA	
Amphibians	420 MYA	
Fish	500 MYA	
Ediacaran Multi-celled Animals	650 MYA	540 MYA
Ediacaran Bacteria	650 MYA	540 MYA
Cyanobacteria	3.5 BYA	

APPENDIX IV:
COMPARISON OF SPECIES
CHARACTERISTICS

Species	Warm-/cold-blooded	Environment	How reproduces	Number of Heart Chambers	Brain and CNS
Fish	Cold	Water exclusively	Female lays eggs, male fertilizes	2	Limited
Amphibian	Cold	Water-land	Female lays eggs, male fertilizes	2	Limited
Reptiles	Cold	Land	Fertilization in female, embryo in shell	3	Reticular
Mammal	Warm	Land	Fertilization in female, live birth	4	Limbic
Primate	Warm	Land	Fertilization in female, live birth	4	Limbic
Human	Warm	Land	Fertilization in female, live birth	4	Neocortex

References for Each Chapter

Prologue

1   Gallup, Home, , (accessed June 6, 2008).

2   Galluo, Evolution-Creationism-Intelligent-Design. (accessed June 6, 2009).

3   emporium.turnpike, mainpts, (accessed June 6, 2008).

4   Thinkexist, once_you_eliminate_the_impossible-whatever/220272

Chapter 1: Electromagnetic Phenomenon Investigations prior to 1897

1   books.google, id=k7Q3AAAAIAAJ

2   Wikipedia, Benjamin_Franklin, Inventions and scientific inquiries (accessed Dec 7, 2008)

3   Wikipedia, Amber (accessed Jul 13, 2008)

4   In2greece, thales (accessed, July 9, 2008)

5   Glasslinks, phoenician (accessed July 10, 2008)

6   Wikipedia Magnetite, (accessed Jul 11, 2008)

7   Dictionary.reference, lodestone

8   Smith, compass2, (accessed July 13, 2008

9   Wikipedia, Giroamo Cardano (accessed July 14, 2008)

10   Wikipedia, De Magnete (accessed July 17, 2008)

11   Rack1, The Magnet (accessed July 11, 2008)

12   Wikipedia Otto von Guericke, (accessed July 17, 2008)

13   Wikipedia C. F. du Fay, Accessed July 17, 2008

14   Wikipedia Electric charge history, ry (accessed Dec 7, 2008)

15   Wikipedia, Leyden_jar, (accessed July 18, 2008)

16   Wikipedia, Electroscope, accessed (July 13, 2008)

17   Physics.kenyon, Electroscope, (accessed July 14, 2008)

18   Wikipedia, Torsion balance (accessed August 4, 2008)

19   Wikipedia, Charles-Augustin de Coulomb (accessed August 4, 2008)

20   Geocities, Sir William Watson, (accesed July 17, 2008)

21   Nndb 742/000091469, , (accessed July 13, 2008)

22   Wikipedia, Luigi_Galvani, (accessed July 13, 2008)

23   Wikipedia, Galvanic cell, (accessed July 18, 2008)

24   Wikipedia, Alessandro Volta, (accessed July 11, 2008)

25   Wikipedia, Voltaic pile (accessed July 11, 2008)

26   Wikipedia, Thomas-Francois Dalibard, (accessed July 14, 2009)

27   Hypertextbook, charge, e/ (accessed July 11, 2008)

28   Biblegateway, Ephesians 2:2, (accessed, Aug 2, 2008)

29   Wikipedia, Hans_Christian_Oersted, sted (accessed Dec 7, 2008)

30   Wikipedia , Michael_Faraday, (accessed July 13, 2008)

31   Wikipedia , William_Sturgeonm (accesed July 18, 2008)

32   Wikipedia, Nicholas_Callan (accessed July 13, 2008)

33   Maxwell, James Clerk. 1865. . *Philosophical Transactions of the Royal Society of London* 155:459–512.

34   Accessexcellence, vision background, (accessed August 6, 2008)

35   Wikipedia, Heinrich Rudolf Hertz, (accessed Dec 7, 2008)

36   Wikipedia, RLC_circuit, (accessed July 10, 2008_

37   Wikipedia, Geissler tube, (accessed, July 14, 2008)

38   Wikipedia, Crookes tube, (accessed July 15, 2008)

39   Wikipedia, G. Johnstone_Stoney, (accessed Jul 20, 2008)

CHAPTER 2: "SOLID" EARTH INVESTIGATIONS PRIOR TO 1897.

1   White, Harvey E.. 1934. *Introduction to Atomic Spectra.* New York: McGraw-Hill.

2   Russell, Bertrand. 1945. *A History of Western Philosophy.* New York: Simon and Shuster.

3   Chapter 1, ref 4

4   Iep, anaximen, (accessed May 20 2008)

5   Wikipedia, Heraclitus. (accessed Jul 7, 2008)

6   Abu.nb.ca, Empedocles, .htm (accessed Jul 7, 2008)

7   Wikipedia, Alchemy, (accessed Dec 7, 2008)

8   Wikipedia, Occult, (accessed Dec 7, 2008)

9   Wikipedia, Philosopher's_Stone, (accessed Dec 7, 2008)

10   Wikipedia, Discovery of the chemical elements, (accessed May 21, 2008)

11   Thefreedictionary, chemistry, (accessed May 21, 2008)

12   Wikipedia, Robert_Boyle, (accessed May 20, 2008)

13   Wikipedia, Jan Baptist van Helmont, (accessed Jan 18, 2009)

14   Holmes, Boynton, ed. 1945. *Beginnings of Modern Science.* Classics Club Edition, Walter J. Black, Roslyn NY

15   Scienceworld.wolfram, Lavoisier, (accessed May 21, 2008)

16   Wikipedia, John Dalton, (accessed Jul, 27, 2008)

17   Wikipedia, Dmitri Mendeleev, (accessed Mar 3, 2008)

18   Chapter 2, reference 4

19   Wikipedia, Thomas_Young,_st) (accessed June 6, 2009)

20   Wikipedia, electromagnetic spectrum, (accessed May 25, 2008)

21 Wikipedia, Balmer series, (accessed July 11, 2008

22 Wikipedia, Johannes_Rydberg, (accessed May 21, 2008)

CHAPTER 3: EMPIRICAL RESOLUTION OF THE "SOLID" EARTH ILLUSION BETWEEN 1897 AND 1911.

1 Wikipedia, J. J. Thomson, disc. cathode rays, (accessed January 2, 2008)

2 Wikipedia, J. J. Thomson, (accessed January 4, 2008)

3 Wikipedia, Wilhelm Conrad Roentgen, (accessed January 3, 2008)

4 Wikipedia, X-ray crystallography, (accessed January 8, 2008)

5 Wikipedia, Arthur Compton, (accessed January 6, 2008)

6 Nobelprize, nobelprize in physics, 1914, (accessed January 9, 2008)

7 hyperphysics.phy-astr.gsu, quantum/bragg, (accessed January 9, 2008)

8 Wikipedia, Bragg diffraction, (accessed January 14, 2008)

9 Wikipedia, Henri Becquerel, (accessed January 16, 2008)

10 Wikipedia, Ernest Marsden, (accessed January 13, 2008)

11 Wikipedia, Ernest Rutherford, , (accessed January 17, 2008)

CHAPTER 4: FINAL RESOLUTION OF THE "SOLID" EARTH ILLUSION, PART I.

1 Wikipedia, Black_body, (accessed January 2, 2008)

2 Wikipedia, Max_Planck, (accessed Dec 8, 2008)

3 Scienceclarified, Photoelectric-Effect, (accessed August 8, 2008)

4 Wikipedia, Einstein: light quanta, (accessed August 8, 2008)

5 Einstein, Albert. 1905. On a heuristic viewpoint concerning the production and transformation of light. *Annalen der Physik* 17:132–148.

6 Wikipedia, Niels_Bohr, (accessed May 25, 2008)

CHAPTER 5: FINAL RESOLUTION OF THE "SOLID" EARTH ILLUSION, PART II.

1   Hyperphysics.phy-astr, de Broglie, (accessed Jan 25, 2009)

2   Davisson, Clinton J., and Lester H. Germer. 1927. Reflection of electrons by a crystal of nickel. Nature 119:558–560.

3   wikipedia, Stern-Gerlach experiment (accessed Sep 13, 2008)

4   W. Heisenberg. 1927. Über den anschaulichen Inhalt der quantentheoretischen Kinematik und Mechanik," trans. J. A. Wheeler and H. Zurek. *Quantum Theory and Measurement* (Princeton University Press, 1983): 62–84.

5   . 1926. the Mechanics of Atoms and Molecules. *Physical. Review*. 28 (6): 1049–1070

6   wikipedia, Quantum_tunneling, (accessed July 20, 2008)

7   wikipedia, Proton [discovered] by Rutherford, (accessed July 25, 2008)

8   Davidparker, [nuclear] twins, (accessed July 24,2008)

9   Laurenvondehsen, Rutherford predicts neutron, (accessed July 30, 2008)

10   wikipedia, Atomic_mass_unit, (accessed Dec 8, 2008)

11   Chemcases, neutron discovery (accessed July 2, 2008)

12   wikipedia, Beta_decay, (accessed July 22, 2008)

13   wikipedia, Nuclear magnetic resonance, (accessed June 29, 2009)

14   Answers, isotope, (accessed July 26, 2008)

15   Colorado, isotopes, (accessed July 27, 2008)

16   Wikipedia, Deuterium, (accessed January 19, 2009)

17   Wwikipedia, Tritium, (accessed January 19, 2009)

18   Wikipedia, Quantum_chromodynamics, s (accessed May 26, 2008)

19   Wikipedia, Mark_Oliphant, nuclear fission, (accessed August 8, 2008)

20   Atomicarchive, Fission2, l (accessed August 8, 2008)

21   Wikipedia, Nuclear_fission, (accessed August 8, 2008)

22   Wikipedia, Binding energy curve, (accessed May 24, 2008)

23   Hyperphysics.phy-astr, alpha particle half-life, (accessed May 24, 2008)

24   Nobelprize. Laureates in 1979, (accessed May 28, 2008)

25   Noblemind, number protons in argon, (accessed May 14, 2008)

26   Wikipedia,. Gordian_Knot, (accessed Mar 1, 2009

CHAPTER 6: SOLVING THE APPARENT MOTION OF THE SUN AND PLANETS AROUND THE EARTH ILLUSION.

1   Scienceworld.wolfram, Aristarchus, (accessed June 14, 2008)

2   Syvum, trigonometry, (accessed Dec 11, 2008)

3   Phy6, Distance to Moon, (accessed Dec 11, 2008)

4   Scienceworld.wolfram, Aristarchus, (accessed August 12, 2008)

5   Wikipedia,. Alexander_the_Great, (accessed Dec 11, 2008)

6   Csep10.phys.utk, aristotle, (accessed Dec 10, 2008)

7   Wikipedia,. Ptolemy, (accessed June 6, 2008)

8   Wikipedia,. Almagest, (accessed June 6, 2008)

9   Wikipedia,. Nicolaus_Copernicus, (accessed June 6, 2008)

10   Wikipedia,. De revolutionibus orbium coelestium, (accessed August 12, 2008)

11   Cnx, Galileo's Telescope, (accessed January 14, 2009)

12   Wikipedia,. Galileo_Galilei, (accessed Dec 12, 2008)

13   Wikipedia,. Tycho_Brahe, (accessed June 6, 2008)

14   Wikipedia,. Johannes_Kepler, (accessed Dec 10, 2008)

15   Adsabs.harvard, SAO/NASA Astrophysics Data System (ADS), (accessed Dec 12, 2008)

16   Newton.cam.ac, Newtons life, (accessed Dec 11 2008)

17   Wikipedia,. Robert_Hook, (accessed Dec 11 2008)

18   Exerpt from Microsoft Encarta (accessed Dec 11 2008)

19   Wikipedia,. Kepler's First Law (accessed Dec 10, 2008)

20   Wikipedia,. Pierre-Simon_Lapla (accessed Dec 11 2008)

21   Wikipedia,. Pierre-Simon Laplace and Napoleon, (accessed Dec 11 2008)

22   Wikipedia,. Theory_of_relativity, (accessed Dec 11 2008)

23   Hawking, Stephen. 1988. *A Brief History of Time.*[New York] Bantam Dell Publishing Group

24   Geocities, galileo, (accessed June 6, 2008)

25   Inventors.about, glass, (accessed Oct 8, 2008)

CHAPTER 7: RESOLVING THE APPARENT SAME MOON AND SUN SIZE AND DISTANCE ILLUSIONS.

1   Juliantrubin, eratosthenes ml (accessed Aug 12, 2008)

2   Wikipedia, Giovanni_Domenico_Cassini, sini (accessed Dec 14, 2008)

3   Physicsforums, Measure Earth-Moon Distance, (accessed Dec 14, 2008l)

4   Astrosociety, mercury orbit, (accessed Dec 14, 2008)

5   Wikipedia, Ole Roemer, (accessed Dec 14, 2008)

CHAPTER 8: RESOLVING THE APPARENT SAME MOON AND SUN SIZE AND DISTANCE ILLUSIONS.

1   Answers, airy disc, Dec 15, 2008)

2   Thefreedictionary, Cosmological, (accessed Dec 15, 2008)

3   Clark, D.H., and F. R. Stephenson. . . Supernovae: A survey of current research"; *Proceedings of the Advanced Study Institute*: 355–370, Cambridge, England: Dordrecht, D. Reidel Publishing Co

4   Howstuffworks, question94, , (accessed Dec 15, 2008)

5   Hyperphysics.phy-astr.gsu, stellar parallax, (accessed Dec 15, 2008)

6   Wikipedia, John_Goodrick, , (accessed Dec 15, 2008)

7   Wikipedia, Cepheid_variable,

8   Leavitt, Henrietta S. 1908. 1777 Variables in the Magellanic Clouds. *Annals of Harvard College Observatory* LX(IV): 87–110.

9   Leavitt, Henrietta S, contribution to Pickering, Edward C. "1912. Periods of 25 Variable Stars in the Small Magellanic Cloud. *Harvard College Observatory Circular* 173:1–3.

10   Cosmology.berkeley, Brightness-distance discussion, (accessed Dec. 8, 2008)

11   Universeadventure, light-magnitude, (accessed Dec. 8, 2008)

12   Wikipedia, Lowell_Observatory, (accessed Dec 15, 2008)

13   Slipher, Vesto Melvin. 1912. The radial velocity of the Andromeda Nebula. *Lowell Observatory Bulletin* 1:2.56–2.57. See also: Slipher. Spectroscopic Observations of Nebulae. *Popular Astronomy* (1910) 23:21–24.

14   Wikipedia, Alexander_Friedman, Dec 1, 2008)

15   Wikipedia, Edwin_Hubble, Dec 1, 2008)

16   Antwrp.gsfc.nasa, sandage hubble, (accessed Dec 1, 2008)

17   Umich, bigbang, June 16, 2008)

18   Wikipedia, George_Gamow, June 16, 2008)

19   bell-labs, cosmology, (accessed October 8, 2008)

20   Hyperphysics.phy-astr.gsu, bbang, (accessed October 7, 2008)

21   space.wikia, Wilkinson Microwave Anisotropy Probe, (accessed October 7, 2008)

22   **Adrian Cho,** A Singular Conundrum: How Odd Is Our Universe? 2007 News Focus, Science 28 SepetemberVOL 317

23   Adrian Cho, 2006, Long-Awaited Data Sharpen Picture of Universe's Birth, *Science*, VOL: 311 24

24   Wikipedia, Star_formation, **(accessed** Dec 15, 2008)

25   Firstgalaxies.ucolick, cosmos, **(accessed** Dec 15, 2008

26   Wikipedia, Milky_Way, **(accessed** Dec 15, 2008

27   Map.gsfc.nasa, rel_stars, (**accessed** Dec 15, 2008

28   Robert Jastrow. 1967. *Red Giants and White Dwarfs*. New York: Harper and Row.

29   Cowan, Ron,An orb, dubbed MOA-2007-BLG-192-Lb and weighing only 3.3 times Earth is found, … the 300th extra solar planet found to date, Science News, July 5, 2008, pp 16-25

30   Planetary Science Institute (PSI), planet formation simulation, (accessed June 22, 2008)

31   J. C. B. Papaloizou 2008 Planetary System Formation, Science VOL 321

32   Gotquestions, big-bang-theory, (**accessed Dec 5, 2008**)

CHAPTER 9: RESOLVING THE APPARENTLY UNCHANGING PHYSICAL FEATURES OF THE EARTH ILLUSION UP TO 1850.

1   Yourdictionary, algebra, (**accessed** Dec 16, 2008)

2   Wikipedia, Geology, (**accessed** Dec 2, 2008)

3   Salam, Abdus. 1987. Islam and Science. In *Ideals and Realities: Selected Essays of Abdus Salam*, by C. H. Lai. 2nd ed. Singapore: World Scientific.

4   Wikipedia, Avicenna (**accessed** August 11, 2008)

5   Winchester, Simon. 2001. *The Map the Changed the World*. New York:Harper Collins

6   Emporium.turnpike, mainpts (**accessed** Dec 17, 2008)

7   Mb-soft, genesis, (**accessed** September 24, 2008)

8   Wikipedia, Homer, (**accessed** September 25, 2008)

9   Emporium.turnpike, mainpts, (**accessed** Dec 17, 2008)

10   Wikipedia, James_Ussher, (**accessed** September 24, 2008)

11   Answersingenesis, catastrophic-plate-tectonics, (accessed June 1, 2008)

12   Christiananswers, aig-c001, (**accessed** June 1, 2008)

13   Nwcreation, fossils, **(accessed** June 1, 2008)

14   Christiananswers, floodwater, tml **(accessed** September 25, 2008)

15   Christiananswers, aig-c010, **(accessed** September 25, 2008)

16   Christiananswers, floodwater tml **(accessed** September 25, 2008)

17   Christiananswers, (accessed August 11, 2008)

18   News and Views Volume 57, Number 2, **(accessed** June 2005)

19   Blodget, A. 2005. Results of a Survey of Archaeologists on the Biblical Flood. *News & Views* 57 (2)

20   Pantheon, deucalion, **(accessed** Sep 20, 2008)

21   In2greece, deucalion, (accessed Sep 20, 2008)

22   Pbs, megaflood, **(accessed** Sep 20, 2008)

23   Vulcan.wr.usgs, description lake missoula, (accessed Sep 20, 2008)

24   Pardee, Montana Joseph Thomas. 1942. Unusual currents in glacial Lake Missoula. *Geological Society of America Bulletin* 53 (11):1569–99.

25   Query.nytimes Iceland zooms in on signs of unrest, DE7DA 143FF934A25752C0A9609C8B63&sec=&spon=&pagewa nted=all **(accessed** Sep 20, 2008)

26   Waitt, Richard B. 1985. "Case for periodic, colossal joekulhlaups from Pleistocene glacial Lake Missoula" *Geological Society of America Bulletin* 96 (10):1271–86.

27   Uwsp, Scablands, (accessed Sep 20, 2008)

28   Frazier, J.B. 2003. Massive ice age floods are largely unmarked—move afoot to mark Missoula Floods with interpretive flood pathway across four states. *Yakima (WA) Herald-Republic,* **(accessed** October 5).

29   Naturaltreasure, genfossils, **(accessed** Dec 17, 2008)

30   Wikipedia, neous_generation **(accessed** September 21, 2008)

31   Rick Steves travel information series on PBS, Milan and Lake Como, dates various

32   Mayer, A. 2001. *The First Fossil Hunters: Paleontology in Greek and Roman Times.* New Jersey: Princeton University Press.

33   *Fabulous fables formed from facts,* Stephen Haine's. 2005 Amazon.com book review of Mayor, A, The First Fossil Hunters:

34   Wikipedia, Leonardo as observer.2C scientist and inventor, (accessed May 31, 2008)

35   Wikipedia, Leonardo_da_Vinci, **(accessed** Sep 25, 2008)

36   Gould, Stephen Jay. 1998. *Leonardo and his mountain of clams.* New York: Harmony Books.

37   Wikipedia, Leonardo_da_Vinci) **(accessed** August 11, 2008)

38   Wikipedia, Neptunism **(accessed** Oct 3, 2008)

39   Wikipedia, Siberian_Traps, **(accessed** Dec 16, 2008)

40   Wikipedia, Plutonism, **(accessed** Oct 3, 2008)

41   Wikipedia, Catastrophism, **(accessed** Oct 3, 2008)

42   Wikipedia, Luis_Walter_Alvarez, **(accessed** Oct 3, 2008)

43   Geolsoc features, (accessed September 28, 2008)

44   Wikipedia, Churchill, Oxfordshire, (accessed September 28, 2008)

45   Wikipedia, William Smith (geologist), (accessed September 28, 2008)

46   James Hutton, biography, **(accessed** September 29, 2008)

47   James Hutton, "James Hutton the Farmer," (accessed September 29, 2008)

48   Strangescience, hutton, (accessed September 29, 2008)

49   James Hutton.org., "Return to Slighhouses and Farm Improvement." (accessed September 29, 2008)

50   Infoplease, people/A0839343, (accessed September 29, 2008)

51 James Hutton.org.,The Theory of the Earth. Retrieved on 2008-04-11. (accessed September 29, 2008)

52 Amnh, "James Hutton: The Founder of Modern Geology." (accessed September 29, 2008)

53 Wikipedia, Jean_de_Charpentier, **(accessed** September 25, 2008)

54 Wikipedia, Karl_Friedrich_Schimper,er **(accessed** September 25, 2008)

55 Wikipedia, Bill_Bryson, **(accessed** Dec 17, 2008)

56 Wikipedia, Short History of Nearly Everything, (accessed September 29, 2008)

57 Wikipedia, Louis_Agassiz, f-4 **(accessed** September 30, 2008)

58 fr.wikisource, Agassiz, Louis. 1840. *Études sur les glaciers. Neuchâtel.* Digital book on Wikisource, (accessed February 25, 2008).

59 Wikipedia, William_Buckland, **(accessed** October 2, 2008)

60 Wikipedia, Roderick_Murchison. **(accessed** October 2, 2008)

61 Wikipedia, Adam_Sedgwick, **(accessed** October 2, 2008)

62 Wikipedia, Silurian, **(accessed** October 2, 2008)

63 Geo.ucalgary, time scale (accessed December 16, 2008)

64 Jersey.uoregon, geoQuerry38, (accessed October 6, 2008)

65 Wikipedia, Charles_Lyell. **(accessed** October 2, 2008)

66 Wikipedia, Uniformitarianism, **(accessed** October 15, 2008)

67 Penelope.uchicago, serapeum, (accessed December, 16, 2008)

68 Wikipedia, Ammianus Marcellinus, **(accessed** *December 16, 2008)*

69 Victorianweb, lyell, **(accessed** August 11, 2008)

70   Lyell, Charles, 1830 *"Principles of Geology 1st vol. 1st edition*, Jan. , London

71   Mnsu, lyell charles, (accessed August 11, 2008)l

72   Adams, Frank D. 1938. *The Birth and Development of the Geological Sciences.* Dover Publications, Inc

73   Bailey, Edward. 1962. Charles Lyell. Nelson, London. referenced in f-czscax_0-0 **(accessed** October 3, 2008)

74   Wikipedia, Charles_Lyell, **(accessed** December, 16, 2008+ .

CHAPTER 10: RESOLVING THE UNCHANGING PHYSICAL FEATURES OF THE EARTH ILLUSION AFTER 1850.

1   Wikipedia, History_of_the_Azore, **(accessed** December 17, 2008)

2   Ashe, Thomas. 1813. History of the Azores, or. Western islands. Oxford University Oxford, U.K.

3   Mariner, ageofex [age of exploration] **(accessed** October 7, 2008)

4   Ucmp.berkeley, wegener, **(accessed** October 7, 2008)

5   Wikipedia, Trilobite, **(accessed** October 7, 2008)

6   Answers, arthur-holmes, **(accessed** October 15, 2008)

7   Thefreedictionary, tectonics, **(accessed** October 15, 2008)

8   Jersey.uoregon, geoQuerry75 [Granite specific gravity], (accessed October 15, 2008)

9   Tamu, chap 04 [Earth's physical propeties], (accessed October 7, 2008)

10   Wikipedia, Harry_Hammond_Hess, **(accessed** August 11, 2008)

11   Waterencyclopedia, Submarines-and-Submersibles, (accessed October 24, 2008)

12   Infoplease, A0814954, [Deep Sea Drilling Project], **(accessed** August 11, 2008)

13 Platetectonics, page_15, [Fault description], **(accessed** Dec 17, 2008)

14 Wikipedia, Subduction, **(accessed** October 7, 2008)

15 Wikipedia, Orogenic **(accessed** October 7, 2008)

16 Geocities, prehistory, ml **(accessed** October 7,2008)

17 Wikipedia, Pacific_Plate, **(accessed** October 7,2008)

18 Sierrahistorical, geology tml **(accessed** October 7,2008)

19 Wikipedia, Seismology, **(accessed** October 6, 2008

20 Infoplease, A0861016 [seismology development], **(accessed** October 7, 2008)

21 Wikipedia, John_Winthrop, 779) **(accessed** October 7, 2008)

22 Enotes, richter-charles-s-f **(accessed** October 7, 2008)

23 Wikipedia, Seismic_wave, **(accessed** October 9, 2008)

24 Pubs.usgs, interior, **(accessed** November 2, 2008)

25 Uh, epi144, [Earth's age by John H, Lienhard], **(accessed December 17, 2008)**

26 Gavin, S., J. Conn, and S. P. Karrer.2008"The Age of the Sun: Kelvin vs. Darwin,", Physics Department, Wayne State University, Detroit, MI, 48201

27 Hyperphysics.phy-astr, clkroc [clocks in the rocks], (accessed November 17, 2008)

28 Mysite, holmes, **(accessed** October 8, 2008)

29 Talkorigins, faq-age-of-earth, tml **(accessed** October 8, 2008)

30 Sciencedirect, [Earth's age, meteorite dating, **(accessed June 2, 2008)**

31 Austarnet, as old as time plimer, (accessed June 2, 2008)

32 Wikipedia, Accelerator mass spectrometry, (accessed May 12, 2009)

33 Manseau, Peter. 2009. Review of *Dating the shroud of Turin*. *Wall Street Journal*, April 11.

34   Astro, age [of universe dating methods by Edward L. Wright] **(accessed** January 27, 2009)

35   Dalrymple, G. Brent. U. S. Geological Survey. "How Old is the Earth, A reply to Creationism" in "Evolutionists confront Creationists", paper presented at the Proceedings of the 63rd Annual Meeting of the Pacific Division, AAAS, April 30, 1984

CHAPTER 11: RESOLVING THE APPARENTLY UNCHANGING ANIMAL AND PLANT STRUCTURES ILLUSION. AFTER DARWIN, EXPLAINING NATURAL SELECTION

1   Singer, Charles. 1931. *A short history of biology.* Oxford University Press, Oxford, U.K.

2   Emily Kearns. 1996. Animals, knowledge about. . 3rd ed., 92, Oxford University Press, Oxford, U.K

3   Chapter 9, reference 32.

4   Buffon.cnrs, [Compte de]Buffon, (accessed December 18, 2008)

5   Ucmp.berkeley, Compte deBuffon, **(accessed June 13, 2008)**

6   Wikipedia, Jean-Baptiste_Lamarck, **(accessed October 10, 2008)**

7   Gould, Stephen Jay. 2002. . Harvard: Belknap Harvard, 170–197.

8   Coleman, William L. 1977. *Biology in the Nineteenth Century: Problems of Form, Function, and Transformation.* Cambridge: , 1–2.

9   Ucmp, cuvier, **(accessed October 10, 2008)**

10   Victorianweb, cuvier, **(accessed October 10, 2008)**

11   Wikipedia, Louis Agassiz, (accessed October 9, 2008)

12   Ucmp, haeckel, **(accessed October 10, 2008)**

13   Wikipedia, Charles_Darwin, **(accessed June 3, 2008)**

14   Wikipedia, The_Origin_of_Species, **(accessed June 4, 2008)**

15 Wikipedia, Beagle **(accessed October 10, 2008)**

16 gct [Galapagos Conservation Trust], Galapagos Islands introduction, (Accessed October 14, 2008)

17 Wikipedia, John_Gould, **(accessed October 10, 2008)**

18 lilt.ilstu, application of Darwinian ideas to creativity, (accessed October 10, 2008)

19 Wikipedia, Thomas_Malthus, **(accessed October 10, 2008)**

20 Marcus, Gary. 2009. Forget About Survival of the Fittest. *Wall Street Journal*, February 11.

21 *Lyell, Charles, Geological Evidences of the Antiquity of Man* 1 vol. 1st edition, Feb. 1863. London: John Murray.

22 Info regarding Darwin in ask.com (Accessed Dec 20, 2008)

23 Wikipedia, Alfred_Russel_Wallace, (Accessed Dec 20, 2008)

24 Wrigtht, Tom, Alfred Russel Wallace's Fans Gear Up for a Darwinian Struggle, *Wall Street Journal*. 2008. December 20

CHAPTER 12: RESOLVING THE APPARENTLY UNCHANGING ANIMAL AND PLANT STRUCTURES ILLUSION AFTER DARWIN, EXPLAINING NATURAL SELECTION.

1 Accessexcellence, Gregor Mendel, (accessed October 12, 2008)

2 Wikipedia, Carl_Correns, (Accessed Dece 20, 2008)

3 Wikipedia, Erich Tschermak von Seysenegg, (accessed December 20, 2008)

4 Wikipedia, Hugo_De_Vries, (Accessed December 20, 2008)

5 Microbiology, anthony van leeuwenhoek, (accessed October 8, 2008)

6 Robert Hook names basic biological unit the cell, The American Naturalist, Vol.73 pgs 517-537, cited in Wikipedia, Cell theory, (accessed October 9, 2008)

7 Wikipedia, Theodor_Schwann, (Accessed December 21, 2008)

8   Wikipedia, Matthias_Jakob_Schleiden, den (Accessed Decemb er 21, 2008)

9   Schleiden, Matthias Jakob 1839,"Contributions to Phytogenesis", cited in (Accessed October 9, 2008)

10   Porpax.bio.miami, cell, (accessed December 21, 2008)

11   Wikipedia, Cell_theory, (Accessed October 9, 2008)

12   Wikipedia, Walther_Flemming, (Accessed October 10, 2008)

13   Wikipedia, Meiosis, (Accessed October 10, 2008)

14   Wikipedia, Edouard_Van_Beneden, (Accessed October 11, 2008)

15   Wikipedia, Thomas_Hunt_Morgan, (Accessed October 11, 2008)

16   Wikipedia, Friedrich_Miescher,

17   Microbiologyprocedure, history-of-nucleic-acids, (accessed October 10, 2008)

18   Fischer, E. 1899. *Berichte der Deutschen Chemischen Gesellschaft 32*: 2550

19   Wikipedia, Richard_Altmann, (Accessed October 12, 2008)

20   Wikipedia, Base_pair, December 22, 2008)

21   Answers, albrecht-kossel, (Accessed October 13, 2008)

22   Wikipedia, Phoebus_Levene, (Accessed October 12, 2008)

23   Levene, P. 1919. . *J Biol Chem* 40 (2): 415–24.

24   Wikipedia, Robert_Feulgen, (Accessed October 12, 2008)

25   Wikipedia, Frederick_Griffith, (Accessed October 12, 2008)

26   Wikipedia, William_Astbury, (Accessed October 12, 2008)

27   Avery, O., C. MacLeod, and M. McCarty. 1944. *nature of the substance inducing transformation of pneumococcal types. Inductions of transformation by a desoxyribonucleic acid fraction isolated from pneumococcus type III*. J Exp Med 79 (2): 137–158.

28   Wikipedia, Oswald_Theodore_Avery, (Accessed October 12, 2008)

29   Hershey, A., and M. Chase. 1952. protein and nucleic acid in growth of bacteriophage. J *Gen Physiol* 36 (1): 39–56.

30   Porpax.bio.miami, DNAdiscovery, (accessed October 16, 2008)

31   Watson, J. D., and F. H. C. Crick. 1953. . *Nature* 171:737–738.

32   Watson, James D. 1968. *The Double Helix: A Personal Account of the Discovery of the Structure of DNA.* New York: Touchstone.

33   Brown, A Reader's Guide to The Double Helix, e.htm (Accessed October 16, 2008)l

34   Wikipedia, Protein_structure, (Accessed October 19, 2008)

35   Newscientist, amino-acid-found-in-deep-space, (accessed October 19, 2008)

36   Wikipedia, Base_pair, (Accessed October 19, 2008)

37   Wikipedia, Crick, Brenner et al. experiment, (accessed October 19, 2008)

38   Wikipedia, Codon, (October 19, 2008)

39   Wikipedia, Ribosome, (Accessed October 19, 2008)

40   Publications.nigms.nih, How Genes Work, (accessed October 19, 2008)

41   Wikipedia, Phillip_Allen_Sharp, (Accessed October 19, 2008)

42   Wikipedia, Exon, (Accessed Ocotber 19, 2008)

43   Wikipedia, Epigenetics, (Accessed February 10, 2009)

44   Venter, J. C., et al. 2001. *Science* 291:1304-1351.

45   Wikipedia, Meiosis, (Accessed Ocitober 20, 2008)

46   Wikipedia, Genetic_recombination, (Accessed October 20, 2008)

CHAPTER 13: RESOLVING THE APPARENTLY UNCHANGING FEATURES OF THE EARTH ILLUSION— HOW THE EARTH WAS REALLY CREATED.

1   Wikipedia, Zircon, (accessed October 21, 2008

2   Valley, John W., et al 2002. . *Geology* 30:351–354.

3   Eurekalert, ancient mineral shows early Earth climate tough on continents, (accessed October 21, 2008)

4   Pmel.noaa, pillows, (accessed October 21, 2008)

5   Furnes, H., et al. 2004. Early Life Recorded in Archean Pillow Lavas, *Science* (April 23): 304.

6   lpl.arizona.edu, CampinsDrake2006, (accessed October 21, 2008)

7   Wikipedia, Origin_of_water_on_Earth, rth (accessed October 21, 2008)

8   Morbidelli, A., et al. (2000) Source regions and timescales for the
        delivery of water to the Earth 35, Vol 35, 1309–1329

9   Space.newscientist, Earth's water brewed at home, not in space, (accessed October 21, 2008)

10   Wikipedia, Shark Bay, Western Australia, (accessed November 9, 2008)

11   Emc.maricopa, Cell Division: Binary Fission and Mitosis, (accessed November 15, 2008)

12   Cst.cmich, basalt , (accessed November 15, 2008)

13   Livingston, D. E., E. E.Brown, and C. Malcolm. 1974. Rb-Sr whole rock isochron ages for "older" Precambrian plutonic and metamorphic rocks of the Grand Canyon, Arizona. *Geological Society of America Abstracts with Programs* 6 (7): 848.

14   Goaustralia.about, stromatolites, .htm (accessed October 22, 2008)

15   Wikipedia, Banded iron formation, (accessed October 22, 2008)

16   Perkins, (2009) Plate tectonics got an early start: the chemistry of minerals preserved in Australian rocks suggests tectonic activity for Earth's earliest eon, *Science News,*

17   Wikipedia, Paleomagnetism, (accessed October 24, 2008(

18   Wikipedia, Magnetic, (accessed October 24, 2008)

19   Peripatus.gen, Rodinia, tml (accessed October 25, 2008)

20   Wikipedia, Neoproterozoic, (accessed October 25, 2008)

21   Wikipedia, Cryogenianm (accessed November 8, 2008)

22   Ucmp, vendian, (accessed October 25, 2008)

23   Emetz, Alexander et al (2004) Geological Framework of the Volhyn Copper Fields... Asgp [Annales Societatis Geologorum Poloniae], referenced in, asgp.pl/2004/74 3/257-265, (accessed October 25, 2008)

24   McMenamin, Mark A., and Dianna L. McMenamin. (1990) The Rifting of Rodina. *The Emergence of Animals*. Columbia University Press, New York, referenced in Wikipedia, Mirovia, (accessed October 25, 2008)

25   Geocities, Origins/transitions, (accessed October 25, 2008)

26   Pbs, The Cambrian Explosion, (accessed November 15, 2008)

27   Wikipedia, Deuterostomes, (Accessed November 10, 2008)

28   Buckland, W. 1841. *Geology and Mineralogy Considered with Reference to Natural Theology*. Lea & Blanchard publication location unjnown.

29   Darwin, C. 1859. *On the Origin of Species by Natural Selection*. John Murray, London.

30   Wikipedia, Prehistoric fish, (accessed October 28, 2008)

31   Wikipedia, Devonian, (accessed October 31, 2008)

32   Answerbag, Do fish drink water?, (accessed November 8, 2008

33   Daeschler, Edward B., Neil H. Shubin, and Farish A. Jenkins, Jr.(2006) and the evolution of the tetrapod body plan. 440:757–763.

34   , Jennifer A. 2005. , *Scientific American*, November 21.

35   Wikipedia, Tiktaalik, (accessed October 31, 2008(

36   Shubin, Neil. 2008. *Your Inner Fish*. Pantheon Books, New York

37   Wikipedia, Neil_Shubim (accessed October 1,2008)

38   Backyardnature, amphibs, (accessed October 28, 2008)

39   Wikipedia, Carboniferous, (accessed October 31, 2008)

40   Wikipedia, Amniote, (accessed October 31, 2008)

41   Wikipedia, Edwin_Stephen_Goodrich, h (accessed October 31, 2008)

42   Wikipedia, Reptile, (accessed October 28, 2008)

43   Wikipedia, Archaeothyris, (accessed Novenber 11, 2008)

44   Crystalinks, reptilianbrain, (accessed Novenber 20, 2008)

45   Wikipedia, Permian-Triassic extinction event, (accessed October 28, 2008)

46   Wikipedia, Roderick_Murchison,   (accessed Novenber 2, 2008)

47   Wikipedia, Permian, (accessed Novenber 2, 2008)

48   Wikipedia, Archosaur, (accessed Novenber 2, 2008)

49   Wikipedia, Cynodonm (accessed Novenber 2, 2008)

50   Wang, Jin YG, et al., Pattern of Marine Mass Extinction Near the Permian–Triassic Boundary in South China. *Science* 289 (5478): 432–436.

51   Ayres, Yolanda. What Really Killed the Dinosaurs. BBC Horizon, Science & Nature TV Program. (accessed Novenber 2, 2008)

52   Wikipedia, Triassic-Jurassic(accessed November 2, 2008)

53   Monographie des Bunten Sandsteins, Muschelkalks und Keupers, und die Verbindung dieser Gebilde zu einer Formation (Stuttgart-Tübingen: Cotta), 1834

54   Wikipedia, Pangaea, (accessed November 2, 2008)

55   Pope, K.O., A.C. Ocampo, G.L. Kinsland, and R. Smith. 1996. Surface expression of the Chicxulub crater. *Geology* 24 (6): 527–530.

56   Wikipedia, Triassic, (accessed November 2, 2008)

57   Wikipedia, Archosaur, (accessed November 2, 2008)

58   Wikipedia, Dinosaur, (accessed November 1, 2008)

59   Wang, S.C., and P. Dodson. 2006. Estimating the Diversity of Dinosaurs. *Proceedings of the National Academy of Sciences USA* 103 (37): 13601–13605.

60   Bakker, R. T., and P. Galton. 1974. Dinosaur monophyly and a new class of vertebrates. *Nature* 248:168–172.

61   Alvarez, L. W., et al. 1980. Extraterrestrial cause for the Cretaceous-Tertiary extinction. *Science* 208:1095–1108.

62   Wikipedia, Deccan Traps, (accessed November 3, 2008)

63   Wikipedia, Mammal, (accessed November 3, 2008)

64   Wikipedia, Limbic system, (accessed November 20, 2008)

65   Wikipedia, Eocene, (accessed November 3, 2008)

66   Stehlen, H.G. 1910. Remarques sur les faunules de Mammifères des couches eocenes et oligocenes du Bassin de Paris. *Bulletin de la Société Géologique de France,* "4 (9): pp 488-520."

67   Diamond, Jared. 1997. *Guns, Germs and Steel, the fates of human societies.* New York: W.W Norton & Co.

68   Goodman, M., et al. 1990. Primate evolution at the DNA level and a classification of hominoids. *Journal of Molecular Evolution* 30:260–266.

69   News24, Human-ape split 'pushed back', (accessed November 3, 2008)

70   Physorg, Scientists narrow time limits for human, chimp split, (accessed November 20, 2008)

71   Wikipedia, Australopithecus afarensis, (accessed November 3, 2008)

72   Pbs, Johanson finds 3.2 million-year-old Lucy, 1974, u.html (accessed November 3, 2008)

73   Wikipedia, Homo_habilis, (accessed November 3, 2008)

74   Wikipedia, Homo_erectus, (accessed November 3, 2008)

75   Zach Zorich First European, Atapuerca, Spain *Archeology* Volume 62 Number 1, Jan/Feb 2009

76   Boehm, Christopher. 1999. *Hierarchy in the forest: the evolution of egalitarian behavior.* Cambridge: Harvard University Press.

77   Hesman Saey, Tina. August 7, 2008. Neandertal mitochondrial DNA deciphered. (accessed November 6, 2008)

78   Pennisi, Elizabeth. 2009. Tales of Prehistoric Human Genome. *Science Magazine* (Feb. 13):866.

79   Wikipedia, Neandertal, (accessed November 6, 2008)

80   Wikipedia, Cro-Magnon, (accessed November 6, 2008)

81   Wikipedia, Cerebral_cortex, (accessed November 20, 2008)

82   Dictionary.com Unabridged (v 1.1). Random House, Inc. 23 Sep. 2008.

83   Soultré Musée ,*Introduction to Prehistory*, Dept de Préhistoire, France

84   Wikipedia, Acheulian, (accessed November 11, 2008)

85   Bar-Yosef, O., and A. Belfer-Cohen. 2001. From Africa to Eurasia—Early Dispersals, Quaternary International 75:19–28.

86   Scarre, C., ed. 2005. *The Human Past.* London: Thames and Hudson.

87   Wikipedia, Mousterian, (accessed November 5, 2008)

88   Wikipedia, Chatelperronia (accessed November 5, 2008)

89   Wikipedia, Aurignacian, (accessed November 6, 2008)

90   Wikipedia, Gravettian, (accessed November 7, 2008)

91   Wikipedia, Burin, (accessed November 7, 2008)

92   Wikipedia, Solutrean, (accessed November 7, 2008)

93   Wikipedia, Magdalenian, (accessed November 7, 2008)

94   Wikipedia, Mesolithic, (accessed November 7, 2008)

95   Wikipedia, Neolithic, (accessed November 7, 2008)

96   Archaeology.about, chalcolithic, (accessed November 7, 2008)

97   Wikipedia, Fertile_Crescent, (accessed November 7, 2008)

98   Thefreedictionary, Hominidae,   (accessed November 7, 2008)

99   Ardry, Robert. 1961. *African Genesis, A Personal Investigation into the Animal Origins and Nature of Man.* New York: Dell Publishing Co.

100   Ardry, Robert. 1966. *Territorial Imperative, A Personal Inquiry into the Animal Origins of Property and Nations.* New York: Athenaeum Press,

CHAPTER 14: WHY THREE EXPLANATIONS OF CREATION?

1   Restak, Richard M., M.D. 1979. *The Brain, the last frontier.* New York: Warner Books,.

2   Talkorigins africanus, africanus (accessed June 6, 2008)

3   Wikipedia, Edward_Tylor,   (accessed June 4, 2008)

4   Themystica, animism, (accessed June 4, 2008)

5   la Barre, Westen. 1970. *The Ghost Dance: Origins of Religion.* New York: Dell Publications.

6   Wikipedia, Shamanism,   (accessed June 20, 2008)

7   Wikipedia, Helladic_period,   (accessed June 20, 2008)

8   Metmuseum, Heilbrunn Timeline of art History, m (accessed June 15, 2008)

9   Pantheon, master of animals, (accessed June 20, 2008)

10   Anthro.palomar, homo,   (accessed June 4, 2008)

11   Mnsu, beliefs, June 16, 2008)

12   Thefreedictionary, creationism,   (accessed June 17, 2008)

13   Starr, Chester G. 1991. *A History of the ancient world.* Oxford, Oxford University Press

14   Avesta, Zoroastrianism and the Avesta   (accessed June 6, 2008)

15   Sacred-texts, Hinduism,   (accessed June 6, 2008)

16   Wikipedia, Eleusinian_Mysteries,   (accessed July 2, 2008)

17   Wikipedia, Golden_age, (accessed July 2, 2008)

18   Wikipedia, Titan_mythology, (accessed July 2, 2008)

19   Wikipedia, Cronus, (accessed July 2, 2008)

20   Wikipedia, Rhea_mythology, (accessed July 2, 2008)

21   Wikipedia, Hades, (accessed July 2, 2008)

22   Wikipedia, Demeter, (accessed July 2, 2008)

23   Wikipedia, Persephone, (accessed July 2, 2008)

24   Wikipedia, Pomegranate, (accessed July 2, 2008)

25   Wikipedia, Triptolemus, (accessed July 2, 2008)

26   1stmuse, history of Alexander, (accessed July 3, 2008)

27   Wikipedia, History_of_Rome, (accessed November 9, 2008)

28   Unrv, major-roman-god-list, hp (accessed November 9, 2008)

29   Dictionary.reference, Monotheism, (accessed November 9, 2008)

30   Wikipedia, Zoroastrianism History, y (accessed November 9, 2008)

31   Wikipedia, Mithraism, (accessed November 16, 2008)

32   Wikipedia, Hebrew_Bible, (accessed November 9, 2008)

33   Wikipedia, Abrahamic Religiom, (accessed November 9, 2008)

34   Wikipedia, Martin Hengel, (accessed November 10, 2008)

35   Wikipedia, Dionysus, (accessed November 10, 2008)

36   Wikipedia, Parallels with Dionysus and Christianty (accessed November 10, 2008)

37   Sellers, Frances Stead. 2005. Origin of Intelligent Design. *Washington Post*, March 19.

38   Wikipedia, William Dembski, (accessed May 12, 2009)

39   Wikipedia, Michael Behe, (accessed May 12, 2009)

40   Wikipedia, The Design Inference, (accessed May 12, 2009)

41   Wikipedia, Specified complexity, (accessed May 12, 2009)

42   Wikipedia, Irreducible complexity, (accessed May 12, 2009)

43 Intelligentdesignnetwork, The intelligent design network, (accessed June 2, 2008)

44 Yourdictionary, anthropolgy, (accessed November 12, 2008)

45 Wikipedia, Existence of God, (accessed November 12 , 2008)

46 Wikipedia, William Paley (accessed November 12, 2008)

47 Wikipedia, Watchmaker analogy, (accessed November 12, 2008)

48 Miller, Kenneth R. 2008. *Only a Theory: Evolution and the Battle for America's Soul.* New York:Viking Penguin

49 Lepeletier, Thomas. 2007, *Darwin Hérétique Léteral retour du créationime,* Paris:editions du seuil, ,

CHAPTER 15: RELIGIOUS REACTION TO EVOLUTION IN THE UNITED STATES.

1 Victorianweb, darwin autobiography, (accessed November 24, 2008)

2 Law.umkc, fundamentalism [Putting Evolution on the Defensive], (accessed November 24, 2008)

3 Neatby, William Blair. 1902. *A History of the Plymouth Brethren.* 2nd ed. London: Hodder and Stoughton.

4 Wikipedia, John_Nelson_Darby, (accessed November 24, 2008)

5 Wikipedia, Dispensationalism, (accessed November 24, 2008)

6 Ordnet.princeton, dispensation on the Web, (accessed November 24, 2008)

7. Wikipedia, Premillennialism, (accessed November 24, 2008)

8 Yourdictionary, eschatology, (accessed November 24, 2008)

9 Walvoordhistory, John F. Walvoord, d.html (accessed November 24, 2008)

10 Wikipedia, Dwight_L._Moody, (accessed November 24, 2008)

11   Wikipedia, Moody_Church, (accessed November 24, 2008)

12   Wikipedia, William_Bell_Riley,   (accessed November 24, 2008)

13   Wikipedia, John_Roach_Straton,   (accessed November 24, 2008)

14   Wikipedia, Butler_Act, (accessed November 24, 2008)

15   Law.umkc, scopes trial, (accessed November 30, 2008)

16   Wikipedia, Scopes_Trial, (accessed November 24, 2008)

17   Shipley, Maynard. 1927. *The War On Modern Science*. New York: Alfred A Kopf.

18   Servintfree, A History of Public Education in the United States, (accessed November 24, 2008)

19   Wikipedia, Kitzmiller v. Dover Area School District, (accessed November 20, 2008

20   Russ, Steve. 2008. Review of *A Challenge Standing on Shaky Clay*. *Science* 322 (3): 47.

21   Olympian, Washington News article, Sat, May 27, 2007

22   Wikipedia, Manichaeism. (accessed November 24, 2008)

23   Paul Kurtz, ed. 2003. *Science and Religion, Are They Compatible?* Amherst: Prometheus Books.

24   Wikipedia, Left Behind (series)

25   Leftbehind, left behind

CHAPTER 16: WHY DO PEOPLE STILL BELIEVE?

1   Bloom, Paul, et al. Childhood Origin of Adult Resistance to Science. *Science Magazine* 316: 996

2   Miller, J. D., E. C. Scott, and S. Okamoto. 2006. SCIENCE COMMUNICATION: Public Acceptance of Evolution *Science* 313: 765

3   Couzin, Jennifer. 2008. Crossing the Divide. *Science* 319 (accessed February 22).

4   News & Views .2005. *Results of a Survey of Archaeologists on the Biblical Flood* , 57: 2,

5   Arlan Blodgett, personal communication

6   Hitchens, Christopher. 2007. *god is not Great.* New York: Grand Central Publishing.

7   Ldolphin, Why I became a YEC, Curt Sewell (accessed November 19, 2008)

8   Ldolphin, [Curt] sewell [Carbon-14 Dating Shows that the Earth is Young], _(accessed November 19, 2008)

9   Wikipedia, Radiocarbon, (accessed November 19, 2008)

10   Travis, John, .2009. On the Origin of The Immune System, *Science*, 324: 580(accessed November 19, 2008)

CHAPTER 17: ANSWERS/EXPLANATIONS IN EVOLUTION.

1   Thefreedictionary, belief, (accessed November 19, 2008)

2   merriam-webster, faith (accessed November 19, 2008)

3   dictionary.reference browse, faith, (accessed November 22, 2008)

4   psychology.about, structuralism, (accessed November 22, 2008)

5   Biblegateway, Romans 5, Verse 9, (accessed November 22, 2008)

6   Phrases, meanings, nasty, brutish and short, (accessed February 13, 2009)

7   Wikipedia, Thomas_Hobbes, (accessed February 13, 2009)

8   Yourdictionary, metabolism, (accessed November 24, 2008)

9   Wikipedia, Menstrual_cycle, (accessed February 17, 2009)

10   Wikipedia, Life_expectancy,

11   Rick Warren. 2007. *The purpose driven life*, Peabody, Mass:Zondervan.

12   Managers-net, frequencydistribution, html (accessed Feb 9, 2009)

13   Herrnstein, R. J., and Charles Murray. 1994. *The Bell Curve – Intelligence and Class Structure in American Life.* New York: Simon & Shuster.

14   Rosner, Bernard. 1990. *Fundamentals of Biostatistics.* Boston: PWS-Kent Publishing Co.

15   Michell, John. 1996. *Who Wrote Shakespeare?* London: Thames and Hudson

16   Holden, Constance. 2008. Parsing the Genetics of Behavior. *Science* 322 (5903): 892–895.

17   Epigenome, Histones and hardware, (accessed Feb 9, 2009)

18   Crystalinks, reptilianbrain, (accessed November 20, 2008)

19   Restak, Richard M., M.D. 1979. *The Brain, the last frontier.* New York: Warner Books.

20   Wikipedia, Limbic system, (accessed November 20, 2008)

21   Biology.about, hippocampus, (accessed November 20, 2008)

22   Wikipedia, Cerebral_cortex, dup (accessed November 20, 2008)

23   MacLean, Paul D. 1969. A triune concept of brain and behavior. Ontario Mental Health Foundation, *U of Toronto Press*

24   Edward Fitzgerald, Translator.1942. *The Rubáiyát of Omar Khayyám,* Roslyn NY:W. J. Black, ,

25   Wikipedia, Apoptosis, (accessed February 15, 2009)

26   Ask.yahoo, number of cells in human body, (accessed February 15, 2009)

27   Kurzweill, R., and T. Grossman. 2004. *Fantastic Voyage, Live Long Enough to Live Forever.* New York: Rodale, Inc

28   Kagan, Donald. 1975. *On the Origins of War, and the preservation of peace.* New York: Double Day Books.

29   Edge, A History of Violence, (accessed February 10, 2009)

30   Joffe, Joseph. 2009. Obama's Popularity Doesn't Mean Much Abroad, as ever, countries have interests, not friends. *Wall Street Journal*, April 18.

31   globalchange.umich.edu, globalchange, (accessed November 24, 2008)

32   Wikipedia, Club_of_Rome, (accessed November 24, 2008)

33   Meadows, D. H., and D. L.Meadows. 1972. *The Limits to Growth*. New York:Potomac Associates

34   Esciencenews, new studies predict record land grab (accessed February 16, 2009)

35   news.bbc.co.uk, science/nature/7865332 , (accessed February 20, 2009)

36   Miller, Geoffrey, et al. 2007. Tip earnings by lap dancers: economic evidence for human estrus?. *Evolution and Human Behaviour* no. 28:375–381.

CHAPTER 18: CONSEQUENCES OF THE SCIENCE-RELIGION CONFLICT.

1   Johnson, Avery. 2009. U. S. Court Rejects Vaccine Connection to Autism. *Wall Street Journal*, February 13.

2   STEVEN NOVELLA .2007 Vaccines & Autism: Myths Myths and Misconceptions, *Skeptical Inquirer* Nov/Dec

3   Know-vaccines, exemptionFAQ, (accessed February 15, 2009)

4   cdc.gov, measles vaccine (accessed February 15, 2009)

5   Nyhan, Paul. 2009. Diseases many come back du to vaccine worries. *Olympian*, March 17.

6   Cancer.about, hpvsymptoms, (accessed February 16, 2009)

7   atheism.about, religious-objections-to-hpv-vaccine, (accessed February 15, 2009)

8   Cbc. prayer-heart-surgery, (accessed February 16, 2009)

9   Wikipedia, Coronary artery bypass surgery, (accessed February 16, 2009)

10 Imrie, Robert, "Dad saw sickness as a test of faith," *The Olympian*, July 26, 2009, pg A 12

11 Tanner, Lindsey.2009. Belief in divine intervention weighs heavy in decisions about continuing life support, *Wall Street Journal*, Aug 19.

12 Rabin, Roni Caryn. 2009. Study links faith and end-of-life treatment. *SF Chronicle*, March 18.

13 Stein, Rob. 2008. Pharmacies that avoid contraceptives growing. *Washington Post*, June 16.

14 Yahoo, af pope africa, (accessed May 21, 2009)

15 Usaid, condomfactsheet, (accessed May 21, 2009)

16 Simon, Stephen. 2009. Texas Opens Classroom Door for Evolution Doubts. *Wall Street Journal*, March 28.

17 Lashawnbarber, Texas School Board Votes on Evolution Challenge, (accessed April 2, 2009)

18 Wood, Lawrence Confronting the Anti-Evolution Attack on Public Education, paper presented at the AAAS, Pacific Division 88th Annual Meeting

NOTES

PROLOGUE

[1] From the Gallup website: "The Gallup organization has studied human nature and behavior for more than 70 years. Gallup's reputation for delivering relevant, timely, and visionary research on what people around the world think and feel is the cornerstone of the organization. Gallup employs many of the world's leading scientists in management, economics, psychology, and sociology, and our consultants assist leaders in identifying and monitoring behavioral economic indicators worldwide. Gallup consultants help organizations boost organic growth by increasing customer engagement and maximizing employee productivity through measurement tools, coursework, and strategic advisory services. Gallup's 2,000 professionals deliver services at client organizations, through the Web, at Gallup University's campuses, and in 40 offices around the world." [1]

[2] These six questions are part of two-line memory aid, which are often invoked to assist investigations:

Six little friends have I

What, When, Where, Who, How and Why

[3] To emphasize the difference between answers that answer *what*, *when*, *where*, *who*, *how*, and *why* the words will be italicized were appropriate.

[4] This is, in a small way, similar to the approach British mathematician **Sir Andrew John Wiles (1953–)** took in solving one of mathematics most enduring and famous puzzles—Fermat's Last Theorem. Wiles

labored for many years developing supporting proofs which ultimately led to the final proof [http://en.wikipedia.org/wiki/Wiles's_proof_of_Fermat's_Last_Theorem].

Fermat's Last Theorem states that equation, $x^n + y^n = z^n$ has no non-zero integer solutions for x, y and z when $n > 2$. In 1630, **Pierre de Fermat (1601–1665)** tantalizingly wrote in the margin of book: "I have discovered a truly remarkable proof which this margin is too small to contain" [http://www.gap-system.org/~history/HistTopics/Fermat's_last_theorem.html].

No one had ever discovered Fermat's proof and the proof of Fermat's Last Theorem has bedeviled mathematicians ever since Fermat died.

After many unsuccessful attempts in the intervening years Wiles finally devised a proof; however, the proof rested upon a number of preliminary proofs, especially the proof of an arcane piece of mathematics termed the "Taniyama-Shimura (T-S) conjecture for semistable elliptic curves." After six years of almost completely secret effort, Wilkes, certain of his proof of the T-S conjecture, presented his proof of it June 21–23, 1993, to an overflowing crowd of mathematicians at the Isaac Newton Institute for Mathematical Studies.

During the three days in which Wiles presented his proof of the T-S conjecture, the mathematicians maintained a respectful silence. At the conclusion of his presentation, almost as an afterthought, Wiles hesitated and then stated, "And of course, this proves Fermat's Last Theorem," at which point the audience, who had known what was coming, rose shouting and applauding the solution to one mathematics' most famous puzzles.

[5] In fact, answering the questions addressed by this book is equivalent to solving a giant puzzle without aid of a guiding picture.

[6] Johannes Kepler's determination that the equation for a planet revolving around the sun is an ellipse is an excellent example of an empirical explanation.).

Kepler's empirical determination of the equation of a planet's orbit is characteristic of most empirically derived explanations of phenomenon in that they lack understanding of the underlying cause of the phenomenon. Kepler's equation was quite accurate, but neither Kepler nor anyone else could explain why the orbit was an ellipse. As also detailed in chapter 6, the answer was provided a number of years later by Sir Isaac Newton who determined the underlying causes.

[7] I will rarely employ the word "theory" in this book since, unfortunately, the word has gotten a bad reputation. As pointed out in Kenneth R Miller's book, *Only a Theory: Evolution and the Battle for America's Soul* ,Viking Penguin, 2008, One often hears, "Oh, it's just a theory," implying it's just a guess, a hunch.

Accordingly, the word theory will generally be avoided. I will instead employ its equivalent, "explanation." Thus, we could employ the phrase "Darwin's explanation of evolution" rather than "Darwin's theory of evolution," since this is what he actually intended—an explanation of evolution, based upon his research and what was known at the time.

However, it will be occasionally convenient to employ the terms "theory" or "theoretical," and in these cases, the terms should be interpreted as explanations based upon an understanding of fundamental principles.

[8] The acronym E/M is formed from electromagnetic; however, this book will employ the similar acronym E&M for electricity and magnetism.

CHAPTER 1: ELECTROMAGNETIC PHENOMENON INVESTIGATIONS PRIOR TO 1897

[1] Following current dating convention, I will employ the term Before Common Era (BCE) and Common Era (CE) and to reference events before or after the "zero" of time.

[2] Presumably, Cardano distinguished between the observation of phenomena that occurred without human intervention, such as lightning, and experiments in which a phenomenon is deliberately produced, such as rubbing amber with fur. For the purpose of this book, the two are equivalent—any observation requires an explanation.

[3] We now know that the electrical current that passes along a nerve is generated by the interaction of sodium and potassium ions.

[4] It is one of the unfortunate facts of the universe we inhabit that the behavior of most phenomena is described by second order differential equations—a reason for the emphasis on mathematics in science education.

[5] As will be discussed in chapter 7, the velocity of light was determined in 1680 by Danish astronomer Olaf Roëmer to be 186,000 miles per second, thus Maxwell had an experimental value to compare with.

CHAPTER 2: "SOLID" EARTH INVESTIGATIONS PRIOR TO 1897.

[1] As Bertram Russell comments in his excellent *History of Western Philosophy*:

> In all of history, nothing is so surprising or so difficult to account for as the sudden rise of civilization in Greece. Much of what makes civilization had already existed for thousands of years in Egypt and in Mesopotamia, and had spread thence to neighboring countries. But certain elements had been lacking until the Greeks supplied them. What the Greeks achieved in art and literature is familiar to everybody, but what they did in the purely intellectual realm is even more exceptional. They invented mathematics (Arithmetic and some geometry had existed among the Egyptians and Babylonians, but mainly in the form of rules of thumb – deductive reasoning from general premises was a Greek innovation) and science and philosophy; they first wrote history as opposed to mere annals; they speculated freely about the nature of the world and the ends of life [2].

As will be discussed in greater detail in chapter 14, the development of proper explanations evolved from the mists of superstition and ignorance; hence it is not surprising that investigators such as Paracelsus working in the fifteenth century straddled two worlds: mystical and practical. In fact, we have those today who straddle these two worlds (e.g., in 2006, Francis Collins, the celebrated leader of the governments Human Genome Project and who wrote this book Collins. F. 2006. The Language of God: A Scientist Presents Evidence for Belief, New York:*Simon and Shuster.*

Collins considers scientific discoveries an "opportunity to worship." In his book, Collins examines and subsequently rejects creationism and

ID. His own belief system is Theistic Evolution, which he prefers to term *Biologos* (see http://en.wikipedia.org/wiki/BioLogos).

Presumably, persons like Dr. Collin's view science as an attempt to understand "God's Creation." Collins comments, "I take the view that God, in His wisdom, used evolution as His creative scheme. I don't see why that's such a bad idea. That's pretty amazingly creative on His part."

Fortunately, Dr. Collins seems to be in the minority. I find it intriguing that a person of Collin's intellectual stature fails to see that invoking God adds nothing to the explanation of evolution.

CHAPTER 3: EMPIRICAL RESOLUTION OF THE "SOLID" EARTH ILLUSION BETWEEN 1897 AND 1911

[1] Interestingly, Becquerel chose the first three letters of the Greek alphabet rather than a Roman alphabet as Roentgen had.

CHAPTER 5: FINAL RESOLUTION OF THE "SOLID" EARTH ILLUSION, PART II

[1] The legend of Alexander the Great cutting the Gordian Knot is often used as a metaphor for an intractable problem solved by a bold stroke ("cutting the Gordian knot"). While probably apocryphal, in the legend while on his conquest of Persia, Alexander slices through a particularly difficult knot in the town of Gordium with his sword rather than attempt to untie it. According to the legend, whoever untied the knot would become King of Persia, a feat Alexander accomplished [25].

CHAPTER 6: SOLVING THE APPARENT MOTION OF THE SUN AND PLANETS AROUND THE EARTH ILLUSION

[1] There is a significant difference between the visual image presented by the planets and the stars. The planets are close enough so that the eye can "resolve" the image of a planet (i.e., we can actually "see" the disc formed by a planet). This is particularly true of the brightest object in the night sky besides the moon, the planet Venus. However, we can also "see" the other planets visible to the human eye, especially Jupiter and Mars. As will be discussed in chapter 13, the stars are so

distant that neither the human eye, nor any earth-based telescope for that matter, can resolve the star's image. What we "see" is an optical phenomenon known as an Airy disc.

[2] The impact on our ability to understand ourselves and the world around us that was provided by the seemingly simple invention of the glass lens cannot be overstated. This invention led to both the microscope, which revealed microscopic world too small to be seen with the unaided eye as well as to the almost simultaneous invention of the telescope, which revealed the unbelievable extent of the universe much too distant to be seen with the unaided eye.

Glass was probably discovered accidentally by using high quartz (silica, the main ingredient in glass) rocks as a fireplace. Glass has been known since the Bronze Age [25]. It is reasonable to suspect that the ability of curved glass to create a "lens" was discovered soon after because the oldest lens artifact, the Nimrud lens, is over three thousand years old, dating back to ancient Assyria [25].

The ancient Greek playwright Aristophanes describes the focusing of the sun's rays by a lens to produce fire in his play *The Clouds* (424 BCE). It is the earliest written mention of a lens [25].

Widespread use of lenses in spectacles began, probably in Italy in the 1200s.

[3] Strictly speaking, the sun and planets revolve about the focal point of the ellipse, but the sun is so much bigger than the planets that the focal point is inside the sun, hence the apparent revolution of planets around the sun.

[4] It is an inconvenient truth that most of the fundamental equations that describe our world are differential equations.

[5] This does seem logical. When we walk or ride in car, we are moving in three spatial dimensions, plus time. We leave place (X) at time (t) and arrive at place (Y) at time (t) plus travel time.

CHAPTER 7: RESOLVING THE APPARENT SAME MOON AND SUN SIZE AND DISTANCE ILLUSIONS

[1] This apparent reduced size with distance is also an illusion, of course.

CHAPTER 8: RESOLVING THE APPARENT SAME MOON AND SUN SIZE AND DISTANCE ILLUSIONS

[1] The variation in angle that occurs when measuring the angles to an object beyond reach and the angles are measured from two different locations. It is termed "parallax" and is often used interchangeably with triangulation.

[2] It is ironic that both John Goodricke and Henrietta Leavitt were both deaf, Goodricke due to scarlet fever in early childhood and Leavett due to an illness contracted after graduating from Radcliff.

[3] In ref [13], Slipher reports on making the first Doppler measurement on September 17, 1912. In his report, Slipher writes: "The magnitude of this velocity, which is the greatest hitherto observed, raises the question whether the velocity-like displacement might not be due to some other cause, but I believe we have at present no other interpretation for it." Three years later, Slipher wrote a review in the journal *Popular Astronomy*, in which he states, "The early discovery that the great Andromeda spiral had the quite exceptional velocity of - 300 km(/s) showed the means then available, capable of investigating not only the spectra of the spirals but their velocities as well" [13]. Slipher reported the velocities for fifteen spiral nebula spread across the entire celestial sphere, all but three had observable "positive" (that is recessional) velocities.

[4] Ignasi Ribas of the Catalonia Institute for Space Studies in Bellaterra, Spain, and his colleagues used light, velocity, and temperature measurements to calculate the true luminosity of a binary star in the galaxy in which the two stars eclipse one another every orbit. They compared their findings to the observed brightness of the stars. This enabled them to calculate Andromeda's distance as 2.52 million light-years, compared to earlier estimates of 2.2 to 2.9 million light-years. The results appear in *Astrophysical Journal Letters* [33].

[5] A recent paper in *Science* magazine clarifies the formation of the first stars via a computer simulation that demonstrates that the first stars formed were tiny "protostars" which eventually became the first real stars [protostar formation in early universe, Science, aug 1, 08, 669].

[6] The phenomenon we call "heat" is caused by the motion of molecules. As air gets hotter, the air molecules move faster. When we

turn on an electric stove, the molecules in the heating coils begin to move very rapidly as the coil heats up.

[7] Since the first extra solar planet was detected in 1992 by the solar wobble induced by the planet, over 300 extra solar planets have been found [34].

CHAPTER 9: RESOLVING THE APPARENTLY UNCHANGING PHYSICAL FEATURES OF THE EARTH ILLUSION UP TO 1850

[1] Interestingly, with the exception of fossils and in a manner reminiscent of the early investigations into the solid earth illusion, investigation into the question of whether the earth was not unchanging as it appeared proceeded along two relatively independent paths: the physical features of the earth and the animal and plant structures. These two paths joined when the true explanation of Earth's creation was fully developed.

[2] As Simon Winchester points out in his excellent discussion of William Smith's contributions, echoing comments made at the beginning of this book:

> "Few outside the world of the rigid Christian fundamentalists today accept the strict inter-predation of James Ussher's arithmetic, which he explained in his monumental work of 1658, *Annalis Veteris et Novi Testamenti*, But nonetheless a 1991 survey showed that fully 100 million Americans still believed that "God created man pretty much in his own image at one time during the last ten thousand years," and anecdotal evidence now suggests that this number is climbing. This might suggest that aspects of the religious climate into which William Smith was born-and that he was to help start changing--are now starting to return." [5] pg 15

[3] I am not sure whether Usher was the first to attempt an estimate of Earth's age, but he is the first person in the historical record.

[4] Winchester's book is not only an excellent biography, it is a window into the intellectual atmosphere of the eighteenth and nineteenth

centuries, pointing out the controversies that existed between the established explanation of ourselves and the world we inhabit based upon 2000 years of biblical interpretation and the correct explanation which was at complete odds with the biblical explanation.

[5] There is a fascinating chronology of Noah's adventure in the ark in this Web site [9].

[6] In the time of Leonardo, many formed their names from a first name and the town in which they were born. *Di* in Italian translates as "from," hence Leonardo di Vinci means Leonardo from [the town of] Vinci.

[7] The Codex Leicester derives its name from Robert Dudley, First Earl of Leicester, who purchased it in 1717 []. That the book is important can be judged by the dollar amount paid by the most recent purchaser—Bill Gates, $30.8 million.

[8] In making these observations of separated but related strata, Leonardo was anticipating the "Father of Stratiography," William Smith by 200 years. However, Leonardo's secretive style of writing prevented others from benefiting from his observations.

[9] Stratum (from the Latin for "a covering" or "blanket") is a layer of rock or soil with internally consistent characteristics that distinguish it from other layers above or below it. Stratiography is the branch of geology that investigates the stratification or creation of a stratum of rock layers. Stratiography is primarily used in the study of sedimentary rocks, rocks formed by the accumulation of sediments, and layered volcanic rocks which are rocks formed by layers of volcanic eruption []. As will be discussed in some detail, examination of rock layers provides important information regarding the history of the earth, the development of life on Earth, and the occasional disasters that have occurred such as the asteroid collision that resulted in the extinction of the dinosaurs.

As mentioned above, rock layers have been studied since the time of Persian geologist Ibn Sina (981–1037 CE), also known by his Latinized name Avicenna []. Ibn Sina was the first to outline the law of superposition of strata. This "law" was later formulated more clearly in the seventeenth century by the Danish scientist Nicolas Steno (1638–1686), a pioneer in both anatomy and geology [108].

Stratiography, considerably improved but essentially the same as developed by Smith, plays an important role in modern industries such as petroleum and limestone quarrying. Both exploration and exploitation of economic materials require predictions of the subsurface stratiography [].

[10] Churchill is a small and picturesque Cotswold village in northwest Oxfordshire, Great Britain, about three miles southwest of Chipping Norton and 77 miles from London. [ch 9, 44]

[11] As will be discussed later, rock strata contain a chronological record of the development of life on Earth. A simple table showing the age relationship of various animal species is presented in appendix IV. Note that this table is arranged in the manner in which fossils would be found, with the most recent species (human) placed on top and oldest at the bottom. One of the most striking example rock formations matching this table is the "Giant Stair Case" in western United States. It begins with the Grand Canyon, which has the oldest rocks, and proceeds to Bryce Canyon, which has the youngest rocks. Almost 2 billion years of life is exposed.

[12] Berwickshire or the County of Berwick was situated on the southeast border with England, and it was a "registration county," a temporary administrative unit used in census results from 1851 to 1911. It was also part of the Scottish Borders Council.

[13] The Royal Society of Edinburgh began as the Philosophical Society, founded in 1783. The Philosophical Society was transformed into The Royal Society of Edinburgh in 1783; hence the RSE predates the RSL by several years [109].

[14] Over time, geologists needed a method for organizing rock groups that are found in many places on Earth. The groups were usually named for the location, such as Devon for the Devonian, or for the initial humans who inhabited the area, such as the Welsh tribe of the Ordovices for the Ordovician.

[15] Greywacke is a gray earthy rock that is a variety of sandstone generally characterized by its hardness, dark color, and poorly-sorted, angular grains of material [http://www.answers.com/topic/greywacke].

[16] Presumably, Lyell, et al. were not conversant with Leonardo's demonstrations, rejecting the idea of the biblical flood.

CHAPTER 10: RESOLVING THE UNCHANGING PHYSICAL FEATURES OF THE EARTH ILLUSION AFTER 1850

[1] A wave is termed elastic when it passes through a medium such as the earth, and the medium is not permanently deformed.

[2] The *P* in P-wave stands for pressure, and the *S* in S-wave stands for shear or perhaps secondary because S-waves travel more slowly through the earth and thus arrive at a seismometer after the P-wave. P-waves are longitudinal in that they oscillate in the direction of wave travel, while S-waves are transverse and oscillate at right angles to the direction of wave travel.

CHAPTER 11: RESOLVING THE APPARENTLY UNCHANGING ANIMAL AND PLANT STRUCTURES ILLUSION UP TO DARWIN

[1] Apparently the long, relatively monotonous, and lonely voyage with no one to talk with became too much for the previous captain, a mistake Fitzroy was not to repeat.

CHAPTER 12: RESOLVING THE APPARENTLY UNCHANGING ANIMAL AND PLANT STRUCTURES ILLUSION AFTER DARWIN, EXPLAINING NATURAL SELECTION

[1] In a similar manner to Leonardo, whose research was lost due to his secretive ways, Mendel's research was lost due to his obscure existence.

[2] From the German spelling, "chromosome" was coined 1888 by German anatomist Wilhelm von Waldeyer-Hartz (1836–1921). Also from the Greek *khroma* (color) and *soma* (body). It was called that because the structures contain a substance that stains readily with basic dyes.

[3] The cells that make up all of the body's cells except the reproductive cells are termed "soma cells." Reproductive cells are termed "gametes," from ancient Greek. Female reproductive cells are termed "gamete," and male reproductive cells are termed "gametes."

[4] The term "protein" was coined by Jöns Jakob Berzelius in 1838 [http://en.wikipedia.org/wiki/J%C3%B6ns_Jakob_Berzelius];

however, it was long known that protein is the primary constituent of all cells. The discoveries which eventually led to our current understanding of proteins began in 1728 with studies by the Italian scholar Jacopo Beccari. They are many and complex and will not be discussed in this book because they ancillary to our purpose. However, reference [] provides an excellent summary.

[5] The fragment deoxy means one less oxygen atom.

[6] If three bases are sufficient to encode the amino acids, one might ask why there are four. The obvious answer is that four are required to construct the DNA "ladder." In a similar manner, three legs for "four-legged" animals would be sufficient but would certainly be awkward and destroy the obvious symmetry of four legs.

CHAPTER 13: RESOLVING THE APPARENTLY UNCHANGING FEATURES OF THE EARTH ILLUSION— HOW THE EARTH WAS REALLY CREATED

[1] See appendix II for the relationship between the Silurian and Devonian periods.

[2] Tetrapod is the scientific term for a four-legged animal. *Tetra* means four and *pod* means leg [101].

[3] One of the principal criticisms of evolution by adherents of creationism and ID is the "lack of transition animals." Clearly Tiktaalik is one, and others are being discovered.

[4] As will be revealed, the human brain has three parts: (1) a reticular formation; (2) a limbic which forms a cap over the reticular formation; and (3) an outer layer neocortex. The reticular formation is often termed the "reptilian" formation as it first appears in the fossil record with the reptiles.

[5] The increasing ability to detect objects such as comets, asteroids, and meteoroids in the vicinity of the earth has engendered much interest is these objects and growing concern that a collision with a large object might occur in the relatively near future. Per reference [http://neo.jpl.nasa.gov/], "Near-Earth Objects (NEOs) are comets and asteroids that have been nudged by the gravitational attraction of nearby planets into orbits that allow them to enter the Earth's neighborhood."

[6] The extinction event that ended the dinosaur rule is often termed the K-T event, where the German word for Cretaceous, *Kretazeisch*, is employed. This is similar to the use of the term EKG for electro-cardio gram, where the K corresponds to the German *Kardio*. Anything to add confusion to an already complex subject.

[7] The term "limbic" was coined by Paul MacLean due to its shape, which reminded him of one of the moon's phases.

[8] The dictionary [http://www.yourdictionary.com/culture] defines culture as "the ideas, customs, skills, arts, etc. of a people or group that are transferred, communicated, or passed along, as in or to succeeding generations."

[9] Breaking the "Stone Ages" code, *Lithic* is Latin for stone, *paleo* means old, *meso* means middle, and *neo* means new; hence Paleolithic means "Old Stone (Age)." Time is implied in Paleolithic but must be made specific when translating to English. The rest of course are middle Stone Age and New Stone Age.

[10] Archeologists denote a "type site" as a location that is considered the model of a particular archeological culture. A type site contains an assemblage of artifacts that are typical of that culture [http://archaeology.about.com/od/sitetypes/Site Types in Archaeology.htm].

[11] When we complete the resolution of the "why three explanations?" question, we will see that animism, which is probably the first religion, can be traced back at least 200 thousand years, although the Aurignacians were probably practicing a form of animism.

[12] More recent references to Crô Magnon appear to prefer the term "early modern humans" [80].

CHAPTER 14: WHY THREE EXPLANATIONS OF CREATION?

[1] The relation of the mind to the body has, of course, been the subject of much discussion, much of it a waste of time. All the evidence to date demonstrates that the mind is a product of electro-chemical intersections between the neurons in the brain [1].

[2] Regarding this matter, the noted historian Chester Starr commented, "To understand any era of the past one must be able to penetrate into the minds of its inhabitants. This is an ever-challenging, yet extremely

difficult task, to which the historian should bring sympathetic imagination and a wide knowledge of the passions of man" [13].

[3] Rather than discuss essentially vague terms such as "spirit" or "god," I prefer the word "supernatural" since this term is usually defined as [41]:

- of or relating to existence outside the natural world;

- attributed to a power that seems to violate or go beyond natural forces;

- of or relating to the immediate exercise of divine power; miraculous.

[4] The belief that gods who "inhabit the shy" have special power evolved into the belief in "bird shamans" who wielded great power in Stone Age cultures. This belief carried over into Greek mythology.

5] The Mycenaean is named for the Bronze Age fortress Mycenae, which dominated the Greek peninsula from about 1600–1100 BCE [42].

[6] For an interesting description of the Greek pantheon and the path from Chaos to Zeus, see [43]. There links to Chaos which state that "In Greek mythology, Chaos or Khaos is the original state of existence from which the first gods appeared. In other words, the dark void of space is made from a mixture of what the Ancient Greeks considered the four elements: earth, air, water and fire."

[7] The number twelve, corresponding to the twelve full moons in a year, plays a significant role in many religious and superstition concepts (e.g., the twelve signs of the Zodiac, the twelve apostles, etc.).

[8] The belief that gods who "inhabit the sky" have special power carried over into Greek mythology.

[9] The Ghost Dance was developed by the declining Native American populations as reprisal for the harsh treatment given by the European settlers who were decimating the Native Americans. Ghost dancers believed the dance would bring back dead warriors who would conquer the white man. Of course, the belief was false.

[10] For an excellent discussion of the human "manufacturing process," *Coming to Life, How Genes Drive Development* by Nobel Laureate Christine Nüsselein-Volhard is one of the best [44].

[11] Le Peletier's title translates as *Darwin the Heretic, the eternal return of creationism* In his book, Peletier discusses "*l'absurditie de l'*Intelligent Design" (the absurdity of intelligent design).

CHAPTER 15: RELIGIOUS REACTION TO EVOLUTION IN THE UNITED STATES

[1] While religious reactions to science have varied in different parts of the world, my interest is the reaction in the United States as evidence indicates that the most severe reaction has occurred here. Witness the election of President George W. Bush largely due to the efforts of religious fundamentalists or see chapter 16, "Consequences of the Science-Religion Conflict."

[2] *Left Behind* is a series of sixteen best-selling novels by Tim LaHaye and Jerry Jenkins, which address a Christian dispensationalist, End of the World viewpoint [21],[24],[25]. Response to the book has been phenomenal, with total series sales exceeding 65 million copies. On their Web site, LaHaye and Jenkins state, "We believe that while no one knows the day or the hour when Christ will return, we have more reason to believe He could come in our lifetime than any generation before us" [http://www.leftbehind.com].

[3] While the anti-evolution crusade is often considered a Southern phenomenon, its two foremost leaders, Riley and his associate, Dr. John Roach Straton, were both born in Indiana [12]; [13].

[4] Richard Dawkins is one of the more outspoken scientists regarding creationism, intelligent design, and religion in general. One of Dawkins's more controversial publications is *The God Delusion*, Houghton Mifflin Co., NY 2006. I find nothing in Dawkins's book to verify that Steven Fuller's assertion that "[Dawkins] arguably owes more to 18th-century secular theodicy than to Darwin's own 19th-century anti-theodicy" [20]. Consider the following passage from Dawkin's book:

> The God of the Old Testament is arguably the most unpleasant character in all fiction: jealous and proud of it; a petty, unjust, unforgiving control-freak; a vindictive, bloodthirsty ethnic cleanser; a misogynistic, homophobic, racist, infanticidal, genocidal, fili-cidal, pestilential, megalomaniacal, sadomasochistic, capriciously malevolent bully.... Thomas Jefferson

-better read -was of a similar opinion [to Winston Churchill]: 'The Christian God is a being of terrific character -cruel, vindictive, capricious and unjust.' (p. 31)

CHAPTER 16: WHY DO PEOPLE STILL BELIEVE?

[1] One of the reasons I emphasized the biblical flood in chapter 5 was the great hold the flood had on persons such as Geoffrey and Blodget, which will be discussed later.

[2] "Scientific Creationism" is the ultimate oxymoron. Creationism's basic belief in an omnipotent, all-knowing God is diametrically opposite of science, which has no place for the supernatural.

CHAPTER 17: ANSWERS/EXPLANATIONS IN EVOLUTION

[1] It might be argued that part of the science-religion conflict can be traced to the fact that religious explanations were considered "the last word" for thousands of years until this "upstart" science began to question this dominance only a few hundred years ago. In many cases, religious leaders have refused to acknowledge science's right to exist.

[2] In modern times, alternate methods like invitro fertilization have been perfected, but as this is a relatively recent development, it will not effect this discussion.

[3] It is ironic that we have so much difficulty with this basic need. Perhaps, as will shown shortly, it's because we don't need sex to survive. Many consider sex "dirty" (e.g., a "dirty joke" or "dirty story" invariably involves sex). Depiction of humans engaged in sex is considered "filth" and proscribed (except for mild simulations) for television and most movies, even though there are many Web sites from which hundreds of films depicting sex can observed.

[4] Of course, those animals who did not meet this criterion are no longer with us.

[5] The fact that approximately thirty minutes is required for the "full feeling" to kick in partially accounts for the fact that people who eat fast tend to overeat. Eating slowly allows time for the full feeling to become effective, which (should) end our eating at that time.

[6] The complexity of the hormonal interactions is beyond the scope of this book but can be found in [8].

[7] The dictionary [http://www.thefreedictionary.com/intention] defines purpose as "a result or effect that is intended or desired; an intention," and intension is defined as "a course of action that one intends to follow."

Some synonyms of "purpose" are: goal, end, aim, object, objective.

Since purpose and intension intertwine, perhaps when we talk of having a purpose, we may really mean achieving a goal or an end result that has value.

[8] Regarding the relationship between sex and territory, Ardry observes, "few men have died for sex, but millions have died in defense of Territory." [ch. 13, ref. 100]

[9] Climate change has been suggested as a principal cause for the Neandertal's demise, however, in an article on this Web site [36] we find the following conclusion:

> Our findings suggest that there was no single climatic event that caused the extinction of the Neandertals," said Katerina Harvati of the Max Planck Institute for Evolutionary Anthropology in Leipzig, Germany. "At most, cold was just a contributing factor in their demise", she added.

[10] I wrote this phrase a number of months ago and was pleasantly surprised to find during a routine Web query regarding the population problem an article titled, "Population: The elephant in the room" [34], written by psychologist Dr. John Feeney, who became an ardent environmentalist specializing in the population control problem [35].

CHAPTER 18: CONSEQUENCES OF THE SCIENCE-RELIGION CONFLICT

[1] As discussed earlier, one of the fundamental reasons for the science-religion conflict is the existence of two school systems: (1) the church and family, and (2) public schools.

INDEX

Mount Wilson, California observatory, 103
Multicelled animals, appear in Ediacaran period, 238
Multicelled animals, requirements for, 239
Murchison, Sir Roderick Impey, first geological time scale, 156
MYA, millions of years ago, 167 and other pages

Natural Selection by reproductive genetic mixing, fully illustrated, 225
Natural Selection Process, partially illustrated, 192
Neptunism and Abraham Werner, incorrect explanation of earth, 146
Neutron, 59
Newton, Sir Isaac, 81, 82
no reason not to believe, xxi, 121, 174
Nuclear decay and radioactivity, 64
Nuclear decay, illustrations of the effects, 65
Nuclear fission, splitting of nucleus, 62
Nuclear fusion, fusing of nuclei, 62
Nuclear fusion, energy source of sun, 162, 375
Nuclear fusion, first demonstrated, 62
Nuclear strong force, 61
Nucleic acid isolated, Altmann Richard , 1890, 208
Nucleic acid: 2 pyrimidines and 2 purines, discovered by Albrecht Kossel, 209
Nuclein, discovered by Johan Miescher, 282
Nucleons, 59, 61, 62
Nucleus Spins, origin of natural magnetism, 62

Objective of book, primary, xv
Objective of book, supporting objectives, xx
Oliphant, Sir Mark, 62
On the Origin of Species by Means of Natural Selection, pub. 1859, 195
On the Revolutions of the Celestial Spheres, pub. 1543, 78
Oxygen, discovery and naming, 25

Paley, William, watch implies a designer, 284